To our daughters, Stacy and Stefani

CONTENTS

University of
Hertfordshire

Writing

3rd EDITION

PROCESS AND PRODUCT

SHARON J. GERSON
DeVry Institute of Technology

STEVEN M. GERSON
Johnson County Community College

Prentice Hall
Upper Saddle River, New Jersey 07458

Library of Congress Cataloging-in-Publication Data

Gerson, Sharon J.,
   Technical writing: process and product/Sharon J. Gerson, Steven M. Gerson.—3rd ed.
      p.   cm.
   Includes bibliographical references and index.
   ISBN 0-13-020871-X
   1. English language—Technical English.   2. English language—Rhetoric.
   3. Technical writing.   I. Gerson, Steven M., 1948-
II. Title.
PE1475.G47   1999
808'.0666—dc21   99-14593
                                                              CIP

Editor-in-Chief: Leah Jewell
Editorial Assistant: Patricia Castiglione
AVP/Director of Production and Manufacturing: Barbara Kittle
Managing Editor: Bonnie Biller
Project Manager: Maureen Richardson
Manufacturing Manager: Nick Sklitsis
Prepress and Manufacturing Buyer: Mary Ann Gloriande
Creative Design Director: Leslie Osher
Interior Design: Circa 86
Cover Design: Maria Lange
Cover Art: Theo Rudnak/SIS, Inc. "Hand in Cloud with Paper, Pencil"
Supervisor of Production Services: John Jordan
Electronic Page Layout: Annie Bartell
Electronic Art Creation: Annie Bartell
Marketing Manager: Brandy Dawson

This book was set in 10.5/12 Sabon Roman by Annie Bartell and was printed
and bound by Courier Companies. Inc. The cover was printed by Phoenix Color Corp.

© 2000 by Prentice-Hall, Inc.
Upper Saddle River, NJ 07458

Printed in the United States of America
10 9 8 7 6 5 4 3 2

ISBN 0-13-020871-X

Prentice-Hall International (UK) Limited, London
Prentice-Hall of Australia Pty. Limited, Sydney
Prentice-Hall Canada Inc., Toronto
Prentice-Hall Hispanoamericana, S.A., Mexico
Prentice-Hall of India Private Limited, New Delhi
Prentice-Hall of Japan, Inc., Tokyo
Pearson Education Asia Pte. Ltd., Singapore
Editora Prentice-Hall do Brasil, Ltda., Rio de Janeiro

CHAPTER 12 INSTRUCTIONS AND USER'S MANUALS 276

Unit 6 • Report Strategies

CHAPTER 13 RESEARCH 320

PREFACE

Welcome to the third edition of *Technical Writing: Process and Product*. In this improved version of our textbook, we have accomplished two goals. First, we have reorganized the book for easier access. Second, we have updated it, especially by focusing on the ways in which electronic technology affects technical writing. In response to technological changes, this textbook provides a comprehensive overview of the types of technical writing you'll encounter on the job—into the twenty-first century.

ORGANIZATION

We have divided this third edition into seven units that provide detailed information on distinct aspects of technical writing. Here's what you'll find in each unit.

Unit 1: Defining Technical Writing

In this section of the book, we lay the foundation. Chapter 1, "An Introduction to Technical Writing," explains how technical writing is similar to and differs from other types of written communication. This chapter also discusses the purposes of technical writing, describes why technical writing is important in the workplace, and focuses on the role teamwork plays in today's corporate environment.

Chapter 2, "Producing the Product," explains why a three-step process—prewriting, writing, and rewriting—will improve your correspondence. We discuss the rationale for the writing process and supply an example of this process in practice. Note that proportions shown in margins of the example memos, letters, and reports are not actual, but adjusted to fit the space limitations of this book. Follow actual dimensions mentioned in Chapter 6 or your instructor's direction.

Chapter 3 provides a detailed discussion of the five major objectives in successful technical writing: clarity, conciseness, accuracy, organization, and ethics.

Chapter 4 explains why audience recognition and involvement are essential to effective technical writing. This chapter teaches you how to communicate with different audiences and focuses on the importance of defining your terms, accommodating multiculturalism, avoiding sexist language, and personalizing your writing.

Unit 2: Correspondence

How do you get a job? Once you have found employment, what types of writing will you need to do on the job? This unit helps you answer these questions. Chapter 5 explains how to write memos and differentiates this type of correspondence from letters and e-mail. Chapter 6 provides criteria and examples for many different types of letters.

Chapter 7, "The Job Search," explains how to look for jobs and how to structure the correspondence (letters of application, résumés, and follow-up letters) you will need to write during your job search.

Unit 3: Visual Appeal

Effective technical communication is visual as well as verbal. We clarify the relationship between words and graphics in this unit of our textbook. Chapter 8 explains the importance of document design. We discuss various ways to create an effective page layout. Chapter 9 expands on this topic by focusing on graphics, including tables and figures. Both chapters provide numerous examples to help you improve the visual appeal of your documents.

Unit 4: Electronic Communication

Technical writing has undergone dramatic changes in the last decade because of new technology. Chapter 10, "E-Mail, Online Help, and Web Sites," highlights the impact of electronic communication on writing in the workplace. In this chapter, we explain how the Internet, intranets, and extranets affect communication locally and globally. We provide criteria for and examples of successful e-mail messages, online help screens, and Web sites.

Unit 5: Technical Applications

Memos, letters, and e-mail are written daily in every business. These types of technical writing allow co-workers to "talk" to each other as well as to vendors, clients, supervisors, subordinates, board members, and stockholders.

Chapters 11 and 12 present different types of technical writing that usually accompany manufactured products and corporate services. Chapter 11 discusses how to write technical descriptions—specifications for mechanisms, tools, and/or pieces of equipment. Chapter 12 explains how to write the instructions and manuals that tell customers how to use or operate the equipment they've purchased. Both chapters include criteria, samples, and process logs to help you communicate effectively.

Unit 6: Report Strategies

As the title of this unit implies, it discusses a wide variety of reports. Chapter 13 provides information on research techniques, including suggestions for conducting and documenting online research. Chapter 14 teaches you how to write a summary.

Chapter 15 discusses the similarities and differences among trip reports, progress reports, lab reports, feasibility reports, and incident reports. Chapter 16 focuses on proposals. Finally, Chapter 17 discusses oral presentations, differentiating between types of oral reports and providing criteria for effective public speaking.

Unit 7: Handbook

Chapter 18 in the final unit lays out the conventions of grammar, punctuation, mechanics, and spelling—all of which are mandatory for successful technical writing.

NEW FEATURES

Many new features in this third edition help you navigate the textbook, write more effectively, and understand how electronic technology has affected technical communication.

Navigating the Textbook

- *Introductory overviews.* Each chapter begins with an overview of its content. Thus, when you begin a new chapter, you will have a road map to help you navigate the text.
- *Chapter highlights.* Each chapter ends with a list of its highlights. This list summarizes the chapter's content, reminding you of what you've read.
- *Overall organization.* As already noted, the book has been reorganized. In response to our reviewers' requests, we have moved the topics of memos, letters, and the job search nearer the beginning of the book. We have also moved the chapters on document design and graphics so that they follow the chapters on correspondence but precede those on electronic communication, technical applications, and reports. Our goal was to create more logical and useful groupings. These changes are further clarified through the unit headings.

Writing Effectively

- *Case studies.* Additional case studies give you a context for practicing the different types of technical communication.
- *New examples.* Throughout the text, we have added many examples to clarify what good writing entails.
- *Definitions.* In response to our reviewers' suggestions, we have added explanations of how and why to define acronyms, abbreviations, and high-tech terminology.
- *Usability testing.* Our new discussions and criteria checklists for usability testing will help you determine the effectiveness of your written instructions and user's manuals.

Electronic Communication

What defines technical writing at the beginning of the twenty-first century, more than any other characteristic, is technology. Many of the additions to our third edition reflect recent technological advances.

- *The Internet, intranets, and extranets.* We define these terms and explain how they affect technical communication.
- *E-mail, online help, and Web sites.* We provide examples, criteria, samples, and end-of-chapter activities to help you write effective electronic communication.
- *Online résumés.* In the past, you wrote paper-bound résumés. In the future, your résumés might be online. We explain how to write résumés in either ASCII or HTML format.

- *Online research.* Libraries still exist, but card catalogs are rare. Most research will be accomplished online. We show you how to navigate the World Wide Web for your research.
- *Internet graphics.* The World Wide Web gives you the opportunity to download countless graphics. We show you how to download and use images you find on the Web while avoiding plagiarism.

These new features make this third edition inclusive and up to date, providing students, instructors, technical writers, and business employees strategies for and models of effective technical communication.

We would like to thank the following reviewers for their contributions to this book: Lori Jo Oswald, Green River Community College; James G. Allen, College of Du Page; John R. Nelson, Jr., University of Massachusetts; Jan Coulson, Oklahoma State University-Okmulgee; Alexander Friedlander, Drexel University.

We would especially like to thank Maureen Richardson for her tireless efforts, patience, and creativity in helping us bring out this third edition.

Sharon J. Gerson
Steven M. Gerson

CHAPTER 1

OUTLINE

An Introduction to Technical Writing

DISTINCTIONS

What is technical writing? What makes it unique? These questions aren't easy to answer. Technical writing is diverse and includes many different types of correspondence, written by different types of people, in different professions, for different reasons. However, one way to distinguish technical writing is to compare it to other types of writing. Once we've done so, the definition of technical writing becomes clear.

There are many types of writing, including fiction, expressive writing, expository writing, persuasive writing, and technical writing.

Fiction includes poetry, short stories, plays, and novels. The authors might draw on experience to create their text, but the writing is imaginative. Creative writing often employs figurative word usage, imaginative imagery, dialect, symbolism, and fictitious characters to present a message.

Expressive writing records a subjective, emotional response to a personal experience (an incident, a place, a person, a phase, etc.). Journal and diary entries are expressive. The goal of expressive writing is to express one's feelings through description and narration. Students in composition classes often write expressive essays, drawing on narration and description to develop ideas.

Expository writing analyzes a topic objectively. Most essays in composition classes are expository. The goal of expository writing is to explain, and thereby reveal your knowledge of, a subject. Once you explain the topic, however, you don't expect a response from the reader.

Persuasive writing, in some ways, combines the emotionalism of expressive writing with the analytical traits of expository writing. Editorials are a good example of persuasive writing. The goal of persuasive writing is to sway your audience's emotional attitude toward a topic.

An audience is expected to read fiction, expressive essays, expository essays, and persuasive writing such as editorials. However, the writer and reader are often detached or disconnected, either by time or location. We read Shakespeare's fiction and Anne Frank's diary, for example, years after their conception. Teachers read student expository essays in an office, hours or days after the essays have been composed. When we read magazines or newspaper editorials and other persuasive texts, we're in a setting or time distant from the writer. Although the writers might have had a particular audience in mind while composing and might have directed comments toward these readers' needs, the writer and the reader are not involved in a dialogue. The writer expresses an idea, and no action is required; no input is expected; no follow-up is mandated. Writers of fiction, expressive writers, expository writers, and even persuasive writers say, "Here's how I feel," "Here's what I think," or "Here's what I believe." It's up to the reader to do with the information as he or she pleases.

Technical writing is different. Technical writing requires give-and-take, a dialogue, a follow-up and/or input and action. When you write a memo, for example, you expect a reaction. Maybe you want a suggestion, a job completed, or problems solved. If you write a letter, you expect to make a sale or to receive an answer to an inquiry. When you write instructions, you know that someone will follow the steps you've outlined. In fictional writing as well as expressive, expository, and persuasive writing, the authors often write in a vacuum. They do not necessarily receive any immediate feedback on their writing from their readers. When you write technical correspondence, however, the detachment between reader and writer, often evident in the other types of writing, does not exist. Instead, the purpose of technical writing is to link you and your boss, you and clients, you and vendors, and you and co-workers. Technical writing creates action. When you write successful technical correspondence, someone on the other end responds.

PURPOSES

Now that we know how technical writing differs from other types of writing, the next question is, "Why do we write technical correspondence?" Technical writing can accomplish many purposes.

For example, automotive technicians document mechanical problems, electronic engineering technicians write maintenance procedures, and computer technicians write instructions for user manuals. However, technical writing is not written just by technicians. Technical writing is also produced by investment counselors who document assets and by marketing specialists who promote a service or product. Businesspeople report on job-related travel, weekly sales calls, meeting minutes, and quality circle activities. Some technical writing documents; some technical writing sells.

The purpose of technical writing is often determined by audience, which then affects the tone of the correspondence. In fact, understanding the interrelationship among *purpose, audience,* and *tone* is essential to answering the question, "Why do we write technical correspondence?"

Let's say that you're an engineer writing a monthly status report. You write this report every month. Your report always goes to the same person, your immediate supervisor. The supervisor reads it and then puts the report in a file for future reference. With this ongoing activity in mind, you know that the purpose of your report is to document. Your goal is to "dump data" and nothing else. You also know that your audience is highly technical, fluent in technical terms and abbreviations. Given your purpose and your audience's level of understanding, you write a report which has a dry, objective, impersonal tone. Thus, the *purpose* of your report and your sense of *audience* create *tone.*

Status Report
PURPOSE (document) + AUDIENCE (technical supervisor)
= TONE (objective/impersonal)

This is a common technical writing situation, but it's not the only environment in which technical writing is generated. Let's say that you're writing a memo, and your purpose is to get action. That sounds simple enough, doesn't it? Writing, however, is never that easy. Who is your audience and what tone is required? Are you writing up or down? If you're writing up to a supervisor or manager (audience), then you're writing to *request* action, *propose* action, or *recommend* action (purpose). If you're writing down to a subordinate (audience), then you're writing to *direct* action (purpose). Once you recognize audience and reassess your purpose, your tone is affected. The tone for a request, proposal, or recommendation written up to a manager is positive, congenial, and polite. The tone for correspondence written down might be more authoritative.

In summary, your purpose in technical writing is determined by your audience's needs. This sense of audience then affects tone. Within these parameters, the purposes of technical writing are wide-ranging.

IMPORTANCE

We've explained what technical writing is and what purposes it serves. The final question is, "Is it important?" The answer is yes.

Technical writing is a significant factor of your work experience for several reasons.

1. Technical writing conducts business. Technical writing is not a frill or an occasional endeavor. It is a major component of the work environment. Through technical correspondence, employees (a) maintain good customer-client relations (follow-up letters), (b) ensure that work is accomplished on time (directive memos), (c) provide documentation that work has been completed (status reports), and (d) generate income (sales letters). Technical writing also (a) keeps machinery working (maintenance instructions), (b) ensures that the correct equipment is purchased (technical descriptions and specifications), (c) gives bosses the information they need for persuasive briefings (summaries), (d) gets you a job (résumés), and (e) informs the world about your company's product (sites on the World Wide Web).

2. *Technical writing takes time.* In addition to serving a valuable purpose in a company, technical writing is important because it is time-consuming. On average, employees spend approximately 20 percent of their work time writing memos, letters, or reports. Twenty percent of a workweek equals one full eight-hour day. You might as well say that every Thursday, for example, you'll just write—nothing else. Of course, that's not how it works. In one week, you spend five minutes writing here, 30 minutes there, an hour here, but it all adds up to a significant amount of time.

Twenty percent is a base figure. Corporate supervisors with whom we've worked say that they not only spend their time writing, but they also spend additional time reviewing and revising the writing composed by their subordinates.

3. *Technical writing costs money.* Time is money. If you spend 20 percent of your work time writing, then your company is paying 20 percent of your salary for your writing skills. Twenty percent of a 40-hour workweeks equals 8 hours—one workday. Imagine how your boss would evaluate your work if you called in sick every Wednesday—you'd be looking for a new job. If you write poorly, yet you spend 20 percent of your workweek doing substandard work, wasting 8 hours per week writing flawed correspondence, then your boss has a right to be concerned. Your time spent writing is part of your salary and part of your boss's work expenditures.

Today a letter costs about $12 to $15 to produce (paper, postage, and your time). In the government sector, this price is tripled, since chains of command require that additional readers sign the correspondence.

Good writing has additional monetary values other than your salary or the cost of correspondence. If you write a clear directive, the reader can get right on the job. A confusing memo, however, wastes time. The reader must ponder the meaning and finally ask you for further explanation. The lost time is unproductive and costs the company. Similarly, if you write a good proposal, you generate income. If the proposal is flawed and unacceptable, your company loses money. Writing isn't just part of your salary; good writing helps pay your wages.

4. *Technical writing is an extension of your interpersonal communication skills.* When you write a memo, letter, or report, you're not just conveying technical information. You're revealing something about yourself to your readers. If you write well, you're telling your audience that you can think logically and communicate your thoughts clearly. If, on the other hand, you write poorly, you give your readers a completely different picture of yourself as a worker. You reveal that you can neither

think clearly nor communicate your thoughts effectively. Technical writing is an extension of your interpersonal communication skills at work, and co-workers will judge your competence based on the effectiveness of your correspondence.

For example, a man with a master's degree in criminal justice enrolled in a technical writing class to improve his writing skills. When asked why, he said, "My supervisors are dissatisfied with my writing. In fact, they've given me ninety days to improve my communication abilities. If I can't, then I'm out of a job." On the other hand, a former student got a job in aviation as a technician. One day he wrote a problem/solution report. The report was well received and was passed around the company as an example of good writing. Management was so impressed with his writing skills that he was promoted to a supervisory level where his communication skills could benefit the company.

Good technical writing can accomplish more than just getting the job done. A well-constructed memo, letter, or report reveals to your readers not only that you know your technical field of expertise but also that you know how to communicate your knowledge thoroughly, accurately, clearly, and concisely. Through good technical writing, you reveal to your audience that you can tell people what to do and can motivate them to do it.

TEAMWORK

BUSINESS AND INDUSTRY EXPECTATIONS

Employers require teamwork from their prospective new hires. Employers aren't just asking for engineering, architectural, accounting, electronic, drafting, and/or telecommunication skills. These degree program objectives are a given. Any student who graduates with a degree from such programs has expertise, presumably, in his or her discipline. Here are the bigger questions employers ask: "What else can you offer us?" "Why should we hire you versus someone else with the same degree?" "How can you help our company?" The successful job candidate must provide more than degree-program expertise. That's where teamwork becomes important.

TOTAL QUALITY MANAGEMENT AND CONTINUOUS QUALITY IMPROVEMENT

In today's workplace, employees have new responsibilities. They aren't required to practice merely their primary job responsibilities—to be good accountants, engineers, architects, or biomedical technicians. They must also participate in maintaining their corporations' essential goal—quality. Thus, most companies stress total quality management (TQM) or continuous quality improvement (CQI). These corporate philosophies stress that companies waste time and money fixing yesterday's problems. The way to stay ahead of the competition and to maintain customer satisfaction is to produce a quality product. This is best achieved by anticipating problems and avoiding them before they occur. Companies have concluded that employees from different areas of expertise, working together in teams, communicating their multidimensional desires for success, can best achieve corporate quality.

TEAM PROJECTS

In school and in business, often you will write in teams. Businesses depend on project teams to ensure quality. Teamwork, a primary tool of the TQM philosophy, helps a company save money by having numerous employees work together on corporate projects.

In fact, teamwork is perceived as so essential that many undergraduate and graduate program directors have mandated team projects in their curriculum. One such director confirmed the importance of teamwork, after having surveyed a cross section of business and industry: " 'The results revealed that more than technical skills, employees need skills that allow them to interact and relate effectively with others.' " Another program director went further, asserting that " 'people don't lose jobs because they don't know how to do things. It's because they fail at relationships' " (Bacon).

Many instruction manuals, proposals, and Web sites are team written. Usually, technical writers work with engineers and graphic artists, as well as with corporate employees in legal, delivery, production, sales, and management. These collaborative team projects even extend beyond the company. Technical writers also work in collaboration with subcontractors and subconsultants from other corporations. The collaborative efforts include communicating with companies in other cities and countries through teleconferences and e-mail. Rarely is a technical writer solely responsible for a work-related activity. TQM and CQI require teamwork—people working together to achieve corporate success.

STRATEGIES FOR SUCCESSFUL COLLABORATION

To successfully collaborate, follow this TQM strategy.

- *Develop your team.* Who will work together? You have two choices. Either select team members by discipline (engineers with engineers, accountants with accountants, automotive technicians with automotive technicians, computer specialists with computer specialists, and so on), or choose diverse team members with different skills (teams composed of an engineer, an accountant, a graphic artist, a marketing specialist, etc.). The first team profile draws strength from individuals with like skills; the second team benefits from multiple perspectives.

- *Choose a team leader.* Either your teacher or your boss selects a team leader, or your group might choose the person best suited to this role. Once this person is chosen, then the delegation of authority begins. Your team leader might do the following:

  1. *Assign duties.* Who will research the material, who will write the various parts of the text, who will interview people, and/or who will proofread and revise?

  2. *Create schedules (project milestones).* When will the group meet, when will research and writing be due, when are revisions required, and/or when is the finished project due?

  3. *Encourage group participation.* How will conflicts among group members be resolved, and/or how will equal participation be achieved?

- *Determine your team goal.* Once the teams are formed, brainstorm an area needing improvement, a corporate goal. Decide what project to work on. For example, maybe the group decides to improve a company's work environment, to increase sales, or to improve a company's benefits package.

- *Identify the problem.* If the group decides that some area needs improvement, then there must be a problem. The key to improvement is identifying the problems impeding the group goal. Perhaps corporate profits are sagging because machines don't work well. Maybe people are uncomfortable, dissatisfied, or overworked. Where's the problem? To improve a company's work environment, the group first must identify the problems which diminish the environment.

- *Analyze the problem's causes.* Once the group has identified problems impeding a goal, next discover the causes of these problems. For example, let's say the group goal is to improve a company's work environment. The team has identified two problems: machine malfunctions and employee dissatisfaction. What's causing these problems? Perhaps machines are malfunctioning due to static electricity in the air, excessively hot and humid room temperatures, or outdated equipment. Regarding personnel dissatisfaction, perhaps the employees have difficulties with child care, distant parking, benefits packages, or ergonomics like heating, cooling, and office space. The team members can't propose solutions to problems until they analyze the problems' causes.

- *Determine potential improvements.* Now it's time to propose the solutions to the problems. Focus on how these potential improvements will be achieved (strategies, timetables, milestones, personnel involved, costs incurred, facilities impacted) and how the company will benefit (improved sales, better employee morale, lower cost of business, satisfied customer base).

- *Verify the suggested solutions.* The team must verify whether the suggested solutions will work. This might require research, interviews, test cases, scenarios, trial runs, or computer modeling.

- *Write the text.* With the above legwork accomplished, the team collaboratively writes the correspondence. This entails revising and printing the finished copy.

Any group activity is challenging to control. Team members don't show up for class; one student monopolizes the activity while another student snoozes; students exert varying amounts of enthusiasm and ability; student personalities clash. Varying levels of involvement and group dynamics can be a headache. However, the benefits outweigh the deficits.

Team projects allow students to help and to learn from each other. The team projects also require that students plan; delegate responsibilities; communicate their attitudes toward word choice, tone, organization, development, and grammar; and learn to work with other people. Often students say that it's easier just to do it themselves rather than confer and compromise. They're right; it is easier. But it's unrealistic. The workplace requires team skills. Employers want workers who can work together, not those who must learn to interact. Teamwork and collaborative writing require communication to solve problems. These are valuable skills that will benefit you in your workplace.

CHAPTER HIGHLIGHTS

1. Technical writing is different from imaginative fiction.
2. Technical writing is like expository writing because both involve analysis.
3. Technical writing is like persuasive writing because the technical writer often tries to convince a reader to act.
4. Technical writers must consider the audience, the tone, and the purpose of every document they write.
5. Technical writing is an important part of many people's everyday work life. It can take up as much as 20 percent of a typical workweek.
6. Technical writing costs a company both time and money, so employees must strive to write well.
7. Your writing at work reflects your abilities to communicate effectively.
8. Employees often write in teams on the job.
9. An effective team leader will facilitate group work.
10. Following a specific total quality management (TQM) strategy will help your team succeed.

ACTIVITIES

1. Visit one or more companies in your career field and interview employees to see how much time they spend writing, what types of documents they write, and how often they work with others in project teams.
2. Invite guest speakers from corporations to discuss their writing activities at work.
3. Research the Internet for technical writing job opportunities to determine what skills are required, what types of writing are performed, and what types of industries employ technical writers.
4. Visit the Society for Technical Communication (STC) Web site to learn about its membership. See which industries employ technical writers, and determine these writers' job responsibilities. Also, learn which colleges and universities have programs in technical writing and what the programs entail. What else can you learn about technical writing from the STC Web site?
5. Research major publications of technical communication, such as *Intercom, Technical Communication*, and *The Journal of Scientific and Technical Communication*. What topics do the articles in these journals focus on?

CHAPTER

2

Producing the Product

You now know what technical writing is and why it's important. You know that technical writing is a major part of your daily work experience, that it takes time to construct the correspondence, and that your writing has an impact on those around you. A well-written memo, letter, report, or e-mail message gets the job done and makes you look good. Poorly written correspondence wastes time and creates a negative image of you and your company.

But recognizing the importance of technical writing does not ensure that your correspondence will be well written. How do you effectively write the memo, letter, or report? How do you successfully produce the finished product?

For some people, producing effective correspondence simply involves getting out a piece of paper or turning on a computer and leaping in. Good things just miraculously occur for them. This is *not* the case for most writers. For most of us, when we get out that piece of paper or turn on the computer, one of two things happens. Either we stare at the blank page or screen until beads of sweat form on our foreheads and we break out in hives because of the dreaded blank page syndrome, or we wander about aimlessly for several pages. In the first instance, we suffer from writer's block, which prevents us from placing even one word on the page or screen. In the second situation, we fill up the page or screen with words, but we never convey a logical thought.

What we need is a solution to these problems. The writer who suffers from blank page syndrome needs help generating information. The writer who wanders aimlessly needs help organizing, formatting, and revising the content. Each of these writers needs a methodical, step-by-step, sequential, easy-to-follow technique for writing effective correspondence.

Professional writers and writing teachers have developed just such a technique. To produce successful technical writing, you need to approach writing as a process.

THE WRITING PROCESS: AN OVERVIEW

The process approach to writing requires the following sequence:

1. Prewrite. Before you can write your technical document, you must have something to say. Prewriting allows you to spend quality time, *prior* to writing the correspondence, generating information. In prewriting, you (a) *determine objectives* (the motivation, rationale, or purpose for writing), (b) *gather data* (the content for your correspondence), and (c) *consider audience* (who will read your correspondence).

2. Write. Once you've gathered your data and determined your objectives, the next step is to state them. You need to draft your document. To do so, you should (a) *organize* the draft according to some logical sequence which your readers can follow easily and (b) *format* the content to allow for ease of access.

3. Rewrite. The final step, and one that is essential to successful writing, is to rewrite your draft. This step requires that you *revise* the rough draft. Revision allows you to perfect your memo, letter, or report so you can be proud of your final product.

To write successfully, you should subdivide your writing activity into these three steps. Doing so will lessen the anxiety caused by writing. Instead of feeling overwhelmed by the task at hand, you'll be able to approach any writing situation one step at a time.

THE RATIONALE FOR PROCESS

Subdivide your writing time as shown in Table 2.1. If you spend eight hours per week writing, you should allocate approximately two of those hours for *prewriting*. Those two hours would not fall in one block, of course. You'd spend five minutes here, thirty minutes there, depending on the scope of your current writing project. Two more hours would be spent drafting the document in the *writing* stage of the process. Finally, you'd spend four hours in the course of your work week *rewriting*.

Why employ this process and abide by our time allocations? Is the sequence really necessary? Why not just leap in and hope for the best? We believe writing as a process is important for the following reasons.

TABLE **2.1** The Writing Process		
PREWRITING (25%)	**WRITING (25%)**	**REWRITING (50%)**

• Determine objectives
• Gather data
• Recognize audience
	• Organize
	• Format
		• Revise (45%)
		    Add
		    Delete
		    Simplify
		    Move
		    Reformat
		    Enhance
		    Correct
		• Proofread (5%)

PREWRITING

Prewriting allows you to *plan* your technical writing. If you don't know where you're going in the correspondence, you'll never get there, and your reader will not get there with you. Through prewriting, you can *determine your objectives.* Before you write the document, you need to know why you are writing and what you hope to achieve. Is your motivation *external* or *internal*? *External motivation* means that someone else has requested the correspondence. Your boss, for example, expects you to write a monthly status report, a performance appraisal of your subordinate, or a memo suggesting solutions to a current problem. Perhaps the writing has been requested by a vendor (a letter documenting need dates), a customer (a letter responding to a complaint), or a potential client (a proposal of services to be rendered).

Internal motivation means that you have decided to write on your own accord. You need information to perform your job more effectively, so you write a letter of inquiry. You need to meet with colleagues to plan a job, so you write an e-mail message calling a meeting and setting an agenda. Perhaps you recognize a problem in your work environment, so you create a questionnaire and transmit it via the company intranet. Then, analyzing your findings, you write a report documenting the problem and asking for help from your boss.

Recognizing the motivation for your correspondence makes a difference. This allows you to provide the appropriate tone and scope of detail in your writing. In contrast, failure to assess motivation can cause communication breakdowns.

What do you hope to achieve in your writing? What is your goal, your objective, your purpose? When you write, maybe you're only trying to document (to place information on the page for filing purposes). However, this assumption is too limited. Documentation is not the only purpose for writing. Are you merely trying to inform (FYI correspondence—"This is just for your information"), or do you want to achieve more than this? Do you want to sell or motivate? In Chapter 1, we noted that a monthly status report might provide toneless documentation. However, a motivational e-mail message must be persuasive, a Web site must create a friendly link between the company and the customer, and a sales letter must sell.

To write effectively, you must understand your objectives. If you don't know why you are writing, how will your reader grasp your intentions?

Once you know why you are writing, the next step is deciding what to say. You have to *gather data*. The page remains blank until you fill it with data. Your correspondence, therefore, will consider personnel, dates, actions required, locations, costs, methods for implementing suggestions, and so forth. As the writer, it's your obligation to flesh out the detail, because until you tell your readers what you want to tell them, they don't know. You can't communicate your content until you've gathered your data.

The final step in prewriting is to *consider your audience*. How many people will read your correspondence? What are their levels of understanding? Before you can determine what to say and how much detail is necessary, you need to answer these questions. (We discuss audience thoroughly in Chapter 4.)

By prewriting, you plan your correspondence. You provide a road map for yourself (your objectives), you ensure that your writing is developed (the data), and you write according to your reader's needs (audience). Prewriting helps you write effectively—proper planning prevents poor performance.

Prewriting Techniques

How do you plan your correspondence? How do you prewrite to determine objectives and gather data? Is there one technique which works in all situations? Many prewriting techniques help writers tackle different types of technical correspondence. A primary function of this text is to provide you with these various techniques and show you how they work best with different types of writing activities. Some of these prewriting techniques include the following:

- *Answering the reporter's questions.* By answering *who, what, when, where, why,* and *how,* you create the content of your correspondence.
- *Mind mapping.* Envision a wheel. At the center is your topic. Radiating from this center, like spokes of the wheel, are different ideas about the topic. Mind mapping allows you to look at your topic from multiple perspectives.
- *Brainstorming/listing.* Performing either individually or with a group, you can randomly suggest ideas (brainstorming) and then make a list of these suggestions.
- *Branching.* This helps you discern the components of a topic. Your main topic is like the trunk of a tree. Each component of the topic represents a separate branch.
- *Flowcharting.* By graphically depicting the steps in a procedure, you can ensure that a chronological organization is maintained and that no steps are omitted.
- *Outlining.* This traditional method of gathering and organizing information allows you to break a topic into major and minor components.

- *Storyboarding*. This pictorial sketch of each page or screen lets you see what your document will look like.

Each of these methods will not appeal to everyone and will not work for every type of correspondence. However, by becoming familiar with them, you'll be able to approach any type of writing assignment with more confidence. No matter what type of writing you must perform, no matter what profession you're in, and no matter what your company's standards are, you'll have prewriting techniques at your command that will help you overcome the blank page syndrome.

WRITING

Writing lets you *package* your data. Once you've gathered your data, determined your objectives, and recognized your audience, the next step is writing the document. You need to package it (the draft) in such a way that your readers can follow your train of thought readily and can easily access your data. Writing the draft lets you *organize* your thoughts in some logical, easy-to-follow sequence. Writers usually know where they're going, but readers don't have this same insight. When readers pick up your document, they can read only one line at a time. They know what you're saying at the moment, but they don't know what your goals are. They can only hope that in your writing, you'll lead them along logically and not get them lost in back alleys of unnecessary data or dead-end arguments.

To avoid leading your readers astray, you need to organize your thoughts. As with prewriting, you have many organizational options. In Chapter 3, we discuss organizing according to (a) space (spatial organization), (b) chronology, (c) importance, (d) comparison/contrast, and (e) problem/solution.

These organizational methods are not exclusive. Many of them can be used simultaneously within a memo, letter, or report to help your reader follow your train of thought.

You also must *format* your text to allow for ease of access. In addition to organizing your ideas, you need to consider how the text looks on the page. If you give your readers a massive wall of words, they'll file your document for future reading and look for the nearest exit. An unbroken page of text is not reader-friendly. To invite your readers into the document, to make them want to read the memo, letter, or report, you need to highlight key points and break up monotonous-looking text. You need to ensure that your information is accessible. (See Chapter 8 for more on formatting.)

REWRITING

Rewriting lets you *perfect* your writing. After you've prewritten (to overcome the blank page syndrome) and written your draft, your final step is to rewrite. There aren't any good writers, only good rewriters. People who write effective documents know that doing so requires a second or third write. Good writers fine-tune, hone, sculpt, and polish their drafts to make sure that their final versions are perfect. To rewrite, you need to *revise*, *revise*, and *revise* again. Revision requires that you look over your draft and do the following:

- *Add* any missing detail for clarity.
- *Delete* dead words and phrases for conciseness.
- *Simplify* unnecessarily complex words and phrases to allow for easier understanding.
- *Move* information around (cut and paste) to ensure that your most important ideas are emphasized.
- *Reformat* (using highlighting techniques) to ensure reader-friendly ease of access.
- *Enhance* the tone and style of the text.
- *Correct* any errors to ensure accurate grammar and content.

We discuss each of these points in greater detail throughout the text.

Revision is possibly the most important stage in the writing process. If you prewrite effectively (gathering your data, determining your objectives, and recognizing your audience) and write an effective draft, you're off to a great start. However, if you then fail to rewrite your text, you run the risk of having wasted the time you spend prewriting and writing. Rewriting is the stage in which you make sure that everything is just right. Failure to do so not only can cause confusion for your readers but can also destroy your credibility.

The process approach to writing, when adhered to, can help you write successfully—in any work environment or writing situation. In fact, the greatest benefit of process is that it's generic. Process is not geared to any one profession or type of correspondence. No author of a technical writing book can anticipate exactly where you'll work, what type of document you'll be required to write, or what your supervisors will expect in your writing. However, we *can* give you a methodology for tackling any writing activity. Writing as a process will help you write any kind of technical document, for any boss, in any work situation.

THE PROCESS IN PRACTICE

Following is a letter produced using the process approach to writing. The document was produced in the workplace by a senior transportation analyst for an international cosmetics firm. He had to write a problem/solution follow-up letter to a sales representative.

PREWRITING

A senior transportation analyst received a phone call from a disgruntled sales representative. The sales rep had not received a shipment of goods on time, and the shipment was incomplete. While talking to the sales rep, the analyst jotted down notes as shown in Figure 2.1 (using the listing method of prewriting).

In addition to listing, the transportation analyst used another prewriting technique—reporter's questions. The note tells us *who* the sales rep is (Beth); *what* her Social Security number, phone number, and sales area are; *what* her problem is (late and missing goods); *where* the shipment originated (Denver); *how* much was

365-6532

Beth Fox
449-87-7247
Milwaukee

1. two weeks ago

2. June $300 short

3. Troy $700 ordered

4. Denver

5. split order

FIGURE 2.1 Listing

ordered ($700); and *when* the shipment was due (two weeks ago). By jotting down this list, the analyst is gathering data.

After concluding his discussion with Beth, the analyst contacted his manager to decide what to do next. This time, the analyst wrote down a list of objectives, as determined by his manager, as shown in Figure 2.2.

The list again answers the reporter's questions: *what* to do (write a letter), *who* gets a copy (manager), *what* to focus on in the letter (we understand your problem; here's an alternative), and *why* to pursue the alternative (better control of shipment).

With data gathered and objectives determined, the analyst was ready to write.

WRITING

First, the analyst wrote a rough draft (revising it as he wrote), as shown in Figure 2.3. In this draft, the analyst made subtle changes by adding new detail and delet-

Send letter to sales rep.

Send copy to manager.

In letter

    discuss problem encountered.

    show alternative method of shipment
    for better control.

Call manager for further help if needed.

FIGURE 2.2 Listing Objectives

Dear Beth —

I appreciate you notifying me of the delay

in delivery of your order. Your orders

are coded to ship via Allied shipping.

                              for Milwaukee
Allied's stated shipping level is

             As you've noticed,
next day. However, because of Allied's

          because of
sorting system, the way Allied ships

                              a multiple
packages, it is possible for an

  carton   To avoid
  order to become split. Alternate

this problem, an alternative   a
delivery is possible through an

delivery service we use in your

district. To do this we need an

      delivery
alternative address. Please

let me know should you need additional

      have any questions or
If you need additional information,

please feel free to contact me.

FIGURE 2.3 Rough Draft

ing unnecessary words. However, he was unsatisfied with this draft, so he tried again (see Figure 2.4).

As is evident from the first two drafts (Figures 2.3 and 2.4), the analyst took the word *rough* seriously. When you draft, don't worry about errors or how the correspondence looks. It's meant to be rough, to free you from worry about making errors. You can correct errors when you revise.

Once the analyst drafted the letter, he typed a clean copy for his manager's approval (Figure 2.5). At this point, the manager added a dateline and added content

~~Thank you for your letter regarding~~

~~the split deliveries.~~

Thank you for letting me know about the

split deliveries of your campaign 19

last week
orders. Our talk ~~lets me~~ gives me

an opportunity not only to expalin

the situation ~~problem~~ situation but also

¶
to offer help. Here's the way ~~the~~

Allied   Allied will deliver to your home.
situation works. Allied sorts packages
However,
individually rather than as a group.

That is, even though we send your

packages
~~orders~~ to them as a unit, ~~they~~

all under your name, they, however,

by complete
load their trucks not ~~according to~~

cartons
order but just as individual ~~boxes.~~

Because of this,
~~Thus, because some,~~ occassionally

one carton ends up ~~one~~ on one truck

carton
while the other is shipped

separately. You ~~then~~ received

such a split order.

FIGURE 2.4 Second Rough Draft

Mrs. Beth Fox
6078 Browntree
Milwaukee, WI 53131

Dear Mrs. Fox:

Thanks for letting me know about the split delivery of your Campaign 19 order. Our talk last week gives me an opportunity not only to explain the situation but also to offer help.

Here's the way Allied Shipping works. Allied will deliver to your home; however, Allied sorts packages individually rather than as a group. That is, even though we send your packages as a unit (all under your name), they load their trucks *not by complete order, but just as individual cartons.* Because of this, occasionally, one carton ends up on one truck with another carton shipped separately. You received such a split order. This is an inherent flaw in their system.

Because we understand this problem, we have an alternative delivery service for you. Here is our option. Free of charge, you can have your order delivered by our delivery agent, who does not split orders. Our agent, however, will only deliver within a designated area. All we need from you is an alternative address of a friend or relative in the designated delivery area.

I realize that neither of these options is perfect. Still, I wanted to share them with you. Your district manager now can help you decide which option is best for you.

Sincerely,

David L. Porter
Senior Transportation Analyst

FIGURE 2.5 Third Draft

Mrs. Beth Fox
6078 Browntree
Milwaukee, WI 53131

Dateline?

Dear Mrs. Fox:

Thanks for letting me know about the split delivery of your Campaign 19 order. Our talk last week gives me an opportunity not only to explain the situation but also to offer help.

Here's the way Allied Shipping works. Allied will deliver to your home; however, Allied sorts packages individually rather than as a group. That is, even though we send your packages as a unit (all under your name), they load their trucks *not by complete order, but just as individual cartons*. Because of this, occasionally, one carton ends up on one truck with another carton shipped separately. You received such a split order. This is an inherent flaw in their system.

However, Allied constantly works with us to eliminate these service failures.

whose system has better control of orders

Because we understand this problem, we have an alternative delivery service for you. Here is our option. Free of charge, you can have your order delivered by our delivery agent, who does not split orders. Our agent, however, will only deliver within a designated area. All we need from you is an alternate address of a friend or relative in the designated delivery area. *If you would like this service, we would need an alternative delivery address in Milwaukee.*

I realize that neither of these options is perfect. Still, I wanted to share them with you. Your district manager now can help you decide which option is best for you.

Should you be unable to establish a different delivery address, we will still work with Allied to ensure that you receive home delivery of your complete orders.

Sincerely,

David L. Porter
Senior Transportation Analyst

FIGURE 2.6 Revised Draft

CAREFREE COSMETICS
83rd and Preen
Kansas City, MO 64141

September 21, 2000

Mrs. Beth Fox
6078 Browntree
Milwaukee, WI 53131

Great Job, David

Dear Mrs. Fox:

Thanks for letting me know about the split delivery of your Campaign 19 order. Our talk last week gives me an opportunity not only to explain the situation but also to offer help.

Allied will deliver to your home; however, Allied sorts packages individually rather than as a group. Even though we send your packages as a unit (all under your name), they load their trucks not by complete order but by individual cartons. Because of this, occasionally one carton ends up on one truck with another carton shipped separately. You received such a split order. This is an inherent flaw in their system; however, Allied constantly works with us to eliminate these service failures.

Because we understand this problem, we have an alternative delivery service for you. Free of charge, you can have your order delivered by our delivery agent, whose system has better control of orders. Our agent, however, only delivers within a designated area. If you would like this service, we would need an alternate delivery address in Milwaukee.

Should you be unable to establish a different delivery address, we will still work with Allied to ensure that you receive home delivery of your complete orders.

Sincerely,

David L. Porter

David L. Porter
Senior Transportation Analyst
pc: R. H. Handley

FIGURE 2.7 Finished Letter

to the second paragraph. He deleted wordiness in the second and third paragraphs. By deleting the entire fourth paragraph, the manager enhanced the tone of the document (see Figure 2.6).

REWRITING

No writing is ever perfect. Every memo, letter, or report can be improved. Note how the manager improved the analyst's typed draft. When the senior transportation analyst received the revised letter from his manager, he typed and mailed the final version (Figure 2.7).

Once the manager received his copy, he wrote the note you see in the letter's top right corner. When you approach writing as a step-by-step process (prewriting, writing, and rewriting), your results usually are positive—and you will receive positive feedback from your supervisors.

Each company you work for over the course of your career will have its own unique approach to writing memos, letters, and reports. Your employers will want you to do it their way. Company requirements vary. Different jobs and fields of employment require different types of correspondence. However, you'll succeed in tackling any writing task if you have a consistent approach to writing. A process approach to writing will allow you to write any correspondence effectively.

CHAPTER HIGHLIGHTS

1. Writing effectively is a challenge for many people. Following the process approach to writing will help you meet this challenge.

2. Prewriting helps you determine your objectives, understand your audience, and gather your data.

3. Prewriting techniques will help you get started. Try answering reporter's questions, mind mapping, brainstorming, branching, flowcharting, outlining, and storyboarding.

4. To begin writing a rough draft, organize your thoughts.

5. Perfect your text by rewriting: adding, deleting, simplifying, moving, reformatting, enhancing, and correcting your documents.

ACTIVITIES

1. To practice prewriting, take one of the following topics. Then, using the suggested prewriting technique, gather data.

   a. *Reporter's questions*. To gather data for your resume, list answers to the reporter's questions for two recent jobs you've held and for your past and present educational experiences.

   b. *Mind mapping*. Create a mind map for your options for obtaining college financial aid.

   c. *Brainstorming/listing*. List five reasons why you have selected your degree program or why you have chosen the school you are attending.

     d. *Branching*. Using branching, categorize the various facets of your life (such as school, work, family, etc.).

     e. *Flowcharting*. Create a flowchart of the steps you followed to register for classes, buy a car, or seek employment.

     f. *Outlining*. Outline your reasons for liking or disliking a current or previous job.

     g. *Storyboarding*. If you have a personal Web site, use storyboarding to graphically depict the various screens. If you don't have such a site, use storyboarding to graphically depict what your site's screens would include.

2. Using the techniques illustrated in this chapter, edit, correct, and rewrite the following flawed memo.

Date: April 3, 2000
To: William Huddleston
From: Julie Schopper
Subject: TRAINING CLASSES

Bill, our recent training budget has increased beyond our projections. We need to solve this problem. My project team has come up with several suggestions, you need to review these and then get back to us with your input. Here's what we have come up with.

We could reduce the number of training classes, fire several trainers, but increase the number of participants allowed per class. Thus we would keep the same amount of income from participants but save a significant amount of money due to the reduction of trainer salaries and benefits. The downside might be less effective training, once the trainer to participant ratio is increased. As another option, we could outsource our training. This way we could fire all our trainers which would mean that we would save money on benefits and salaries, as well as offer the same number of training sessions, which would keep our trainer to participant ratio low.

What do you think. We need your feedback before we can do anything so even if your busy, get on this right away. Please write me as soon as you can.

---

3. Choose the best prewriting techniques for each of the following communication needs, and then justify your choice.

   a. You are planning an oral presentation about benefits and insurance options for newly hired personnel at your company.

   b. You must prepare a short proposal for marketing a new product, highlighting your product's purpose or the problems it will solve, the costs incurred, any new personnel required, facilities needed, and the benefits derived.

   c. You are writing a recommendation report for the purchase of computer printers. You must first decide on criteria, with the help of your project team.

   d. You are writing an instruction for testing electronic equipment on your company's recently purchased voltmeter.

   e. You are writing an incident report about an accident in the company's boiler room, which must include the date, location, time, personnel involved, causes, and financial ramifications.

Objectives in Technical Writing

CLARITY

The ultimate goal of good technical writing is clarity. If you write a memo, letter, or report that is unclear to your readers, which your readers can't understand, then what have you accomplished? You've wasted time. First, your readers don't understand your point or can't follow your train of thought. They must write you a follow-up inquiry to determine your needs. This wastes *their* time. Once you receive the inquiry, you must rewrite your correspondence, trying to clarify your initial intentions. You've now written twice to accomplish the same goal. This wastes *your* time.

To avoid these time-consuming endeavors, write for clarity. But how do you do this?

PROVIDE SPECIFIC DETAIL

One way to achieve clarity is by supplying specific, quantified information. If you write using vague, abstract adjectives or adverbs, such as *some* or *recently*, your readers will interpret these words in different ways. The adverb *recently* will mean thirty minutes ago to one reader, yesterday to another, and last week to a third reader. This adverb, therefore, is not clear. The same applies to an adjective like *some*. You write, "I need some information about the budget." Your readers can only guess at what you mean by *some*. Do you want the desired budget increase for 2000, the budget expenditures for 1998, the allotted budget increase for 1999, the guidelines for implementing a budget increase, the budgeted allotment for travel, or the explanation for the budget decrease for training?

Look at the following example of vague writing caused by imprecise, unclear adjectives. (Vague words are underlined.)

ACTIVITY REPORT DRAFT

Our <u>latest</u> attempt at molding preform protectors has led to <u>some</u> positive results. We spent <u>several</u> hours in Dept. 15 trying different machine settings and techniques. <u>Several</u> good parts were molded using two different sheet thicknesses. Here's a summary of the findings.

First, we tried the <u>thick</u> sheet material. At 240°F, this thickness worked well.

Next, we tried the <u>thinner</u> sheet material. The <u>thinner</u> material is less forgiving, but after a <u>few</u> adjustments we were making good parts. Still, the <u>thin</u> material caused the most handling problems.

The engineer who wrote this report realized that it was unclear. To solve the problem, she rewrote the report, quantifying the vague adjectives.

ACTIVITY REPORT REVISION

During the week of 10/4/00, we spent approximately 12 hours in Dept. 15 trying different machine settings, techniques, and thicknesses to mold preform mold protectors. Here's a report on our findings.

.030″ Thick Sheet

At 240°F, this thickness worked well.

0.15″ Thick Sheet

This material is less forgiving, but after decreasing the heat to 200°F, we could produce good parts. Still, material at .015″ causes handling problems.

Your goal as a technical writer is to communicate clearly. To do so, state your exact meaning through specific, quantified word usage.

ANSWER THE REPORTER'S QUESTIONS

A second way to write clearly is to answer reporter's questions—who, what, when, where, why, and how. The best way we can emphasize the importance of answering these reporter's questions is by sharing with you the following memo, written by a highly placed executive to a newly hired employee.

DATE: November 11, 2000
TO: Mary Jane Post
FROM: Don Goldenbaum
SUBJECT: TECHNICIAN MEETING

Please be prepared to plan a presentation on month-end reports.
Please be sure that your explanations are very detailed. Thanks.

That's the entire memo. The questions are, "What doesn't the newly hired employee know?" "What additional information would that employee need to do the job?" "What needs clarifying?"

Simply, the employee needs answers to reporter's questions. For example, *what* is the subject of the presentation? If the answer is month-end reports, we still lack clarity. *Which* of the 12 month-end reports does Don want Mary Jane to discuss? *Who* is the audience? We read the word *technician* in the subject line, but is Don focusing on automotive technicians, electronic engineering technicians, or dental hygiene technicians? *Why* is this presentation being made? That is, what's the rationale or motivations for the meeting? *When* will the presentation be made, and *how* much detail is "very detailed"? *Where* will Mary Jane and the technicians meet for the presentation? Mary Jane has a right to ask for clarity on one more point: *what* exactly is she supposed to do? Is she being asked to make a presentation, plan a presentation, or merely prepare to plan a presentation?

This memo's lack of clarity, a result of its inability to answer reporter's questions, causes Mary Jane stress, anxiety, and tension. As a new hire, she wants to do her job well, but Don's unclear correspondence hinders her. You must write clearly to avoid causing your reader stress and to help your reader do the job effectively.

In contrast, the memo on the following page achieves clarity by answering reporter's questions.

Becky knows *where* the meetings will be held (Conference Room C); *who* selected her for the committee (her manager); *what* she will accomplish (writing a proposal about changes to the CIA system); *when* the meetings will occur (Friday, Monday, and Wednesday); *why* she's involved (to assess and revise a current system); and *how* the work will be accomplished (the sequence of work during the three days of meetings).

USE EASILY UNDERSTANDABLE WORDS

Another key to clarity is using words which your readers can understand easily. Avoid *obscure words*, and be careful when you use *acronyms*, *abbreviations*, and *jargon*.

Avoid Obscure Words

A good rule of thumb is to *write to express, not to impress; write to communicate, not to confuse.* If your reader must use a dictionary, you're not writing clearly.

DATE: September 5, 2000
TO: Becky Stapleton
FROM: Dave Woodring
SUBJECT: CASH IN ADVANCE (CIA) PROCEDURE TASK FORCE

You have been chosen by your manager to be a member of the CIA Task Force. This committee will assess current CIA procedures and revise the system.

The following is your schedule for Task Force activities:

Friday, September 12: 8:30–10:30 a.m.
   Assess the current system and brainstorm new ideas.

Monday, September 15: 8:30–10:30 a.m.
   Review the suggested changes, add new ideas, and test the adjusted system.

Wednesday, September 17: 8:30–10:30 a.m.
   Review the adjusted system and write a proposal confirming your changes.

All meetings will be held in Conference Room C. Thank you for your involvement. If you have any questions, please call me at ext. 1849.

Try to make sense of the following examples of unclear writing.

The following rules are to be used when determining whether or not to duplicate messages:

- Do not duplicate nonduplicatable messages.
- A message is considered nonduplicatable if it has already been duplicated.

Your job duties will be to assure that distributed application modifications will execute without abnormal termination through the creation of production JCL system testing.

It's hard to believe, but these examples were written by businesspeople who were trying to communicate *something*. However, the examples are filled with old-fashioned words. The words are too difficult to understand.

Following is a list of difficult, out-of-date terms and the modern alternatives.

Obscure Words	Easy Words
aforementioned	already discussed
initial	first
in lieu of	instead of
accede	agree
as per your request	as you requested
issuance	send
this is to advise you	I'd like you to know
subsequent	later
inasmuch as	because
ascertain	find out
pursuant to	after
forward	mail
cognizant	know
endeavor	try
remittance	pay
disclose	show
attached herewith	attached
pertain to	about
supersede	replace
obtain	get

Impressive writing is correspondence we can understand easily. A modern thrust in technical writing is to write the way you speak ... unless you speak poorly. Try to be casual, almost conversational.

Using Acronyms, Abbreviations, and Jargon

In addition to obscure words, a similar obstacle to readers is created by acronyms, abbreviations, and jargon.

We've all become familiar with common acronyms such as *scuba* (self-contained underwater breathing apparatus), *radar* (radio detecting and ranging), *NASA* (National Aeronautics Space Agency), *FICA* (Federal Insurance Contributions Act), and *MADD* (Mothers Against Drunk Driving)—single words created from the first letters of multiple words. We're comfortable with abbreviations like *FBI* (Federal Bureau of Investigations), *JFK* (John F. Kennedy), *NFL* (National Football League), *IBM* (International Business Machines), and *L.A.* (Los Angeles). Some jargon (in-house language) has become so common that we reject it as a cliché. Baseball jargon is a good example. It's hard to tolerate sportscasters who speak baseball jargon, describing line drives as "frozen ropes" and fast balls as "heaters."

However, more often than not, acronyms, abbreviations, and jargon cause problems, not because they are too common but because no one understands them. Your technical writing loses clarity if you depend on them. You might think your readers understand them, but do they?

Try to guess the meaning of the following abbreviations and jargon.

- *CIA.* Your first guess was Central Intelligence Agency, wasn't it? Wrong! By this abbreviation we mean *cash in advance.*
- *Spaghetti.* Your first guess made you envision a romantic dinner at a candlelit pasta house, didn't it? Wrong! We are using fire prevention jargon. For firefighters, spaghetti is a term for various fire hoses and lines.
- *FIFO.* If you guessed that this is a natural, all-bran cereal, you're wrong, but don't worry. This is a confusing acronym, except for business office personnel, who immediately recognize it as meaning "first in, first out."

You have to decide when to use acronyms, abbreviations, and jargon and how to use them effectively. One simple rule is to define your terms. You can do so either *parenthetically* or in a *glossary.* Rather than just writing *CIA*, write *CIA (Cash in Advance).* Such parenthetical definitions, which are only used once per correspondence, don't take a lot of time and won't offend your readers. Instead, the result will be clarity.

If you use many potentially confusing acronyms or abbreviations, or if you need to use a great deal of technical jargon, then parenthetical definitions might be too cumbersome. In this case, supply a separate glossary. (This point is discussed in greater detail in Chapter 16.) A *glossary* is an alphabetized list of terms, followed by their definitions, as in the following example.

CPA	Certified Public Accountant
FICA	Federal Insurance Contributions Act (Social Security taxes)
Franchise	Official establishment of a corporation's existence
Gross pay	Pay before deductions
Line of credit	Amount of money which can be borrowed
Net pay	Pay after all deductions
Profit and loss statement	Report showing all incomes and expenses for a specified time

USE VERBS IN THE ACTIVE VOICE VERSUS THE PASSIVE VOICE

It has been decided that Joan Smith will head our Metrology Department.

The preceding sentence is written in the *passive voice* (the primary focus of the sentence, *Joan Smith,* is acted on rather than initiating the action). Passive voice causes two problems.

1. Passive constructions are often unclear. After reading the preceding sentence, our question is, *who* decided that Joan Smith will head the department? To solve this problem and to achieve clarity, replace the vague indefinite pronoun *it* with a precise noun: "Kin Norman decided that Joan Smith will head our Metrology Department."

2. Passive constructions are often wordy. Passive sentences always require helping verbs (such as *has been*). When we revise the sentence to read "Kin Norman decided that Joan Smith will head our Metrology Department," the helping verb *has been* disappears.

The revision ("Kin Norman decided that Joan Smith will head our Metrology Department") is written in the *active voice*. When you use the active voice, your subject (*Kin Norman*) initiates the action.

Another common problem with passive voice construction concerns prepositions. Look at the following example.

Overtime is favored by hourly workers.   (passive voice)

Again, this sentence, written in the passive voice, uses a helping verb (*is*), has the doer (*hourly workers*) acted on rather than initiating the action, and also includes a preposition *(by)*. The sentence is wordy. Revised, the sentence reads as follows:

Hourly workers favor overtime.   (active voice)

We've omitted the helping verb *is*; we've deleted the preposition *by*; and the subject, *Hourly workers*, initiates the action. The sentence is less wordy and more precise.

Occasionally, it's appropriate to use the passive voice. When an individual is less important than an inanimate object, passive voice is appropriate. For example:

The new software was oversold by the salesperson who guaranteed ease of use.   (The individual, *salesperson*, is not as important as the inanimate object, *software*.)

You might also use passive voice when the individual is unknown.

The tax preparation software can be learned easily even when the user is new to our CPA firm.   (The individual, *user*, is unnamed.)

CONCISENESS

After clarity, your second major goal in technical writing is conciseness. Conciseness is important for at least two reasons.

First, remember how time-consuming technical writing is in the work environment? American workers spend approximately eight hours per week writing and additional time reading and revising others' writing. Conciseness in writing can help save some of this time. If you write concisely, providing thorough detail in fewer words, you can save yourself time and take up less of your readers' time.

Second, concise writing can aid comprehension. If you dump an enormous number of words on your readers, they might give up before finishing your correspondence or skip and skim so much that they miss a key concept. Wordy writing will lead your readers to think, "Oh no! I'll never be able to finish that. Maybe I can skim through it. I'll probably get enough information that way." Conciseness, on the other hand, makes your writing more appealing to your readers. They'll think, "Oh, that's not too bad; I can read it easily." If they can read your correspondence easily, they will read it with greater interest and involvement. This, of course, will aid their comprehension.

Let's look at some poor writing—writing which is wordy, time-consuming to read, and not easily comprehensible.

> Please prepare to supply a readout of your findings and recommendations to the officer of the Southwest Group at the completion of your study period. As we discussed, the undertaking of this project implies no currently known incidences of impropriety in the Southwest Group, nor is it designed specifically to find any. Rather, it is to assure ourselves of sufficient caution, control, and impartiality when dealing with an area laden with such potential vulnerability. I am confident that we will be better served as a company as a result of this effort.

Is that paragraph easy to understand? No, it's not. Why? What gets in your way? Do you have difficulty following it because you are an outsider and are not aware of the situation that generated it? That's only part of the problem. The reason you have difficulty understanding this paragraph is because it's poorly written. It causes difficulty for two reasons: (a) the paragraph is too long, and (b) the words and sentences in the paragraph are too long.

LIMIT PARAGRAPH LENGTH

The number of lines or words in a paragraph is arbitrary. Obviously, as the writer, you must decide what's best. Some paragraphs, due to the complexity of the subject matter, might require more development. Other paragraphs requiring less development can be shorter.

Nonetheless, an excessively long paragraph is ineffective. In a long paragraph, you force your reader to wade through many words and digest large amounts of information. This hinders comprehension. In contrast, short, manageable paragraphs invite reading and help your reader understand your content.

As a rule of thumb, a paragraph in a technical document should consist of (a) no more than four to six typed lines, or (b) no more than 50 words. Sometimes you can accomplish these goals by cutting your paragraphs in half; find a logical place to stop a paragraph and then start a new one. Even the foregoing difficult example can be improved in this way.

Please prepare to supply a readout of your findings and recommendations to the officer of the Southwest Group at the completion of your study period. As we discussed, the undertaking of this project implies no currently known incidences of impropriety in the Southwest Group, nor is it designed specifically to find any.

Rather, it is to assure ourselves of sufficient caution, control, and impartiality when dealing with an area laden with such potential vulnerability. I am confident that we will be better served as a company as a result of this effort.

The writing is still difficult to understand, but at least it's a bit more manageable. You can read the first paragraph, stop, and consider its implications. Then, once you have grasped its intent, you can read the next paragraph and try to tackle its content. The paragraph break gives you some room to breathe.

LIMIT WORD AND SENTENCE LENGTH

In addition to the length of the example paragraph, the writing is flawed because the paragraph is filled with excessively long words and sentences. This writer has created an impenetrable wall of haze—the writing is foggy. In fact, we can determine how foggy this prose is by assessing it according to Robert Gunning's fog index. Here's his mathematical way of determining how foggy your writing is.

The Fog Index

1. Count the number of words in successive sentences. Once you reach approximately 100 words, divide these words by the number of sentences. This will give you an average number of words per sentence.

2. Now count the number of long words within the sentences you've just reviewed. Long words are those with three or more syllables. You can't count (a) proper names, like Leonardo DaVinci, Christopher Columbus, or Alexander DeToqueville; (b) long words that are created by combining shorter words, such as *chairperson* or *firefighter*; or (c) three-syllable verbs created by *ed* or *es* endings, such as *united* or *arranges*. Discounting these exceptions, count the remaining multisyllabic words. (A good example of a multisyllabic word is the word *mul-ti-syl-lab-ic*.)

3. Finally, to determine the fog index, add the number of words per sentence and the number of long words. Then multiply your total by 0.4.

Given this system, let's see how the original difficult paragraph (page 31) scores. The paragraph is composed of 92 words in 4 sentences. Thus, the average number of words per sentence is 23. The paragraph contains 16 multisyllabic words (*recommendations, officer, completion, period, undertaking, currently, incidences, impropriety, specifically, sufficient, impartiality, area, potential, vulnerability, confident,* and *company*).

$$
\begin{array}{rl}
23 & \text{(words per sentence)} \\
+ 16 & \text{(multisyllabic words)} \\
\hline
39 & \text{(total)}
\end{array}
$$

$$
\begin{array}{r r}
39 & \text{(total)} \\
\times\ .4 & \text{(fog factor)} \\
\hline
15.6 & \text{(fog index)}
\end{array}
$$

What does a fog index of 15.6 mean? Look at Table 3.1. It shows that the paragraph is written at a level midway between college junior and senior, definitely above the danger line.

Why is this level of writing considered dangerous? A fog index of 15.6 enters the danger zone for two reasons.

* Approximately 22 percent of Americans graduate from college. Thus, if you're writing at a college level, you could be alienating approximately 78 percent of your audience.

* Studies show that college graduates read at approximately a tenth-grade level. Thus, even if you're writing to college graduates, you can't assume college-level reading skills.

Given these facts, many businesses ask their employees to write at a sixth- to eighth-grade level. To accomplish this, you would have to strive for an average of approximately 15 words per sentence and no more than 5 multisyllabic words per 100 words.

$$
\begin{array}{r l}
15 & \text{(words per sentence)} \\
+\ 5 & \text{(multisyllabic words)} \\
\hline
20 & \text{(total)}
\end{array}
$$

$$
\begin{array}{r l}
20 & \text{(total)} \\
\times\ .4 & \text{(fog factor)} \\
\hline
8.0 & \text{(fog index level)}
\end{array}
$$

TABLE 3.1   Fog Index and Reading Level

	FOG INDEX	BY GRADE	BY MAGAZINE
	17	College graduate	No popular magazine
	16	College senior	scores this high.
	15	College junior	
	14	College sophomore	
Danger Line	13	College freshman	
	12	High school senior	*Atlantic Monthly*
	11	High school junior	*Time* and *Newsweek*
	10	High school sophomore	*Reader's Digest*
	9	High school freshman	*Good Housekeeping*
	8	Eighth grade	*Ladies' Home Journal*
	7	Seventh grade	Modern romances
	6	Sixth grade	Comics

You can't always avoid multisyllabic words. Scientists would find it impossible to write if they could never use words like *electromagnetism*, *nitroglycerine*, *telemetry*, or *trinitrotolulene*. The purpose of a fog index is to make you aware that long words and sentences create reading problems. Therefore, although you can't always avoid long words, you should be careful when using them. Similarly, you can't always avoid lengthy sentences. However, try not to rely on sentences over 15 words long. Vary your sentence lengths, relying mostly on sentences under 15 words long.

Here are ways to lower a potentially high fog index.

Use the Meat Cleaver Method of Revision

One way to limit the number of words per sentence is to cut the sentence in half or thirds. The following sentence, which contains 44 words, is too long.

> VERSION 1
>
> To maintain proper stock balances of respirators and canister elements and to ensure the identification of physical limitations which may negate an individual's previous fit-test, a GBC-16 Respirator Request and Issue Record will need to be submitted for each respirator requested for use.

If we use the meat cleaver approach, we can make this sentence more concise and easier to understand.

> VERSION 2
>
> Please submit a GBC-16 Respirator Request and Issue Record for each requested respirator. We then can maintain proper respirator and canister element stock balances. We also can identify physical limitations which may negate an individual's previous fit-test.

This sentence, now cut in thirds, is more digestible. Because we have less to swallow whole, we can understand the content more easily.

Avoid Shun Words

In the preceding examples, the original sentence contained 44 words: the revised version is composed of 3 sentences totaling 38 words. Where did the missing 6 words go?

One way to write more concisely is to shun words ending in *-tion* or *-sion*—words ending in a *shun* sound. For example, the original sentence reads "to ensure the identification of physical limitations." To revise this, we simply wrote "identify physical limitations." That's 3 words versus 6 in the original version. Deleting 3 words in this case reduces wordiness by 50 percent. Shun words are almost always unnecessarily wordy.

Let's try another example. Instead of writing "I want you to take into consideration the following," you could write "consider the following." That's 9 words versus 3, a 66.6 percent savings in word count.

Look at the following shun words and their concise versions.

Shun Word	Concise Version
came to the conclu*sion*	concluded (or decided)
with the excep*tion* of	except for
make revi*sions*	revise
investiga*tion* of the	investigate
consider implementa*tion*	implement
utiliza*tion* of	use

Avoid Camouflaged Words

Camouflaged words are similar to shun words. In both instances, a key word is buried in the middle of surrounding words (usually helper verbs or unneeded prepositions). For example, in the phrase *with the exception of*, the key word *except* is camouflaged behind the unneeded *with*, *the*, *-tion*, and *of*. Once we prune away these unneeded words, the key word *except* is left, making the sentence less wordy.

Camouflaged words are common. Here are some examples and their concise versions.

Camouflaged Word	Concise Version
make an *amend*ment to	amend
make an *adjust*ment of	adjust
have a *meet*ing	meet
*thank*ing *you* in advance	thank you
for the purpose of *discuss*ing	discuss
arrive at an *agree*ment	agree
at a *later* moment	later

Avoid the Expletive Pattern

Another way to write more concisely is to avoid the following expletives.

- *there* is, are, was, were, will be
- *it* is, was

Both these expletives *(there* and *it)* lead to wordy sentences. For example, consider the following sentence.

There are three people who will work for Acme.

This sentence can be revised to read, "Three people will work for Acme." The original sentence contains nine words; the revision has six. We've omitted three words by deleting the expletive *there*. Your response to this revision could be "So what, who cares? What's the point of deleting three words?" Deleting three words doesn't seem like much. However, the omission equals a 33 percent savings in word count, helping us achieve conciseness. In one sentence, that might be a minimal achievement, but if you can delete three words from every sentence, the benefits will add up.

The expletive *it* creates similar wordiness, as in the following sentence.

It has been decided that ten engineers will be hired.

If we delete the expletive *it*, the sentence reads, "Ten engineers will be hired." The original sentence contained ten words; the revision has five. We've achieved a 50 percent savings in word count.

Omit Redundancies

Redundancies are words that say the same thing. Conciseness is achieved by saying something once rather than twice. For example, in each of the following instances, the boldface words are redundant.

during **the year of** 2000
(2000 is obviously a year; the words *the year of* are redundant.)

in **the month of** December
(As in the preceding example, *the month of* is redundant; what else is December?)

needless to say
(If it's needless to say, why say it?)

The computer will cost **the sum of** $1,000.
(One thousand dollars *is* a sum.)

the results **so far achieved** prove
(A result, by definition, is something which has been achieved.)

our **regular** monthly status reports require
(Monthly status reports must occur every month; regularity is a prerequisite.)

We collaborated **together** on the project.
(One can't collaborate alone!)

the **other** alternative is to
(Every alternative presumes that some other option exists.)

This is a **new** innovation.
(As opposed to an old innovation?!)

the consensus **of opinion** is to
(The word *consensus* implies opinion.)

Avoid Wordy Phrases

Sentences may be wordy not because you've been redundant or because you've used shun words, camouflaged words, or expletives. Sometimes sentences are wordy simply because you've used wordy phrases.

Here are examples of wordy phrases and their concise revisions.

Wordy Phrase	Concise Revision
in order to purchase	to buy
at a rapid rate	fast (or state the exact speed)
it is evident that	evidently
with regard to	about
in the first place	first
a great number of times	often (or state the number of times)
despite the fact that	although
is of the opinion that	thinks
due to the fact that	because
am in receipt of	received
enclosed please find	enclosed is
as soon as possible	by 11:30 A.M.
in accordance with	according to
in the near future	soon
at this present writing	now
in the likely event that	if
rendered completely inoperative	broken

ACCURACY

Clarity and conciseness are primary objectives of effective technical writing. However, if your writing is clear and concise but incorrect—grammatically or textually—then you've wasted your time and destroyed your credibility. To be effective, your technical writing must be *accurate*.

Accuracy in technical writing requires that you *proofread* your text. The examples of inaccurate technical writing on pages 37-38 are caused by poor proofreading (we've underlined the errors to highlight them).

First City Federal Savings and Loan
1223 Main
Oak Park, Montana

October 12, 2000

Mr. and Mrs. David Harper
2447 N. Purdom
Oak Park, Montana

Dear Mr. and Mrs. <u>Purdom</u>:

Note that the savings and loan incorrectly typed the customer's street rather than the last name.

National Bank
1800 Commerce Street
Houston, TX

September 9, 2000

Adler's Dog and <u>Oat</u> Shop
8893 Southside
Bellaire, TX

Dear <u>Sr.</u>:
In response to your request, your account with us has been <u>close</u> out. We <u>are</u> <u>submitted</u> a check in the amount of $468.72 (your existing balance). If you have any questions, please <u>fill</u> free to <u>conact</u> us.

In addition to all the other errors, it should be "Dog and <u>Cat</u> Shop," of course. All the errors make the writer look incompetent.

To ensure accurate writing, use the following proofreading tips.

1. *Let someone else read it*. We miss errors in our own writing for two reasons. First, we make the error because we don't know any better. Second, we read what we think we wrote, not what we actually wrote. Another reader might help you catch errors.

2. *Use the gestation approach*. Let your correspondence sit for a while. Then, when you read it, you'll be more objective.

3. *Read backwards*. You can't do this for content. You should only read backwards to slow yourself down and to focus on one word at a time to catch typographical errors.

4. *Read one line at a time*. Use a ruler or scroll down your PC to isolate one line of text. Again, this slows you down for proofing.

5. *Read long words syllable by syllable*. How is the word *responsiblity* misspelled? You can catch this error if you read it one syllable at a time (re-spon-si-bl-i-ty).

6. *Use technology*. Computer spell checks are useful for catching most errors. They might miss proper names, homonyms (*their*, *they're*, or *there*) or incorrectly used words such as *device* to mean *devise*.

7. *Check figures, scientific and technical equations, and abbreviations*. If you mean $400,000, don't write $40,000. Double-check any number or calculations. If you mean to say *HCl* (hydrochloric acid), don't write *HC* (a hydrocarbon).

8. *Read it out loud.* Sometimes we can hear errors that we can't see. For example, we know that *a outline* is incorrect. It just sounds wrong. *An outline* sounds better and is correct.

9. *Try scattershot proofing.* Let your eyes roam around the page at random. Sometimes errors look wrong at a glance. If you wander around the page randomly reading, you often can isolate an error just by stumbling on it.

10. *Use a dictionary.* If you're uncertain, look it up.

If you commit errors in your technical writing, your readers will think one of two things about you and your company: (a) they'll conclude that you are stupid, or (b) they'll think that you're lazy. In either situation, you lose. Errors create a negative impression at best; at worst, a typographical error relaying false figures, calculations, amounts, equations, or scientific/medical data can be disastrous.

ORGANIZATION

If you're clear, concise, and accurate, but no one can follow your train of thought because your text rambles, you still haven't communicated effectively. Successful technical writing must also be well organized.

Here's an analogy to explain the importance of organization. Most artists can't just dip a brush in paint and then splatter that paint on canvas. People want to make sense of what they see, and splattered images cause confusion. The same applies to technical writing. As the writer, you can't haphazardly throw words on the page and expect readers to understand you clearly. In contrast, you should order that information on the page logically, allowing your readers to follow your train of thought.

No one method of organization always works. Following are five patterns of organization which you can use to help clarify content.

SPATIAL

If you are writing to describe the parts of a machine or a plot of ground, you might want to organize your text spatially. You would describe what you see as it appears in space—left to right, top to bottom, inside to outside, or clockwise. These spatial sequences help your readers visualize what you see and, therefore, better understand the physical qualities of the subject matter. They can envision the layout of the land you describe or the placement of each component within the machine.

For example, let's say you are a contractor describing how you will refinish a basement. Your text reads as follows:

> At the basement's north wall, I will build a window seat 7' long by 2' wide by 2' high. To the right of this seat, on the east wall, I will build a desk 4' high by 5' long by 3' wide. On the south wall, to the left of the door, I will build an entertainment unit the height of the wall including four, 4' high by 4' wide by 2' deep shelving compartments. The west wall will contain no built-ins. You can use this space to display pictures and to place furniture.

Note how this text is written clockwise, uses points of the compass to orient the reader, and includes the transitional phrases "to the right" and "to the left" to help the reader visualize what you will build. That's spatial organization.

CHRONOLOGICAL

Whereas you would use spatial organization to describe a place, you would use chronology to document time or the steps in an instruction. For example, an emergency medical technician reporting services provided during an emergency call would document those activities chronologically.

> At 1:15 P.M., we arrived at the site and assessed the patient's condition, taking vitals (pulse, respiration, etc.). At 1:17 P.M., after stabilizing the patient, we contacted the hospital and relayed the vitals. By 1:20 P.M., the patient was on an IV drip and en route to the hospital. Our vehicle arrived at the hospital at 1:35 P.M. and hospital staff took over the patient's care.

Chronology also would be used to document steps in an instruction. No times would be provided as in the EMT report. In contrast, the numbered steps would denote the chronological sequence a reader must follow.

IMPORTANCE

Your page of text is like real estate. Certain areas of the page are more important than others—location, location, location. If you bury key data on the bottom of a page, your reader might not see the information. In contrast, content placed approximately one-third from the top of the page and two-thirds from the bottom (eye level) garners more attention. The same applies to a bulletized list of points. Readers will focus their attention on the first several points more than on the last few.

Knowing this, you can decide which ideas you want to emphasize and then place that information on the page accordingly. Organize your ideas by importance. Place the more important ideas above the less important ones.

The following agenda is incorrectly organized.

Agenda

- Miscellaneous ideas
- Questions from the audience
- Refreshments
- Location, date, and time
- Subject matter
- Guest speakers

Which of these points is most important? Certainly not the first two. A better list would be organized by importance, as follows:

Agenda

- Subject matter
- Guest speakers

- Location, date, and time
- Refreshments
- Questions from the audience
- Miscellaneous ideas

COMPARISON/CONTRAST

Many times in business you will need to document options and/or ways in which you surpass a competitor. These require that you organize your text by comparison/contrast. You compare similarities and contrast differences. For example, if you are writing a sales brochure, you might want to present your potential client alternatives regarding services, personnel, timetables, and fee structures. Table 3.2 shows how comparison/contrast provides the client options for cost and features. Each housing option is comparable in that the developer provides a 4-bedroom, 3½-bath home with fully equipped kitchen. However, the homes contrast regarding garage size, deck availability, and basement. These options then affect the cost. The table, organized according to comparison/contrast, helps the reader understand these distinctions.

PROBLEM/SOLUTION

Every proposal and sales letter is problem/solution-oriented. When you write a proposal, for instance, you are proposing a solution to an existing problem. If your proposal focuses on new facilities, your reader's current building must be flawed. If your proposal focuses on new procedures, your reader's current approach to doing business must need improvement. Similarly, if your sales letter promotes a new product, your customers will purchase it only if their current product is inferior.

TABLE 3.2 Housing Costs		
ITEM	**FEATURES**	**COSTS**
The Broadmoor	4 bedrooms, 3½ baths 2-car garage fully equipped kitchen	$180,000
The Aspen	4 bedrooms, 3½ baths finished basement 3-car garage fully equipped kitchen	$190,000
The Regency	4 bedrooms, 3½ baths patio deck finished basement with 1/2 bath 3-car garage fully equipped kitchen	$200,000

To clarify the value of your product or service, therefore, you should emphasize the readers' need (their problem) and show how your product is the solution. Note how the following summary from a proposal is organized according to problem/solution.

SUMMARY

Your city's 20-year-old wastewater treatment plant does not meet EPA requirements for toxic waste removal or ozone depletion regulations. This endangers your community and lessens property values in its neighborhoods.

Anderson and Sons Engineering Company has a national reputation for upgrading wastewater treatment plants. Our staff of qualified engineers will work in partnership with your city's planning commission to modernize your facilities and protect your community's values.

The summary's first paragraph identifies the problem. The second paragraph promotes the solution. The problem/solution organization clarifies the writer's intent.

ETHICS

Here's the scenario. You're a technical writer responsible for producing a maintenance manual. Your boss tells you to include the following sentence.

NOTE: Our product has been tested for defects and safety by trained technicians.

When read literally, this sentence is true. The product has been tested, and the technicians are trained. However, you know that the product has been tested for only 24 hours by technicians trained on site without knowledge of international regulations.

So where's the problem? As a good employee, you're required to write what your boss told you . . . aren't you? Even though the statement isn't completely true, you legally can include it in your manual . . . can't you?

The answer to both questions is no! Actually, you have an ethical responsibility to write the truth. Your customers expect it, and it's in the best interests of your company. Of equal importance, including the sentence in your manual is illegal. Though the sentence is essentially true, it implies something that is false. Readers will assume that the product has been *thoroughly* tested by technicians who have been *correctly* trained. Thus, the sentence deceives the readers. Such comments are "actionable under law" if they lead to false impressions (Wilson WE-68). If you fail to properly disclose information, including dangers, warnings, cautions, or notes like the sentence in question, then your company is legally liable. (We discuss how to correctly write hazard alert messages in Chapter 12.)

Knowing this, however, doesn't make writing easy. Ethical dilemmas exist in corporations. The question is, what should you do when confronted with such problems?"

One way to solve this dilemma is by checking your actions against these three concerns: legal, practical, and ethical. For example, if you plan to write operating instructions for a mechanism, will your text be

1. *Legal*, focusing on liability, negligence, and consumer protection laws?
2. *Practical*, since dishonest technical writing backfires and can cause the company to lose sales or to suffer legal expenses?
3. *Ethical*, written to promote customer welfare and avoid deceiving the end user? (Bremer et al. 76–77)

These are not necessarily three separate issues. Each interacts with the other. Our laws are based on ethics and practical applications.

LEGALITIES

If you're uncertain, that's what lawyers are for. When asked to write text that profits the company but deceives the customer, for instance, you might question where your loyalties lie. After all, the boss pays the bills, but your customers might also be your next-door neighbors. Such conflicts exist and challenge all employees. What do you do? You should trust your instincts and trust the laws. Suffice it to say that laws are written to protect the customer, the company, and you—the employee. If you believe you are being asked to do something illegal that will harm your community, seek legal counsel.

PRACTICALITIES

Though it might appear to be in the best interests of the company to hide potentially damaging information from customers, such is not the case. First, as technical writer, your goal is candor. That means that you must be truthful, stating the facts. It also means that you must not lie, keeping silent about facts that are potentially dangerous (Girill 178–79). Second, practically speaking, the best business approach is *good* business. The ultimate goal of a company is not just making a profit, but making money the right way—"good ethics is good business" (Guy 9). What good is it to earn money from a customer who will never buy from you again, or who will sue for reparation? That's not practical.

ETHICALITIES

Here's the ultimate dilemma. Defining ethical standards has been challenging for professional organizations. As Shirley A. Anderson-Hancock, manager for the Society for Technical Communication (STC) Ethical Guidelines Committee, states, "Everyone who participates in discussions of ethical issues may offer a different perspective, and many viewpoints may be legitimate" (6). Due to the difficulty of clearly defining what is and what is not ethical, the STC struggled for three years before publishing its guidelines in 1995 (see Figure 3.1).

These ethical guidelines, as the STC Ethical Guidelines Committee admits, had to be "sufficiently broad to apply to different working environments for the next several years" (Anderson-Hancock 6).

STC Ethical Guidelines for Technical Communicators

Introduction

As technical communicators, we observe the following ethical guidelines in our professional activities. Their purpose is to help us maintain ethical practices.

Legality

We observe the laws and regulations governing our professional activities in the workplace. We meet the terms and obligations of projects we undertake. We ensure that all terms of our contractual agreements are consistent with the STC Ethical Guidelines.

Honesty

We seek to promote the public good in our activities. To the best of our ability, we provide truthful and accurate communications. We dedicate ourselves to conciseness, clarity, coherence, and creativity, striving to address the needs of those who use our products. We alert our clients and employers when we believe material is ambiguous. Before using another person's work, we obtain permission. In cases where individuals are credited, we attribute authorship only to those who have made an original, substantive contribution. We do not perform work outside our job scope during hours compensated by clients or employers, except with their permission; nor do we use their facilities, equipment, or supplies for personal gain. When we advertise our services, we do so truthfully.

Confidentiality

Respecting the confidentiality of our clients, employers, and professional organizations, we disclose business-sensitive information only with their consent or when legally required. We acquire releases from clients and employers before including their business-sensitive information in our portfolios or before using such material for a different client or employer or for demonstration purposes.

Quality

With the goal of producing high quality work, we negotiate realistic, candid agreements on the schedule, budget, and deliverables with clients and employers in the initial project planning stage. When working on the project, we fulfill our negotiated roles in a timely, responsible manner and meet the stated expectations.

Fairness

We respect cultural variety and other aspects or diversity in our clients, employers, development teams, and audiences. We serve the business interests of our clients and employers, as long as such loyalty does not require us to violate the public good. We avoid conflicts of interest in the fulfillment of our professional responsibilities and activities. If we are aware of a conflict of interest, we disclose it to those concerned and obtain their approval before proceeding.

Professionalism

We seek candid evaluations of our performance from clients and employers. We also provide candid evaluations of communication products and services. We advance the technical communication profession through our integrity, standards, and performance.

(Approved by the STC Board of Directors, April 1995)

FIGURE 3.1

One way to clarify these necessarily broad standards is by looking at the STC Code for Communicators, which reads as follows:

> As a technical communicator, I am the bridge between those who create ideas and those who use them. Because I recognize that the quality of my services directly affects how well ideas are understood, *I am committed to excellence in performance and the highest standards of ethical behavior.*
>
> I value the worth of the ideas I am transmitting and the cost of developing and communicating ideas. I also value the time and effort spent by those who read or see or hear my communication.
>
> I therefore recognize my responsibility to communicate technical information truthfully, clearly, and economically.
>
> My commitment to professional excellence and ethical behavior means that I will
>
> - Use language and visuals with precision.
> - Prefer simple, direct expression of ideas.
> - Satisfy the audience's need for information, not my own need for self-expression.
> - Hold myself responsible for how well my audience understands my message.
> - Respect the work of colleagues, knowing that a communication problem may have more than one solution.
> - Strive continually to improve my professional competence.
> - Promote a climate that encourages the exercise of professional judgment and that attracts talented individuals to careers in technical communication.

Let's expand on each of these points. In doing so, we can achieve a working guide for ethical standards.

1. Use language and visuals with precision. In a recent survey comparing technical writers and teachers of technical writing, we discovered an amazing finding; professional technical writers rate grammar and mechanics higher than teachers do (Gerson and Gerson). On a 5-point scale (5 equaling "very important"), writers rated grammar and mechanics 4.67, whereas teachers rated grammar and mechanics only 3.54. That equals a difference of 1.13, which represents a 22.6 percent divergence of opinion.

Given these numbers, would we be precise in writing "Teachers do not take grammar and mechanics as seriously as writers do"? The numbers accurately depict a difference of opinion, and the 22.6 percent divergence is substantial. However, these figures do not assert that teachers ignore grammar. To say so is imprecise and would constitute an ethical failure to present data accurately. Even though writers are expected to highlight their client's values and downplay their client's shortcomings, technical writers ethically cannot skew numbers to accomplish these goals (Bowman and Walzer). Information must be presented accurately, and writers must use language precisely.

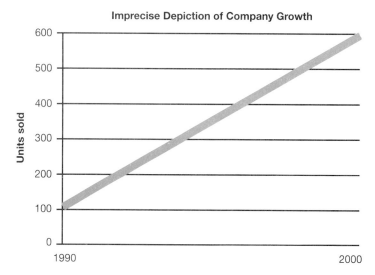

FIGURE 3.2

Precision also is required when you use visuals to convey information. Look at Figures 3.2 and 3.3. Both figures show that XYZ's sales have risen. As with language, the writer is ethically responsible for presenting visual information precisely.

2. Prefer simple, direct expression of ideas. While writing an instructional manual, you might be asked to use legal language to help your company avoid legal problems (Bowman and Walzer). Here's a typically obtuse warranty that legally protects your company:

> Acme's liability for damages from any cause whatsoever, including fundamental breach, arising out of this Statement of Limited Warranty, or for any other claim related to this product, shall be limited to the greater of $10,000 or the amount paid for this product at the time of the original purchase, and shall not apply to claims for personal injury or damages to personal property caused by Acme's negligence, and in no event shall Acme be liable for any damages caused by your failure to perform your responsibilities under this Statement of Limited Warranty, or for loss of profits, lost savings, or other consequential damages, or for any third-party claims.

Can your audience easily grasp this warranty's lengthy sentence structure and difficult-to-understand words? No, the readers will become frustrated and will fail to recognize what is covered under such a warranty. Although you are following your boss's directions, you are not fulfilling one of your ethical requirements to use simple and direct language.

Writing can be legally binding *and* easy to understand. The first requirement does not negate the second. As a successful technical writer who values "the time and effort spent by those who read or see or hear [your] communication," you

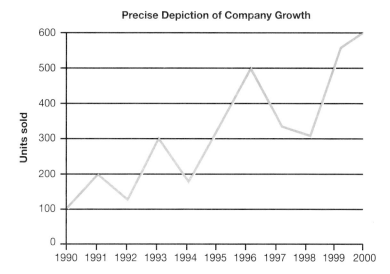

FIGURE 3.3

want to aid communication by using simple words and direct expression of ideas whenever possible.

3. Satisfy the audience's need for information, not my own need for self-expression. The importance of simple language also encompasses the third point in STC's Code for Communicators: satisfying "the audience's need for information, not my own need for self-expression." The previously mentioned warranty might be considered poetic in its sentence structure and sophisticated in its word usage. It does not communicate what the audience needs, however. Such elaborate and convoluted writing will only satisfy one's need for self-expression: that's not the goal of effective technical writing.

Electronic mail (e-mail) presents another opportunity for unethical behavior. Because e-mail is so easy to use, many company employees use it too frequently. They write e-mail for business purposes, but they also use e-mail when writing to family and friends. Employees know not to abuse their company's metered mail by using corporate envelopes and stamps to send in their gas bills or write thank-you notes to Aunt Rose. These same employees, however, will abuse the company's e-mail system (on company time) by writing e-mail messages to relatives coast to coast. They are not satisfying their business colleagues' need for information; instead, they are using company-owned e-mail systems for their own self-expression (Hartman and Nantz 60). That's unethical.

4. Hold myself responsible for how well my audience understands my message. As a technical writer, you must place the audience first. When you write a precise proposal or instruction, using simple words and syntax, you can take credit for helping the reader understand your text. Conversely, you must also accept responsibility if the reader fails to understand.

Ethical standards for successful communication require the writer to always remember the readers—the real people who read manuals to put together their children's toys, who read corporate annual reports to understand how their stocks are doing, and who read proposals to determine whether to purchase a service. An ethical writer remembers that these real people will be frustrated by complex instructions, confused by inaccessible stock reports, and misled by inaccurate proposals. It's unethical to forget that your writing can frustrate people and even endanger them. It's unethical to incorrectly use your writing to mislead or to confuse. In contrast, as an ethical writer, you need to treat these readers like neighbors, people you care about, and then write accordingly. Take the time to check your facts, present your information precisely, and communicate clearly so that your readers—your friends, co-workers, and clients—are safe and satisfied. You are responsible for your message (Barker et al.).

5. Respect the work of colleagues. Each of the ethical considerations already discussed relates to your clients, the readers of your technical communication. This fifth point differs in two ways: first, it relates to co-workers or professional colleagues; second, it highlights the importance of online ethics, the ethical dilemmas presented by electronic communications on the Internet. We can discuss these two points simultaneously.

Much of today's technical writing takes place online, electronically (we discuss electronic communication in Chapter 10). This new venue for technical writing creates three unique reasons for an increased focus on ethics. As the technical writer, you must ethically consider *confidentiality*, *courtesy*, and *copyright* when working with the Internet or any private electronic network.

- *Confidentiality.* The 1974 Privacy Act allows "individuals to control information about themselves and to prevent its use without consent" (Turner 59). The Electronic Communication Privacy Act of 1986, which applies all federal wiretap laws to electronic communication, states that e-mail messages can be disclosed only "with the consent of the senders or recipients" (Turner 60). However, both of these laws can be abused easily on the Internet. First, neither law specifically defines "consent." Data such as your credit records can be accessed without your knowledge by anyone with the right hardware and software. Your confidentiality can be breached easily.

  Second, the 1986 Electronic Communication Privacy Act fails to define "the sender." In the workplace, a company owns the e-mail system, just as it owns other more tangible items such as desks, computers, and file cabinets. Companies have been held liable for electronic messages sent by employees. Thus, many corporations consider the contents of one's e-mail and one's e-mailbox company property, not the property of the employee. With ownership comes the right to inspect an employee's messages. Although you might write an e-mail message assuming that your thoughts are confidential, this message can be monitored without your knowledge (Hartman and Nantz 61).

  These instances might be legal, but are they ethical? Though a company can eavesdrop on your e-mail or access your life's history through databases, that doesn't mean they should. As an employee or corporate manager, you should respect another's right to confidentiality. Ethically, you should avoid the temptation to read someone else's e-mail or to access data about an individual without his or her consent.

- *Courtesy*. As already noted, e-mail is not as private as you might believe. Whatever you write in your e-mail correspondence can be read by others. Given this reality, you should be very careful about what you say in e-mail. Specifically, you don't want to offend co-workers by "flaming" (writing discourteous messages). Before you criticize a co-worker's ideas or ability, and before you castigate your employer, remember that common courtesy, respect for others, is ethical (Adams et al. 328).

- *Copyright*. A final ethical consideration relates to copyright laws. Every English teacher you've ever had has told you to avoid *plagiarism* (stealing another writer's words and ideas). Plagiarism, unfortunately, is an even greater problem on the Internet. The Internet is perhaps the world's largest library without walls; it's almost a universal commons where anyone can set up a soap box and speak (or print) his or her opinion.

    This incredible ability to disseminate massive amounts of information presents problems. If you were an unethical technical writer, you could access data from the Internet and easily print it as your own. After all, it's hard to trace the identity of a writer on the Net, and Internet information can be downloaded by anyone (Adams et al. 328).

    An ethical writer will not fall prey to such temptation. If you haven't written it, give the other author credit. Words are like any other possession. Taking words and ideas without attributing your source through a footnote or parenthetical citation is wrong. You should respect copyright laws.

6. *Strive continually to improve my professional competence. Promote a climate that encourages the exercise of professional judgment.* We've combined these last two tenets from the STC's Code for Communicators since together they sum up all of the preceding ethical considerations. Professional competence, for example, includes one's ability to avoid plagiarizing. Competence should also include professional courtesy, respect for another's right to confidentiality, a sense of responsibility for one's work, and the ability to write clearly and precisely. Ultimately, your competence is dependent upon professional judgment. Do you know the difference between right and wrong? Recognizing the distinction is what ethics is all about.

As a writer, you will always be confronted by a multitude of options, such as loyalty to your company, responsible citizenship, need for a salary, accountability to your client and your co-workers, and personal integrity. You must weigh the issues—ethically, legally, and practically—and then write according to your conscience.

STRATEGIES FOR MAKING ETHICAL DECISIONS

When confronted with ethical challenges, try following these writing strategies (Guy 165).

1. *Define the problem*. Is the dilemma legal, practical, ethical, or a combination of all three?

2. *Determine your audience.* Who will be affected by the problem? Clients, co-workers, management? What is their involvement, what are their individual needs, and what is your responsibility—either to the company or to the community?

3. *Maximize values; minimize problems.* Ethical dilemmas always involve options. Your challenge is to select the option that promotes the greatest worth for all stakeholders involved. You won't be able to avoid all problems. The best you can hope for is to minimize those problems for both your company and your readers while you maximize the benefits for the same stakeholders.

4. *Consider the big picture.* Don't just focus on short-term benefits when making your ethical decisions. Don't just consider how much money the company will make now, how easy the text will be to write now. Focus on long-term consequences as well. Will what you write please your readers so that they will be clients for years to come? Will what you write have a long-term positive impact on the economy or the environment?

5. *Write your text.* Implement the decision by writing your memo, letter, proposal, manual, or report. When you write your text, remember to

   - Use precise language and visuals.
   - Use simple words and sentences.
   - Satisfy the audience's need for information, not your own need for self-expression.
   - Take responsibility for your content, remembering that real people will follow your instructions or make decisions based on your text.
   - Respect your colleagues' confidentiality, be courteous, and abide by copyright laws.
   - Promote professionalism and good judgment.

CHAPTER HIGHLIGHTS

1. If your technical writing is unclear, your reader may misunderstand you and then do a job wrong, damage equipment, and/or contact you for further explanations.

2. Use details to ensure reader understanding. Whenever possible, specify and quantify your information.

3. Answering *who*, *what*, *when*, *where*, *why*, and *how* (the reporter's questions) helps you determine which details to include.

4. For some audiences, you should avoid acronyms, abbreviations, and jargon.

5. Words that are not commonly used (legalisms, old-fashioned terms, etc.) should be avoided.

6. Write to express, not impress—to communicate, not to confuse.

7. Avoid passive voice constructions, which tend to lengthen sentences and confuse readers.

8. Writing concisely helps save you and your readers time.

9. Shorter paragraphs are easier to read, so they hold your reader's attention.

10. When possible, use short, simple words (always considering your reader's level of technical knowledge).

11. Apply readability formulas to determine your text's degree of difficulty.

12. Proofreading is essential to effective technical writing.

13. Well-organized documents are easy to follow.

14. Different organizational patterns—spatial, chronology, importance, comparison/contrast, and problem/solution—can help you explain material.

15. Consider whether or not your technical writing is legal, practical, and ethical.

ACTIVITIES

Specify

The following sentences are vague and imprecise. They will be interpreted differently by different readers. Revise these sentences—replace the vague, impressionistic words with more specific information.

1. We need this information as soon as possible.

2. The machinery will replace a flawed piece of equipment in our department.

3. Failure to purchase this will have a negative impact.

4. Weather problems in the area resulted in damage to the computer systems.

5. The most recent occurrences were caused by insufficient personnel.

6. Fire in the office caused substantial losses.

7. If we can't solve this problem soon, we'll lose a large percentage of our business.

8. The automobile has a smaller turning radius than last year's model.

9. Several employees commended her for her expertise.

10. Make your explanations very detailed.

Avoid Obscure Words

Obscure words make the following sentences difficult to understand. Improve the sentences by revising the difficult words and making them more easily understandable.

1. As Very Large Scale Integration (VLSI) continues to develop, a proliferation of specialized circuit simulators will be utilized.

2. As you requested at the commencement of the year, I am forwarding my regular quarterly missive.

3. Though Randolph was cognizant of his responsibility to advise you of any employment abberations, he failed to abide by this mandate.

4. Herewith is an explanation of our rationale for proferring services, pursuant to your request.

5. Please be advised that the sale constitutes a successful closure.

6. Inasmuch as we have endeavored to determine the causes of the dilemma without success, we are terminating this fact-finding operation immediately.

7. To facilitate the initiation of this activity, we have assigned the ensuing job responsibilities.

8. Can you assist us in ascertaining the causes pertaining to yesterday's mechanism malfunction?

9. I wonder if you would be so kind as to avail yourself of this opportunity to respond accordingly to our questionnaire.

10. In lieu of further discussion, we want to state in the affirmative that what transpired was due to the fact that the vehicle had insufficient braking capabilities to avoid the collision.

Use the Active Voice versus the Passive Voice

Use of the passive voice often leads to vague, wordy sentences. Revise the following sentences by writing them in the active voice.

1. Implementation of this procedure is to be carried out by the metrology department.

2. Benefits derived by attending the conference were twofold.

3. The information was demonstrated and explained in great detail by the training supervisor.

4. Discussions were held with representatives from Allied, who supplied analytical equipment for automatic upgrades.

5. Also attended was the symposium on polymerization.

6. Process control systems for foam encapsulation were reviewed with vendors.

7. Effort should be expended to reduce overtime. Overtime in excess of eight hours should be closely monitored.

8. The reassignment of this activity was the result of changes requested by manufacturing.

9. Installation of the fiber optic networking is estimated to occur early next month.

10. Misapplication of a dry film lubricant has been the primary cause of defectiveness.

Limit Paragraph Length

You can achieve clarity and conciseness if you limit the length of your paragraphs. An excessively long paragraph (beyond eight typed lines) requires too much work for your reader. Revise the following paragraph to make it more reader-friendly.

As you know, we use electronics to process freight and documentation. We are in the process of having terminals placed in the export departments of some of our major customers around the country so they may keep track of all their shipments within our system. I would like to propose a similar tracking mechanism for your company. We could handle all of your export traffic from your locations around the country and monitor these exports with a terminal located in your home office. This could have many advantages for you. You could generate an export invoice in your export department which could be transmitted via the computer to our office. You could trace your shipments more readily. This would allow you to determine rating fees more accurately. Finally, your accounting department would benefit. All in all, your export operations would achieve greater efficiency.

Reduce Word Length

Multisyllabic words can create long sentences. To limit sentence length, limit word length. Find shorter words to replace the following words.

1. advise	6. endeavor	11. prohibit
2. anticipate	7. inconvenience	12. residence
3. ascertain	8. indicate	13. subsequent
4. cooperate	9. initially	14. sufficient
5. determine	10. presently	15. terminate

Reduce Sentence Length

Each of the following sentences is too long. Revise them using the techniques suggested in this chapter: use the meat cleaver approach; avoid shun words, camouflaged words, and expletives; omit redundancies; and delete wordy phrases.

1. In regard to the progress reports, they should be absolutely complete by the fifteenth of each month.

2. I wonder if you would be so kind as to answer a few questions about your proposal.

3. I am in receipt of your memo requesting an increase in pay and am of the opinion that it is not merited at this time due to the fact that you have worked here for only one month.

4. On two different occasions, I have made an investigation of your residence, and I believe that your sump pump might result in damage to your neighbor's adjacent property. I have come to the conclusion that you must take action to rectify this potential dilemma, or your neighbor might seek to sue you in a court of law.

5. In this meeting, our intention is to acquire a familiarization with this equipment so that we might standardize the replacement of obsolete machinery throughout our entire work environment.

6. It is evident that the company's request for electrical equipment to be placed in the laboratory has become rather important inasmuch as this need is prioritized in our CEO's most recent letter.

7. It is anticipated that these changes will lead to a reduction in the failure rate.

8. There is the possibility that we will implement these suggestions early next month.

9. New personnel will be assessed when brought on board and then tested on a yearly basis in order to ensure their continued successful job prowess.

10. If there are any questions that you might have, please feel free to contact me by phone.

Organize

1. *Spatial:*

    a. Using spatial organization, write a paragraph describing your classroom, your office, your work environment, your dorm room, your apartment, or any room in your house.

    b. Using spatial organization, write an advertisement describing the interior of a car, the exterior of a mechanism or tool, a piece of clothing, or a motorcycle.

2. *Chronological:* Organizing your text chronologically, write a report documenting your drive to school or work, your activities accomplished in class or at work, your discoveries at a conference or vacation, or your activities at a sporting event.

3. *Importance:* A fashion merchandizing retailer asked her buyers to purchase a new line of clothing. In her memo, she provided them the following list to help them accomplish their task. Reorganize the list by importance, and justify your decisions.

DATE:	January 15, 2000
TO:	Buyers
FROM:	Nancy Aaronson
SUBJECT:	CLOTHING PURCHASES

It's time again for our spring purchases. This year, let's consider a new line of clothing. When you go to the clothing market, focus on the following:

- Colors
- Materials
- Our customers' buying habits
- Price versus markup potential
- Quantity discounts
- Wholesaler delivery schedules

Good luck. Your purchases at the market are what make our annual sales successful.

4. *Comparison/contrast:* Visit two auto dealerships, two clothing stores, two restaurants, two music shops, two prospective employers, two colleges, etc. Based on your discoveries, write a report using comparison/contrast to make a value judgement. Which of the two cars would you buy, which of the two restaurants would you frequent, and at which of the two music shops would your purchase CDs?

5. *Problem/solution: Case study*

You work for Acme Electronics as an electrical engineer. Carol Haley, your boss, informs you that your department's electronic scales are measuring tolerances inaccurately. You are asked to study the problem and determine solutions. In your study, you find that one scale (ID #1893) is measuring within 90 percent of tolerance; another scale (ID #1887) is measuring within 75 percent of tolerance; a third scale (ID #1890) is measuring within 60 percent of tolerance; a final scale (ID #1885) is measuring within 80 percent of tolerance. Standards suggest that 80 percent is acceptable. To solve this problem, the company could purchase new scales ($2,000 per scale); reduce the vibration on the scales by mounting them to the floor ($1,500 per scale); reduce the vibration around the scales by enclosing the scales in plexiglass boxes ($1,000 per scale).

Write a memo to your boss detailing your findings (the problems) and suggesting the solutions.

Consider Ethics

1. The Society for Technical Communication constantly is trying to redefine its Code of Ethics. As a class, how would you define the word *ethics*? Brainstorm new definitions as they apply to technical communication, and come to a class consensus. Then in small groups, based on your class's definition of ethics,

   - List five or more examples of ethical responsibilities you believe technical writers should have when writing memos, letters, reports, proposals, and/or instructions (other than those already discussed in this chapter).
   - List five or more examples of failures to abide by ethical responsibilities you've seen either in writing or in other types of media (recordings, movies, television, newspapers, magazines, news reporting, etc.).
   - Technical writing is factual, as are newspapers, magazines, and radio and television news reports; recordings, movies, and television are art forms. Do the same ethical considerations apply for all types of media? If there are differences, explain your answer.

2. Bring to class examples of ethically flawed communication. These could include poorly written warranties (which are too difficult to understand) or dangers, warnings and cautions (which do not clearly identify the potential

for harm). You might find misleading annual reports or unethical seeming advertisements. Poor examples could even include graphics which are visually misleading. Then in small groups, rewrite these types of communication or redraw the graphics to make them ethical.

3. Every day, the Internet is presenting ethical dilemmas to lawmakers and to the populace. Congress is currently debating laws to curb unethical practices online. These unethical practices include hacking, pornography, solicitation, infringements on confidentiality, hate mail, inappropriate advertisements, and unauthorized viewing of e-mail.

To update your knowledge of ethics problems online, research any of the topics listed above. Then, present your findings as follows:

- Write an individual or group report on your findings (we discuss reports in Unit 6).
- Give an individual or group oral presentation on your findings (we discuss oral presentations in Chapter 17).
- Write a summary of the article(s) you've researched (we discuss summaries in Chapter 14).

Case Study/Collaborative Writing

The Blue Valley Wastewater Treatment Plant processes water running to and from Frog Creek, a water reservoir that passes through the North Upton community. This water is usually characterized by low alkalinity (generally, <30 mg/l), low hardness (generally, <40 mg/l), and minimal water discoloration. Inorganic fertilizer nutrients (phosphorus and nitrogen) are also generally low with limited algae growth.

Despite the normal low readings, algae-related tastes and odors occur occasionally. Although threshold odors range from 3 to 6, they have risen to 10 in summer months. Alkalinity rises to <50 mg/l, hardness to <60 mg/l, and discoloration intensifies. Taste and odor problems can be controlled by powdered activated carbon (PAC); nutrient-related algae growth can be controlled by filtrated ammonia. Both options are costly.

The odors and tastes are disturbing North Upton residents. The odors are especially bothersome to outdoor enthusiasts who use the trails bordering Frog Creek for biking and hiking. Residents also worry about the impact of increased alkalinity on fish and turtles, many of which are dying, further creating an odor nuisance. Frog Creek is treasured for its wildlife and recreational opportunities.

The Blue Valley Wastewater Treatment Plant is under no legal obligation to solve these problems. The alkalinity, hardness, color, and nutrient readings are all within regulated legal ranges. However, community residents are insistent that their voices be heard and that restorative steps be taken to improve the environment around their homes.

This is an ethical and practical dilemma. In response to this dilemma, divide into small groups to write one of the following documents.

- You are Blue Valley Wastewater Treatment Plant's Director of Public Relations. Write a letter to the City Commission stating your plant's point of view. (We discuss letters in Chapter 6.)
- You are North Upton community's resident representative. Write a letter to the City Commission stating your community's point of view.
- You are an employee for the Blue Valley Wastewater Treatment Plant. Write a memo to the plant director suggesting ways in which the plant could solve this environmental and/or community relations problem. (We discuss memos in Chapter 5.)
- You are the Blue Valley Wastewater Treatment Plant's director. Write a memo to the engineering supervisor (your subordinate) stating how to solve this environmental and/or community relations problem.

These documents could be organized by the problem/solution or comparison/contrast methods. Whichever method you choose, consider the strategies for making ethical decisions, discussed in this chapter.

4

Audience Recognition and Involvement

OBJECTIVES

When you write a memo, letter, or report, someone reads it. That individual or group of readers is your audience. As mentioned in Chapter 1, a primary difference between technical writing and other types of writing is the importance of audience. The audience for other types of writing (including fiction and expressive, expository, and persuasive writing) is not required to act after reading the text. This is not the case in technical writing.

When a technical document is submitted, the reader responds. Your audience either follows a procedure, answers a request, makes a decision, or files the document for future reference. Whatever the case, the writer writes the technical document and expects the reader (the audience) to react. The writer and reader are not removed, as they often are from other types of writing. In technical writing, the writer and reader are fused. They are involved in a business transaction.

Because technical writing links writer and reader, you must consider the importance of audience. To compose effective technical writing, you should achieve (a) audience recognition and (b) audience involvement.

AUDIENCE RECOGNITION

When writing a technical document, ask yourself: Who is the reader? What is his or her level of understanding? What is his or her position in relation to your job title? If you don't know the answers to these questions, your technical writing might miss the mark. Your letter may contain jargon or acronyms the reader won't understand. The tone of the memo may be inappropriate for management (too dictatorial) or for your subordinates (too lax).

On the other hand, successful technical writing achieves audience recognition. If you're going to write well, you need to develop a profile of your audience. The following audience profiles will help you achieve audience recognition.

- High-tech audience
- Low-tech audience
- Lay audience
- Multiple audiences

HIGH-TECH AUDIENCE

High-tech readers work in your field of expertise. They might work directly with you in your department, or they might work in a similar capacity for another company. Wherever they work, they are your colleagues because they share your educational background, work experience, or level of understanding.

If you're a computer programmer, for example, another computer programmer who is working on the same system is your high-tech peer. If you are an environmental engineer working with hazardous wastes, other environmental engineers focusing on the same concerns are your high-tech peers.

Once you recognize that your reader is high tech, what does this tell you? High-tech readers have the following characteristics.

- They are experts in the field you are writing about. If you write an e-mail message to one of your department colleagues about a project you two are working on, your associate is a high-tech peer. If you write a letter to a vendor requesting specifications for a system he or she markets, that reader is a high-tech expert. If you write a journal article geared toward your fellow experts, they are high tech.

- Because their work experience and/or education are comparable to yours, high-tech readers share your level of understanding. Therefore, they'll understand high-tech jargon, acronyms, and abbreviations. You do not have to explain to an electronics technician, for example, what *MHz* means. Defining *megahertz* for this high-tech reader would be unnecessary and even offensive. We recently read a procedure written by a manager to his engineers detailing how to write departmental technical reports. In the memo, the boss specified that the engineers should use "metric SI units in their reports." Because the boss recognized that his audience was high tech, he did not have to define *SI* as "System International."

- High-tech readers require minimal detail regarding standard procedures or scientific, mathematical, or technical theories. Two physicists, when discussing the inflation theory of the universe's creation, would not labor over the importance of quarks and leptons (subatomic particles). Although this jargon is unintelligible to most of us, to high-tech peers no explanations are required.

- High-tech peers read to discover new technical knowledge or for updates regarding the status of a project.

- High-tech readers need little background information regarding a project's history or objectives unless the specific subject matter of the correspondence is new to them. If, for example, you're writing a status report to your first-line supervisor, who has been involved in a project since its inception, then you will not need to flesh out the history of the project. On the other hand, if you have a new supervisor or if you are updating a colleague new to your department, even though these readers are high tech, you will need to provide background data.

If you work in an environment in which you write only to high-tech peers, you are a rare and lucky individual. Writing to high-tech readers is rather easy since you can use acronyms, abbreviations, jargon, complex graphics, and so on. You usually don't have to define terms or provide background information. Writing within a technical environment, however, also requires that you write to low-tech readers.

LOW-TECH AUDIENCE

Low-tech readers include your co-workers in other departments. Low-tech readers also might include your bosses, your subordinates, or your colleagues who work for other companies. For instance, if you're a biomedical equipment technician, the accountant or personnel director or graphic artists in your company are low-tech peers. These individuals have worked around your company's equipment and, therefore, are familiar with your technology. However, they do not understand the intricacies of this technology. Your bosses are often low tech because they no longer work closely with the equipment. Although they might have been technicians at one time, as they moved more and more into management, they moved further and further away from technology. Your subordinates might be low tech because their levels of education and/or work experience are less than yours. Finally, although your colleagues at other companies have your level of education and/or work experience, they could be low tech if they aren't familiar with your company's procedures or in-house jargon, acronyms, and abbreviations.

When you write memos, letters, and reports to low-tech readers, remember that they share the following characteristics.

- Low-tech readers are familiar with the technology you're writing about, but their job responsibilities are peripheral to the subject matter. They either work in another department, manage you, work under your supervision, or work outside your company.

- Because low-tech readers are familiar with your subject matter, they understand *some* abbreviations, jargon, and technical concepts. To ensure that readers understand your content, therefore, define your terms. An abbreviation like *VLSI* can't stand alone. Define it parenthetically: VLSI (very large scale integration). Similarly, technical jargon like *silicon foundry* needs a follow-up explanation. You should write instead, "A silicon foundry, as the name implies, is a factory or manufacturer which casts into silicon an integrated chip based on a customer's specifications." In addition, technical concepts must be defined for low-tech readers. For example, whereas high-tech readers understand the function of pressure transducers, a low-tech reader needs further information, such as "Pressure transducers: Solid-state components sense proximal pressure in the patient tubing circuit. The transducers convert this pressure value into a proportional voltage for the control system."

- Since the low-tech reader is not in your normal writing "loop"—that is, not someone to whom you write often regarding your field of expertise—you need to provide more background information. When you submit a status report to upper-level management, for example, you can't just begin with work accomplished. You need to explain why you are working on the project (objectives, history), who is involved (other personnel), when the project began and its scheduled end date, and how you are accomplishing your goals. Low-tech readers understand the basic concepts of your work, but they have not been involved in it daily. Fill them in on past history.

LAY AUDIENCE

Customers and clients who neither work for your company nor have any knowledge about your field of expertise are your lay audience. If you work in telecommunications for a telephone company, for example, and you write a letter to a client regarding a problem with the company's phone line, your audience is a lay reader. If your field of expertise is biomedical equipment and you write a procedures manual for the patient end user, you're writing to a lay audience. Or if you are an automotive technician writing a service report for a customer, that customer is a lay reader. Although you understand your technology, your reader, who uses the phone or the medical equipment or the car, is not an expert in the field. These readers are using your equipment or require your services, but the technology you are writing about is not within their daily realm of experience.

This makes writing for a lay audience difficult. It's easy to write to a high-tech reader who thoroughly understands the technology you are discussing. However, writing to a low-tech co-worker or a lay audience totally outside your field of expertise is demanding.

To write successfully to a lay audience, remember that these readers share the following characteristics.

MAINTENANCE

Cleaning the Head Section

The heads, capstan, and pinch rollers get dirty easily. If this head section becomes dirty, the high-frequency sound will not be reproduced and the stereo balance will be impaired. This hurts your system's sound quality.

To avoid these problems, clean your system's head section regularly by following these simple steps:

1. Push the STOP/EJECT button to open the cassette door.
2. Dip a cleansing swab into the cleaning fluid.
3. Wipe the heads, capstan and pinch rollers with the swab.
4. Allow 30 seconds to dry.

Note:
- Do not hold screwdrivers, metal objects, or magnets close to the heads.
- When demagnetizing the heads, be sure the unit's POWER switch is in the OFF position.

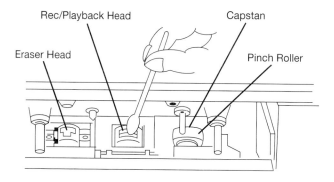

FIGURE 4.1 Cleaning the Head Section

- Lay readers are unfamiliar with your subject matter. They do not understand your technology. Therefore, you should write simply. That's not to say that you should insult your lay reader with a remedial discussion or with a patronizing tone. You must, however, explain your topic clearly. You achieve clarity through precise word usage, depth of detail, and simple graphics.
- Since your lay readers don't understand your technology or work environment, they won't understand any of your in-house jargon, abbreviations, or acronyms. Avoid high-tech terms or define them thoroughly.

• Lay readers will need background information. If you leap into a discussion about a procedure without explaining to your lay readers why they should perform each step, they will not understand the causes or rationale. High-tech and possibly even low-tech readers might not need such explanations, but the lay reader needs you to clarify. For example, look at the maintenance procedure, which is part of a user's manual provided for the purchaser of an audio recorder (Figure 4.1). The introductory paragraph clarifies for the lay reader why he or she should clean the head section, instead of incorrectly assuming that the reader will understand without being told. This procedure also uses a simple graphic to clearly depict what action is required.

On the other hand, look at the procedure for cable preparation from an installation manual geared toward high-tech readers (Figure 4.2). It provides no background about why to perform the action, no clarity about how to perform the steps, and no graphics to help the reader visualize the procedure.

A lay reader would have no idea what is going on here. Why are we performing this act? What's the purpose? What will be achieved? Perhaps these questions are irrelevant for the technician, who knows the background or rationale, but a lay audience will be confused without an introductory overview. In addition, a lay audience will not know what a gasket, a cable dielectric, or a center conductor is. These terms are high-tech jargon which a lay reader will not understand. Finally, what dimensions are we supposed to achieve? The high-tech reader, who is familiar with this operation, knows what is meant by *proper length*, but a lay audience won't. And, when is a clamp seated "properly in gasket"? Again, a lay audience needs these points clarified.

Composing effective technical writing requires that you recognize the difference between high-tech, low-tech, and lay audiences. If you incorrectly assume that all readers are experts in your field, you'll create problems for yourself as well as for your readers. If you write using high-tech terms to low-tech or lay audiences, your readers will be confused and anxious. You'll waste time on the phone clarifying the points that you did not make clear in the technical document.

1. Place nut and gasket over cable and cut jacket to dimension shown.
2. Comb out braid and fold out. Cut cable dielectric to dimension shown. Tin center conductor.
3. Pull braid wires forward and taper toward center conductor. Place clamp over braid and push back against cable jacket.
4. Fold back braid wires as shown, trim to proper length (D), and form over clamp as shown. Solder contact to center conductor.
5. Insert cable and parts into connector body. Make sure sharp edge of clamp seats properly in gasket. Tighten nut.

FIGURE 4.2 Cable Preparation

MULTIPLE AUDIENCES

Correspondence isn't always sent to just one type of audience. Sometimes your correspondence has multiple audiences. Sometimes you write to *all* of the afore-mentioned audiences *simultaneously*.

For example, when writing a report, most people assume that the first-line supervisor will be the only reader. This might not be the case, however. The first-line supervisor could send a copy of your report to the manager, who could then submit the same report to the executive officer. Similarly, your first-line supervisor might send the report to your colleagues or to your subordinates. Or, your report might be sent out to other lateral departments. Figure 4.3 shows the possibilities.

Writing correspondence for multiple readers with different levels of under-standing and different reasons for reading creates a challenge for you. When you add the necessity of using a tone which will be appropriate for all of these varied readers, the writing challenge becomes even greater.

How do you meet such a challenge? The first key to success is recognizing that multiple audiences exist and that they share the following characteristics.

- Your intended audience will not necessarily be your only readers. Others might receive copies of your correspondence.
- Some of the multiple readers will be unfamiliar with the subject matter. You will have to provide background data (objectives, overviews) to clarify the history of the report for these readers. In a short letter, memo, report, or e-mail message, this background information can't be too elaborate. Often, all you do is provide a reference line suggesting where the readers can find out more about the subject matter if they wish—"Reference: Operations

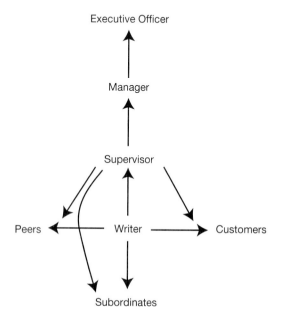

FIGURE 4.3

Procedure 321 dated 9/21/00." In longer reports, background data will appear in the summary or abstract, as well as in the report's introduction.

Summaries, abstracts, introductions, and references are especially valuable when you consider one more fact regarding technical writing: reports, memos, and letters are kept on file. When you write the correspondence initially, you can assume that your reader or readers have knowledge of the subject matter. But months (or years) later when the report is retrieved from the files, will your readers still be familiar with the topic? Will you still have the same readers? Many people, some of whom didn't even work for your company at the time of the original writing, will read your correspondence. These future multiple readers need background information.

- Multiple readers have diverse understandings of your technology. Some will be high tech; others will be low tech. This requires that you define jargon, abbreviations, and acronyms. As mentioned earlier in this chapter, you can define your terms either parenthetically within your text or in a glossary, depending on the length of the correspondence. Short e-mail, memos, or letters allow for only parenthetical definitions. Long reports, Web sites, and manuals allow for glossaries and extended definitions.

- Correspondence geared toward multiple readers must have a matter-of-fact, businesslike tone. You shouldn't be too authoritative, since upper-level management might read the memo, letter, or report. You shouldn't be too ingratiating, since lower-level subordinates might also read the correspondence.

The memo in Figure 4.4 is written to a multiple audience. "To: Distribution" is a common way to direct correspondence to numerous readers. The multiple audience, then, affects the memo's content.

High-tech terms such as *operating procedure, engineering notice*, and *product quality requirements* are followed by parenthetical abbreviations— (OP), (EN), and (PQR). If these abbreviations alone had been presented, certain high-tech readers would have understood them. Other readers in the distribution list, however, might not have known what OP, EN, or PQR meant. To cover all readers in the multiple audience, the writer correctly used both the written-out terms and the abbreviations.

Finally, the tone of the memo is appropriate for readers with different levels of responsibility. Although it contains no direct commands, which might have been offensive to management, the memo is still assertive. The matter-of-fact tone simultaneously suggests to management that these actions will be carried out and informs subordinates to do so.

DEFINING TERMS FOR DIFFERENT AUDIENCE LEVELS

Communicating with audiences at different levels, including high-tech, low-tech, lay, and multiple readers, is challenging because each industry has its own specialized vocabulary. Usually your high-tech readers work most closely with you and share your expertise; colleagues understand your high-tech jargon. Therefore,

DATE: July 7, 2000
TO: Distribution
FROM: Nancy Roediger
SUBJECT: Revision of Operating Procedure (OP) 354 dated
             5/31/97

The reissue of this procedure was the result of extensive changes
requested by Engineering, Manufacturing, and Quality. These
procedural changes will be implemented immediately according
to Engineering Notice (EN) 185.

Some particularly significant changes are as follows:

1. *Substitutions*—An asterisk (*) can no longer be used to identify
   substitutable items in requirement lists. Substitution for items
   called out in work orders will be authorized by Engineering.
   Substitution also will be specified by item numbers rather
   than generic description names.
2. *Product Quality Requirements (PQR)*—These will be included
   in work directions either by stating the requirements or by
   calling out the PQR.
3. *Oral Instructions*—When oral instructions for process
   adjustments are given, the engineer will be present in the
   department.

OP 354 and EN 185 will be reviewed by all managers. Next, the
managers will review the changes with their supervisors to make
sure that each supervisor is aware of his or her responsibilities.
These reviews will occur immediately.

Distribution:

R.L. Peters   H.C. Ramos
M.K. Arias   R.A. Travers
J.D. Mowers   B.D. Smith
S.T. Thomas   J.H. Hunt
P.D. Bingham   L.R. Rochelle
A.R. Daily   S.B. Haas

FIGURE 4.4 Memo for Multiple Audiences

you can use acronyms and abbreviations for these high-tech readers without providing definitions. However, it's your responsibility as a technical writer to define terms that might be unfamiliar to audiences at other levels. You can accomplish this goal by using any or all of the following:

- Words or phrases that define your terms, presented either parenthetically or in a glossary
- Sentences that define your terms, either in a glossary or following the unfamiliar terms
- Extended definitions of one or more paragraphs

Low-tech readers at your company have different areas of expertise. They are unfamiliar with your high-tech acronyms and abbreviations, but they are familiar with your work environment. Therefore, parenthetical definitions or brief definitions in a glossary should suffice. You could define *ATM* parenthetically as (*asynchronous transfer mode*). Then, your low-tech reader won't misconstrue *ATM* as *automatic teller machine*. You could also use a glossary.

HTTPS	Hypertext Transfer Protocol, Secure
TDD	telecommunication device for the deaf
TTY	teletypewriter

Lay readers are further removed from your immediate work world. Customers, members of a city council, end users of your equipment, and vendors have no knowledge of your high-tech jargon and won't be helped if you merely define your terms. These readers need more information. You could provide either a follow-up sentence or an extended definition. If you provide a sentence definition, include the following:

Term + Type + Distinguishing characteristics

The *term* consists of the words, abbreviations, and/or acronyms you are using. The *type* explains what class your term fits into. For example, a "car" is a "thing," but the word *thing* applies to too many types of "things," including bananas, hammers, and computers. You need to classify your term more precisely. What type of thing is a car? "Vehicle" is a more precise classification.

But a motorcycle, a dirigible balloon, and a submarine are also types of vehicles. That's when the *distinguishing characteristics* become important. How does a car differ from other vehicles? Perhaps the distinguishing characteristics of a car would include the facts that cars are land-driven, four-wheeled vehicles. Thus, the definition would read as follows:

A car is a *vehicle* that contains *four wheels* and is *driven on land*.

Term + Type + Distinguishing characteristics

Using a sentence to define *HTTP*, you would write

HyperText Transfer Protocol is a computer access code that provides secure communications on the Internet, an intranet, or an extranet.

When you need to provide an extended definition of a paragraph or more, in addition to providing the term, type, and distinguishing characteristics, also consider including examples, procedures, and descriptions. Look at the following definition of a voltmeter.

The voltmeter is an instrument used to measure voltage. The voltmeter usually consists of a magnet, a moving coil, a resistor, and control springs. Types of voltmeters include the microvoltmeter, millivoltmeter, and kilovoltmeter, which measure voltages with a span of 1 billion to 1. By connecting between the points of a circuit, voltmeters measure potential difference.

There are often ways to define your terms.

- Place the definition in the front matter or an appendix of a document (such as a manual).
- Use endnotes in a report with other researched material.
- Use pop-up screens for online help incorporated into your Internet, intranet, or extranet site. The reader can access the definitions by clicking on "hot buttons" or hypertext links. (See Figure 10.4 in Chapter 10 for more information on electronic communication.)

Table 4.1 shows techniques for communicating with audiences with different levels of technical knowledge. These various techniques for defining terms are neither foolproof nor mandatory. You are always the final judge of how much information to provide for your intended readers. However, remember that if your readers fail to understand your content, then you have failed to communicate. No one will complain if you define your terms, but readers who are uncertain about your meaning either will call you for assistance, wasting your time, or will perform tasks incorrectly, wasting their time. It's better to provide too much information than not enough.

MULTICULTURAL AUDIENCES

In addition to considering audience levels (high tech, low tech, lay, and multiple), you also will be writing to people for whom English is a second language. In today's global economy, your company will market its products or services worldwide. Here's one startling example to prove that point. International sales for U.S.-manufactured computers reached 59 percent in 1991 (King 28). That figure, of course, highlights the profitability of marketing goods internationally. However, a corporation's income isn't the only factor affected by a global market; if you are your company's technical writer, you also are affected. You have to be aware of your expanded audience.

TABLE **4.1**	How to Communicate to Different Audience Levels	
AUDIENCE LEVEL	HOW TO COMMUNICATE	SAMPLE
High-Tech	Use jargon, acronym, or abbreviation alone.	The wastewater is being treated for DBPs.
Low-Tech	Use jargon, acronym, or abbreviation with a parenthetical definition.	The wastewater is being treated for DBPs (disinfection by-products).
Lay	Use jargon, acronym, or abbreviation with a parenthetical definition *and* a brief explanation or extended definition.	The wastewater is being treated for DBPs (disinfection by-products), such as acid, methane, chlorine, and ammonia.

As a technical writer, you now must go beyond merely considering whether your readers are high tech, low tech, or lay. Due to the multicultural makeup of your audience, you must ensure that your writing accommodates language barriers and cultural customs. The classic example of one company's failure to recognize the importance of translation concerns a car that was named Nova. In English, *nova* is defined as a star that spectacularly flares up. In contrast, *no va* in Spanish is translated as "no go," a poor advertisement for an automobile.

To communicate effectively to a multicultural audience, follow these guidelines.

1. *Define acronyms and abbreviations.* Acronyms and abbreviations cause most readers a problem. Although you and your immediate colleagues might understand such high-tech usage, many readers won't. This is especially true when your audience is not a native of the United States.

To avoid communication problems, define your acronyms and abbreviations parenthetically the first time you use them. This applies even to acronyms and abbreviations that most people take for granted. Do you know what *FYI* means? You probably do, but would your readers in Japan, France, Mexico, and China understand this commonly used acronym? Why take the risk? Just define it parenthetically: *FYI (for your information)*.

A current e-mail acronym is *LOL*. When this is placed at the end of a document or sentence, the writer is telling the readers not to take the message seriously. *LOL* is defined as *laughing out loud*. How many readers abroad will recognize this acronym?

Here's another example. Corporate employees often abbreviate the job title *system manager* as *sysmgr*. However, in German, the title *system manager* is called the *system leiter*; in French, it's *le responsable*. The abbreviation *sysmgr.* would make no sense in either of these countries (Swenson WE-193).

2. *Avoid jargon and idioms.* The same dilemma applies to *jargon* and *idioms*, words and phrases that are common expressions in English but which could be meaningless outside our borders. Every day in the United States, we use *on the other hand* as a transitional phrase and *in the black* or *in the red* to denote financial status. What will these idioms mean in a global market?

Similarly, the computer industry says, "The system crashed so we rebooted." A literal translation of this jargon into Chinese, German, or French will only confuse the readers (Swenson WE-194).

3. *Distinguish between nouns and verbs.* Many words in English act as both nouns and verbs. This is especially true with computer terms, such as *file*, *scroll*, *paste*, *code*, and *help*. If your text will be translated, make sure that your reader can tell whether you're using the word as a noun or a verb (Rains).

4. *Watch for cultural biases/expectations.* Your text will include words and graphics. As a technical writer, you need to realize that many colors and images that connote one thing in the United States will have different meanings elsewhere.

For example, the idioms *in the red* and *in the black* will not necessarily communicate your intent when they are translated. Even worse, the colors black and red have different meanings in different cultures. Red in the U.S. connotes danger; therefore, *in the red* suggests a financial problem. In China, however, the word *red* has a positive connotation, which would skew your intended meaning. The word *black* often implies death and danger, yet *in the black* suggests financial stability. Such contradictions could confuse readers in various countries.

Moreover, images like those of an apple, a grim reaper, or a police officer might represent such concepts as wholesomeness, death, and safety in the United States. These same images, however, are not universal. In various countries abroad, the image of a police officer could be perceived as the symbol of totalitarian oppression. The grim reaper would be meaningless to a Hindu or Buddhist. Biblically, apples represent sin, not American apple pie. Hands are an even greater problem. For example, the "thumbs up" gesture is considered obscene in many countries. Even an image as simple as an electrical plug presents multicultural problems, since these plugs vary from country to country.

Animals represent another multicultural challenge. In the United States, we say you're a "turkey" if you make a mistake, but success will make you "soar like an eagle." The same meanings don't translate in other cultures. Take the friendly piggy bank, for example. It represents a perfect image for savings accounts in the United States, but pork is a negative symbol in the Mideast. If you are "cowed" by your competition in the United States, you lose. Cows, in contrast, represent a positive and sacred image in India (Horton 686–93).

5. *Be careful when using slash marks (/).* Our fourth consideration, "Watch for cultural biases/expectations," uses a slash mark. What do we mean by this typographic character? Does the slash mark mean "and," "or" or both "and/or"? The word *and* means "both," but the word *or* means "one or the other," not nec-

essarily both. If your text will be translated to another language, will the transla-tor know what you meant by using a slash mark? To avoid this problem, deter-mine what you want to say, and then say it (Rains).

6. Avoid humor and puns. Humor is not universal. In the United States, we talk about regional humor. If a joke is good in the South but not in the North, how could that same joke be effective overseas? Microsoft's software package *Excel* is promoted by a logo which looks like an *X* superimposed over an *L*. This visual pun works in the United States because we pronounce the letters *X* and *L* just as we would the names of the software package. If your readers are not familiar with English, however, they might miss this clever sound-alike image (Horton 686).

7. Leave space for a translation. In Chapter 10, we discuss the importance of electronic communication. If your technical writing will be conveyed not on paper but on disk or on the Internet, you must consider software's line-length and screen-length restrictions. A page of text could have 56 lines that average 80 char-acters per line. In contrast, a computer screen might allow for only 20 to 30 lines of 60 characters.

Why is this a problem? Won't a page of English text translate to the same length in any language? The answer is no. The word count of a document written in English will expand more than 30 percent when translated into some European languages. For example, the four-letter word *user* has 12 characters in German (Hussey and Homnack RT-46); the simple word *print* is translated as *impression* in French; *file* in Spanish is *archivo*; and *view* becomes, *visualizzare* in Italian (Horton 691). You'll want to leave enough room on your screen to accommodate these words when they are translated.

8. Avoid figurative language. Many of us use sports images to figuratively illustrate our points. We "tackle" a chore; in business, a "good defense is the best offense"; we "huddle" to make decisions; if a sale isn't made, you might have "booted" the job; if a sale is made, you "hit a home run"; if you know you're not going to succeed at a task, you just "bail." Each of these sports images might mean something to native speakers, but will they communicate worldwide? We doubt it. Instead, say what you mean, using precise words (Weiss 14).

Cultural differences and language barriers present the technical writer many challenges. You might even want to ignore these problems as being too time-con-suming. Creating text that can be translated successfully will cost your company time and money—up front. The payoff comes later, however, when you ensure your company's reputation and avoid potential lawsuits. By recognizing the multicultur-al basis of your audience, you can protect your company from unintentional cultural offenses. Similarly, by considering the needs of multicultural readers, you can create text that is easy to follow and understand. This will minimize misreadings which could lead to mechanism failures or physical harm—each of which could cost your company in legal fees.

Recognizing the importance of the global marketplace is smart business and a wise move on the part of the technical writer.

SEXIST LANGUAGE

Many of your readers will be women. This does not constitute a separate audience category. Women readers will be high tech or low, management or subordinate. Thus, you don't need to evaluate a woman's level of understanding or position in the chain of command any differently than you do for readers in general.

Recognize, however, that women constitute over half the workforce. As such, when you write, you should avoid *sexist language*, which is offensive to all readers.

Let's focus specifically on ways in which sexism is expressed and techniques for avoiding this problem. Sexism creates problems through omission, unequal treatment, and stereotyping, as well as through word choice.

OMISSION

When your writing ignores women or refers to them as secondary, you are expressing sexist sentiments. The following are examples of biased comments and their nonsexist alternatives.

Biased	Unbiased
The pioneers crossed the prairie with their women, children, and possessions.	Pioneer families crossed the prairie with their possessions.
Slaves were allowed to marry and have their wives and children live with them.	Slaves were allowed to marry and live with their families.
Radium was discovered by a woman, Marie Curie.	Radium was discovered by Marie Curie.
When setting up his experiment, the researcher must always check for errors.	When setting up experiments, the researcher must always check for errors.
As we acquired scientific knowledge, men began to examine long-held ideas more critically.	As we acquired scientific knowledge, people began to examine long-held ideas more critically.

UNEQUAL TREATMENT

Modifiers that describe women in physical terms not applied to men are patronizing.

Biased	Unbiased
The poor women could no longer go on; the exhausted men . . .	The exhausted men and women could no longer go on.
Mrs. Acton, a statuesque blonde, is Joe Granger's assistant.	Jan Acton is Joe Granger's assistant.

STEREOTYPING

If your writing implies that only men do one kind of job and only women do another kind of job, you are stereotyping. For example, if all management jobs are held by men and all subordinate positions are held by women, this is sexist stereotyping.

Biased	Unbiased
Current tax regulations allow a head of household to deduct for the support of his children.	Current tax regulations allow a head of household to deduct for child support.
The manager is responsible for the productivity of his department; the foreman is responsible for the work of his linemen.	Management is responsible for departmental productivity. Supervisors are responsible for their personnel.
The secretary brought her boss his coffee.	The secretary brought the boss's coffee.
The teacher must be sure her lesson plans are filed.	The teacher must file all lesson plans.

Sexist language disappears when you use pronouns and nouns which treat all people equally.

PRONOUNS

Pronouns such as *he*, *him*, or *his* are masculine. Sometimes you read disclaimers by manufacturers stating that although these masculine pronouns are used, they are not intended to be sexist. They're only used for convenience. This is an unacceptable statement. When *he*, *him*, and *his* are used, a masculine image is created, whether or not such companies want to admit it.

To avoid this sexist image, avoid masculine pronouns. Instead, use the plural, generic *they* or *their*. You also can use *he or she* and *his or her*. *(S)he* is not a good compromise; it's just too odd looking. Sometimes you can solve the problem by omitting all pronouns.

Biased	Unbiased
Sometimes the doctor calls on his patients in their homes.	Sometimes the doctor calls on patients in their homes.
The typical child does his homework after school.	Most children do their homework after school.
A good lawyer will make sure that his clients are aware of their rights.	A good lawyer will make sure that his or her clients are aware of their rights.

NOUNS

Use nouns that are nonsexist. To achieve this, avoid nouns that exclude women and denote that only men are involved.

Biased	Unbiased
mankind	people
manpower	workers, personnel
the common man	the average citizen

wise men	leaders
businessmen	businesspeople
policemen	police officers
firemen	firefighters
foreman	supervisor
chairman	chairperson
stewardess	flight attendant

AUDIENCE INVOLVEMENT

In addition to audience recognition, effective technical writing demands audience involvement. You not only need to know who you're writing to in your technical correspondence (audience recognition), but also you need to involve your readers—lure them into your writing and keep them interested. Achieving audience involvement requires that you strive for (a) personalized tone and (b) reader benefit.

PERSONALIZED TONE

Companies don't write to companies; people write to people. A major thrust in modern technical writing is person-to-person communication. Remember that when you write your memo, letter, report, or procedure, another person will read it. As such, you want to achieve a personalized tone to involve your reader. This personalization can be accomplished in any of the following ways: (a) pronoun usage, (b) names, and (c) contractions.

Pronouns

The best way to personalize correspondence is through pronoun usage. If you omit pronouns in your technical writing, as some old schools of thought advocated, the text will read as if it has been computer generated, devoid of human contact. On the other hand, when you use pronouns in your technical writing, you humanize the text. You reveal that the memo, letter, report, or procedure is written by people, for people.

The generally accepted hierarchy of pronoun usage is as follows:

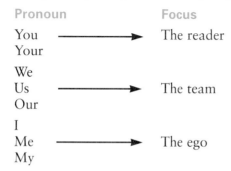

Pronoun		Focus
You Your	⟶	The reader
We Us Our	⟶	The team
I Me My	⟶	The ego

The first group of pronouns, *you/your*, is the most preferred. When you use *you* or *your*, you are speaking directly to your reader(s) on a one-to-one basis. The readers, whoever they are, read the word *you* or *your* and see themselves in the pronoun, envisioning that they are being spoken to, focused on, and singled out. In other words, *you* or *your* appeals to the reader's sense of self-worth. We all like to be thought of as special and worthy of the writer's attention. By focusing on the pronouns *you* or *your* in technical writing, you make your readers feel special and involved in the writing transaction.

The second group of pronouns (*we*, *us*, and *our*) uses team words to connote camaraderie and group involvement. These pronouns are especially valuable when writing to multiple audiences or when writing to subordinates. In either instance, *we*, *us*, or *our* implies to the readers that "we're all in this together." Such a team concept helps motivate by making the readers feel an integral part of the whole.

The third group (*I*, *me*, and *my*) denotes the writer's involvement. These pronouns, however, if overused, can connote egocentricity ("All I care about is *me*, *me*, *me*"). Because of this potential danger, emphasize *you* and *your* and downplay *I*, *me*, and *my*. The rule of thumb is to strive for a two-to-one ratio. For every *I*, *me*, or *my*, double your use of *you* or *your*. We're not saying that you should avoid using first-person pronouns. You can't write without involving yourself through *I* or *my*. We're just saying that an overuse of these first-person pronouns creates an egocentric image, whereas an abundance of *you* orientation achieves audience involvement.

Let's look at an automobile manufacturer's user manual which omits pronouns and, thus, reads as if it's computer generated.

CLAIMS PROCEDURE
To obtain service under the Emissions Performance Warranty, take the vehicle
to the company dealer as soon as possible after it fails an I/M test along with
documentation showing that the vehicle failed an EPA-approved emissions test.

Compare this flat, dry, dehumanized example with the more personalized revision.

CLAIMS PROCEDURE
How do you get service under the Emissions Performance Warranty? To get
service under this warranty, take your car to the dealer as soon as possible
after it has failed an EPA-approved test. Be sure to bring along the document
that shows your car failed the test.

Without any pronouns, the first version has no personality. It's dull and stiff, and it reads as if no human beings are involved. In contrast, when pronouns are added in the second version, the writing involves the reader, who is being spoken to on a person-to-person basis.

Many writers believe that the first version is more professional. However, professionalism doesn't require that you write without personality or deny the existence of your reader. The pronoun-based writing in the second version is more friendly. Friendliness and humanism are positive attributes in technical writing.

The second version also ensures reader involvement. As a professional, you want your readers (customers, clients, colleagues) to be involved, for without these readers no transaction occurs. You end up writing in a vacuum denying your audience, which is not the goal of technical writing.

This desire for reader involvement through pronoun usage is evident in other examples of technical writing. Here are a few lines from a legal contract, which again shows that professional writers are striving to involve readers through pronoun usage.

DEFINITIONS
Throughout this policy, *you* and *your* refer to the "named insured" shown in the Declaration. *We*, *us*, and *our* refer to the company providing the insurance.

Using pronouns takes a conscious effort. Technical writers want to erase the old-fashioned style of writing which denies the reader's existence or depersonalizes the reader through stiff euphemisms such as the "named insured." Instead, writers want to involve the readers and humanize the text. Pronouns soften the technical edge of technical writing.

The following is an example of dry, depersonalized correspondence and a more friendly revision.

Depersonalized	Friendly
Dear Sir:	Dear Mr. Handleman:
With regard to lost policy 123, enclosed is a lost policy form. Complete this form and return it ASAP. Upon receipt, the company will issue a replacement policy.	Your lost policy 123 can be replaced easily. I've enclosed a form to help us replace it for you. All you need to do is fill it out for us. As soon as we get it, we'll send you your replacement policy.

Names

Another way to personalize and achieve audience involvement is through use of names—incorporating the reader's name in your technical writing. By doing so, you create a friendly reading environment in which you speak directly to your reader. Some examples follow.

Note that the reader's first name is used in the first example and the reader's last name in the second example. When do you use first names versus last names? The decision is based on your closeness to or familiarity with your reader. If you know your reader well, have worked with him or her for a while, and know your reader won't be offended, then use the first name. However, if you don't know the reader well, haven't worked with this reader before, or worry that the reader might be offended, then use the surname instead. In either instance, calling a reader by name will involve that reader and personalize your writing.

DATE: June 22, 2000
TO: Steve McMann
FROM: Maureen Pierce
SUBJECT: Cable Purchase Orders

Steve, attached are requisitions for cable replacement for
California. These are held pending our analysis and your approval:

Requisition No.	Amount	Location
1045	$3,126	Waconia
1825	$4,900	Chaska
2561	$3,829	Chanhassen

When you sign the attached requisition forms, Steve, we'll get
right on the purchases.

Contractions

A new rule in technical writing is to write naturally and conversationally, the way
you talk (unless you speak poorly). However, you shouldn't use slang or dialect.

Pfeiffer Consulting

February 13, 2000

Alex Davis
Home Health Care
1238 Roe
Phoenix, AZ 34143

Subject: Proposed Business Writing Seminar

Thank you, Mr. Davis, for inviting us to help your colleagues with
their business writing. We're looking forward to working with you.

Here's a reminder of our agreed arrangements:

Date: February 22, 2000
Time: 8:00 a.m.-5:00 p.m.
Site: Cedar Inn

If we can provide any other information, please let us know.

One way to achieve conversationalism in technical writing is to use contractions (*we're, let's, here's, it's, can't,* etc.). Contractions, although long held in contempt by English teachers, now are accepted in composition classes as well as in technical writing. We all use them every day, and they make technical writing more approachable and more friendly.

The following examples prove our point. The first is from a user manual; the second is from a memo written by a CEO and distributed to an entire company.

> CONNECTING THE DG-40 TO YOUR COMPUTER
>
> Now that your printer is working, it's time to hook it up to your computer. It's best to turn off both the computer and the printer when you hook them up.
>
> Remember that each computer communicates differently with a printer. If your computer communicates through a parallel interface, all you'll need is a cable. If your computer requires another kind of interface, then you'll need an interface board.
>
> If you don't know what a parallel interface is, your computer manual or dealer will tell you.

In addition to the contractions, note that the preceding text achieves audience involvement through pronouns.

> Dear Friends,
>
> I can't tell you how proud I am of the extraordinary effort each of you has put forth toward reaching our goal. As you've probably heard by now, we attained 99.89% of our March 31, 2000, ship schedule.
>
> We did not quite reach our gold-ring goal of 100 percent ship performance, but we are so close that we're going to celebrate as if we did. We are committed to continuous improvement. I'd like to invite you to a special celebration this Tuesday.
>
> Let's take a moment to congratulate ourselves before we get back to work. We've got more units to ship and more gold rings to reach. When we work together as a team, we can make significant achievements.

This letter epitomizes motivation through personalization. The text involves the readers by making them feel a part of the team (through pronouns) and softens the technical edge of the writing (through conversational contractions).

Unfortunately, most antiquated technical writing is impersonal and alienates readers by its stiffness. The foregoing examples break that mold. Most modern technical writing strives to engage readers, not deny them. (At least one exception to this would be lab reports, in which impersonal objectivity is preferred.) Why will contractions help to involve readers? Contractions, which are conversational, help make technical writing less threatening. Often, technical writing reads too legalistically. This style deters most of us. In contrast, technical writing which uses contractions feels more casual and accessible. Contractions appeal to us because they duplicate our normal speech patterns.

READER BENEFIT

A final way to achieve audience involvement is to motivate your readers by giving them what they want or need. We're not saying that you should make false promises. Instead, you should show your audience how they will benefit from your technical writing. You can do this one of two ways: (a) explain the benefit and (b) use positive words and verbs.

Explain the Benefit

Until you tell your readers how they'll benefit, they don't know. Therefore, in your letter, memo, report, or manual, state the benefit clearly. You can do this anywhere, but you're probably wise to place the statement of benefit either early (first paragraph or abstract/summary) or late (last paragraph or conclusion). Placing the benefit early in the writing will interest readers and help ensure that they read on, remaining alert throughout the rest of the document. Placing the benefit at the end could provide a motivational close, leaving readers with a positive impression when they finish.

For example, when you write a procedure, you want your readers to know that by performing the steps appropriately, they will avoid mechanism breakdowns, maintain tolerance requirements, achieve a proper fit, or ensure successful equipment operation. By following the procedures, they will reap a benefit.

The following examples of procedures show how reader benefit is conveyed.

> *INSTRUCTIONS FOR POURED FOUNDATIONS*
> A poured foundation will provide a level surface for mounting both the pump and motor. Carefully aligned equipment will provide you a longer and more easily maintained operation.

> *INSTRUCTIONS FOR DRIVE BELT INSTALLATION*
> When installing your Amox Drive Belt, you need to achieve a belt tension of 3/8″ deflection for a 9-lb. to 13-lb. load applied at the center span. Check tension frequently during the first 30 hours of operation. Correct belt tension is important for maximum belt life in a large drive system.

Memos can also develop reader benefit to involve audiences. Figure 4.5 achieves audience involvement through reader benefit in many ways.

In this instance, the reader benefit is emphasized in the last paragraph. The pronouns *we* and *our* in the last sentence involve the readers in the activity. A team concept is implied. The benefit, however, is evident in the positive words sprinkled throughout the memo—*improved, strengthen, reduce, timely, accurate, happier, pride,* and *contributions*. These words motivate readers by making them perceive a benefit from their actions.

MEMORANDUM

DATE: December 2, 1999
TO: Fred Mittleman
FROM: Bev Adams
SUBJECT: Improved Billing Control Procedures

Three times in the last quarter, one of the billing cycles excluded franchise taxes. This was a result of the CRT input being entered incorrectly.

To strengthen internal controls and reduce the chance of future errors, the following steps will be implemented, effective January 1, 2000.

1. Input	The control screen entry form will be completed by the accounting clerk.
2. Entry	The CRT entry clerk will enter the control screen input and use local print to produce a hard copy of the actual input.
3. Verification	The junior accountant will compare the hard copy to the control screen and initial if correct.
4. Notification	The junior accountant will tell data processing to start the billing process.

The new procedure will provide more timely and accurate billing to our customers. Customer service will be happier with accounting, and we will have more pride in our contributions.

FIGURE 4.5 Memo Achieving Audience Involvement

Use Positive Words and Verbs

In all of the preceding examples geared toward reader benefit, the motivation or value is revealed through positive words or verbs. A sure way to involve your audience is to sprinkle positive words throughout your correspondence or to inject verbs (power writing) into your text. Positive words give your writing a warm glow; verbs give it punch. Positive words and verbs can sell the reader on the benefits of your subject matter.

Positive Words

advantage	effective	happy	profitable	successful
asset	efficient	please	satisfied	thank you
benefit	enjoyable	pleased	succeed	value
confident	favorable	pleasure	success	

To clarify how important such words are, let's look at some examples of negative writing followed by positive revisions.

Negative

- We cannot process your request. You failed to follow the printed instructions.

- The error is your fault. You keep your books incorrectly and cannot complain about our deliveries. If you would cooperate with us, we could solve your problem.

- We have received your letter complaining about our services.

- Your bill is now three weeks overdue. Failure to pay immediately will result in lower credit ratings.

- The invoice you sent was useless by the time it arrived. You wasted my time and money.

Positive

- So that we may process your request rapidly, please fill in line 6 on the printed form.

- To ensure prompt deliveries, let's get together to review our bookkeeping practices. Would next Tuesday be convenient?

- Thank you for writing to us about our services.

- If you're as busy as we are, you've probably misplaced our recent bill (mailed three weeks ago). Please send it in soon to maintain your high credit ratings.

- We're proud of our ability to maintain schedules. But we need your help. When you return invoices by the fifteenth, we save time and you save money.

Making something positive out of something negative is a challenge, but the rewards for doing so are great. If you attack your readers with negatives, you lose. When you involve your readers through positive words, you motivate them to work with you and for you.

Verbs also motivate your readers. The following verbs will motivate your readers to action.

accomplish	establish	organize
achieve	guide	plan
advise	help	prepare
analyze	implement	produce
assess	improve	promote

assist	increase	raise
build	initiate	recommend
conduct	install	reduce
construct	insure (ensure)	reestablish
control	lead	serve
coordinate	maintain	supervise
create	manage	support
develop	monitor	target
direct	negotiate	train
educate	operate	use

The memo in Figure 4.6 uses power verbs to motivate the reader. This memo brings together all of the techniques we've been discussing. It achieves audience involvement by speaking directly to the reader ("Martha"), by using pronouns, by including contractions to soften the technology, by using verbs that give the memo punch, and by employing positive words to motivate the reader to action.

MEMO

DATE: July 22, 2000
TO: Martha Collins
FROM: Bob Kaplan
SUBJECT: Assignment of Contract Administration Project

Martha, thank you for accepting the contract administration project. To accomplish our goals with this project, please proceed according to the following guidelines:

1. Analyze other operating companies to see how they ensure neutrality when administering contracts.
2. Coordinate our consultants to guarantee that they achieve impartiality in contract administration.
3. Develop a standard operating procedure for all contract negotiations.
4. Prepare a report of your findings and recommendations for our advisory team.

Your successful completion of this project, Martha, will help us avoid difficulties in future contract administrations. I'm confident you'll do a great job, and I know the company will be better served as a result of your expertise.

FIGURE 4.6 Memo Using Power Verbs

AUDIENCE CHECKLIST

To communicate effectively with your readers, ask yourself the following questions. Then circle the appropriate responses or fill in the needed information.

✓ What is my audience's level of understanding regarding the subject matter?
- High-tech
- Low-tech
- Lay

✓ Given my audience's level of understanding, have I written accordingly?
- Have I defined my acronyms, abbreviations, and jargon?
- Have I supplied enough background data?
- Have I used the appropriate types of graphics?

✓ Who else might read my correspondence?
- How many people?
- What are their levels of understanding?
  - High-tech
  - Low-tech
  - Lay
- Will my audience be multicultural?

✓ What is my role in relation to my audience?
- Do I work for the reader?
- Does the reader work for me?
- Is the reader a peer?
- Is the reader a client?

✓ What response do I want from my audience? Do I want my audience to act, respond, confirm, consider, decide, or file the information for future reference?

✓ Will my audience act according to my wishes? What is my audience's attitude toward the subject (and me)?
- Negative
- Positive
- Noncommittal
- Uninformed

Is my audience in a position of authority to act according to my wishes (can he or she make the final decision)?

If not, who will make the decision?

What are my audience's personality traits?

- Slow to act
- Eager
- Receptive
- Questioning
- Organized
- Disorganized
- Reticent

✔ Have I motivated my audience to act?

- Have I involved my reader in the correspondence by using pronouns and/or his or her name?
- Have I shown my reader the benefit of the proposal by using positive words and/or verbs?
- Have I considered any questions or objections my audience might have?

✔ Have I considered my audience's preferences regarding style?

- Will my reader accept contractions?
- Is use of a first name appropriate?
- Should I use my reader's last name?

✔ Have I avoided sexist language?

- Have I used *their* or *his or her* to avoid sexist singular pronouns like *his*?
- Have I used generic words such as *police officer* versus sexist words like *policeman*?
- Have I avoided excluding women, writing sentences such as "All present voted to accept the proposal" versus the sexist sentence "All the men voted to accept the proposal"?
- Have I avoided patronizing, writing sentences such as "Mr. Smith and Mrs. Brown wrote the proposal" versus the sexist sentence "Mr. Smith and Judy wrote the proposal"?

CHAPTER HIGHLIGHTS

1. High-tech readers understand acronyms, abbreviations, and jargon. However, not all of your readers will be high tech.

2. Low-tech readers need glossaries or parenthetical definitions of technical terms.

3. Lay readers usually will not understand technical terms. Consider providing these readers with extended definitions.

4. Multiple audiences have different levels of technical knowledge. This fact affects the amount of technical content and terms you should include in your document.

5. The definition of a technical term usually includes its type and distinguishing characteristics but can also include examples, descriptions, and procedures.

6. For multicultural audiences, define terms, avoid jargon and idioms, and consider the context of the words you use. Avoid cultural biases, avoid complicated punctuation, be careful with humor, and allow space for translation.

7. More than 50 percent of your audience will be female, so you should avoid sexist language.

8. To get your audience involved in the text, personalize it by using pronouns, the reader's name, and contractions.

9. Show your reader how he or she will benefit from your message or proposal.

ACTIVITIES

Revision

1. Revise the following sentences to avoid sexist language.

   a. All the software development specialists and their wives attended the conference.

   b. The foremen met to discuss techniques for handling union grievances.

   c. Every technician must keep accurate records for his monthly activity reports.

   d. The president of the corporation, a woman, met with her sales staff.

   e. Throughout the history of mankind, each scientist has tried to make his mark with a discovery of significant intellectual worth.

   f. The chairman of the meeting handled the debate well.

   g. All manmade components were checked for microscopic cracks.

   h. The supervisors have always brought the girls flowers on Mother's Day.

   i. Mr. Smith (the CEO) introduced his staff: Mr. Jones (the vice president), Mr. Brown (the chief engineer), and Judy (the executive secretary).

   j. The policemen and firemen rushed to the scene of the accident.

2. The following sentences are either dull, dry, and impersonal or egocentric. Revise them to achieve audience involvement through personalization, adding pronouns, names, and/or contractions. Strive for a two-to-one ratio, using *you* twice you for every *I*.

   a. The company will require further information before processing this request.

   b. It has been decided that a new procedure must be implemented to avoid further mechanical failures.

    c. The department supervisor wants to extend a heartfelt thanks for the fine efforts expended.

    d. I think you have done a great job. I want you to know that you have surpassed this month's quota by 12 percent. I believe I can speak for the entire department by saying thank you.

    e. If the computer overloads, simultaneously press Reset and Control. Wait for the screen command. If it reads "Data Recovered," continue operations. If it reads "I/O Error," call the computer resource center.

    f. Ampex Corporation announces the opening of a new office in the Fairway Village area.

    g. Telech's new communication system can help meet business telecommunication needs. This new system offers multiparty extensions, call forwarding, call waiting, intercom, and call accounting.

    h. There are three ways to solve the problem: line checks, system checks, and/or random checks.

    i. I want a member from each department to hear my speech so I can receive feedback on what I can do to improve my content and my delivery.

    j. Here is how to number pages. To number from page one, press Page Layout and then the number 1. To number from any other page, press Page Layout and then the required number, whatever it might be.

3. Revise the following sentences to achieve reader benefit, using positive and/or power verbs.

    a. We cannot lay your cable until you sign the attached waiver.

    b. John, don't purchase the wrong program. If we continue to keep inefficient records, our customers will continue to complain.

    c. You have not paid your bill yet. Failure to do so might result in termination of services.

    d. If you incorrectly quote and paraphrase, you will receive an *F* on the assignment.

    e. Send me the requested information by January 12.

    f. Poor input of information delays shipments, which forces us to pay unneeded expenses. These costs eventually come out of your year-end raises.

    g. Thank you for your recent call complaining about our product.

    h. Your team has lost 6 of their last 12 games.

    i. Your memo suggesting an improvement for the system has been rejected. The reconfigurations you suggest are too large for the area specifications. We need you to resubmit if you can solve your problem with calculations.

4. Write definitions as instructed below.

    a. Find examples of definitions provided in computer word processing programs, e-mail packages, Internet dictionaries, and online help screens. Determine whether these examples provide the *term*, its *type*, and its *dis-*

tinguishing characteristics. Are the definitions effective? If so, explain why. If not, rewrite the definitions for clarity.

b. Find examples of definitions in manuals such as your car's user manual, a manual that accompanied your computer, or manuals packaged with your radio/alarm clock, coffeemaker, VCR, or lawn mower. Note where these definitions are: are they placed parenthetically within the text or in a glossary? Next, determine whether the definitions provide the *term*, its *type*, and its *distinguishing characteristics.* Are the definitions effective? If so, explain why. If not, rewrite the definitions for clarity.

c. Find examples of definitions in your textbooks. Where are these definitions placed—parenthetically within the text or in a glossary? Next, determine whether the definitions provide the *term*, its *type*, and its *distinguishing characteristics.* Are the definitions effective? If so, explain why. If not, rewrite the definitions for clarity.

d. Select 5 to 10 terms, abbreviations, and/or acronyms from your area of expertise or interest (telecommunications, accounting, electronic engineering, automotive technology, biomedical records, data processing, etc.). First, define each term using words and phrases. Next, define each term in a sentence, conveying the *term*, its *type*, and its *distinguishing characteristics.* Finally, define each term using a longer paragraph. In these longer definitions, include examples, descriptions, and/or processes.

Team Projects

1. Individually, find examples of writing geared to high-tech, low-tech, and lay audiences. To do so, read professional journals, find procedures and instructions, look at marketing brochures, read trade magazines, and/or ask your colleagues and co-workers for memos, letters, or reports.

   Once you've found these examples, bring them to class. In small groups, discuss your findings to determine whether they are written for high-tech, low-tech, or lay readers. Use the Audience Evaluation Form on page 88 to record your decisions. Share these with all class members.

2. Once you've made your decisions regarding team project 1, select one of the high-tech examples and rewrite it for a low-tech or lay audience. To do so, work with your group to define your high-tech terms, improve the tone through pronouns and positive words, enhance the page layout through highlighting techniques, and add appropriate graphics.

3. In small groups composed of individuals from like majors, list 10 high-tech terms (jargon, acronyms, and/or abbreviations) unique to your degree programs. Then, envisioning a lay audience, parenthetically define and briefly explain these terms.

   To test the success of your communication abilities, orally share these high-tech terms with other students who have different majors. First, state the high-tech term to see if they understand it. If they don't, provide the

AUDIENCE EVALUATION FORM

EXAMPLE NUMBER	1	2	3	4	5
TYPE (CIRCLE ONE)	HIGH LOW LAY	HIGH LOW LAY	HIGH LOW LAY	HIGH LOW LAY	HIGH LOW LAY
Criteria					
Language					
• Abbreviations					
• Acronyms					
• Jargon (defined?)					
Content					
• General					
• Specific					
• Background					
Tone					
• Formal (high)					
• Less formal (low)					
• Least formal (lay)					
Format					
• Highlighting					
• Graphics					
—type					
—complexity					
—color					
—labeling					

parenthetical definition. How much does this help? Do they understand now? If not, add the third step—the brief explanation.

How much information do the readers need before they understand your high-tech terms?

4. In small groups, read the following examples to determine whether they are written to a high-tech or low-tech audience. To justify your answer, con-

sider the use of acronyms, abbreviations, and/or jargon; the presentation of background data; the length of sentences and paragraphs; and the way in which the text has been formatted (how it looks on the page).

EXAMPLE 1

This documentation summarizes the items to be tested and the methodology utilized in the testing routine. The test plan tested four new 8536 Switcher features: OIM channel commands, OIM disconnect commands, OIM dump commands, and OIM log-off messages.

The patch files associated with the base must disable the features; consequently, only testing to ensure that this is operational will be addressed.

Software development personnel, after testing the features, will deliver copies of this report and product acceptance releases to program management, who are then responsible for writing functional specifications and submitting them to administration for manufacturing and marketing.

EXAMPLE 2

Purpose: This document identifies the time to be tested and the methods used for testing. The tests center on the 8536 Switcher (a network consisting of two nodes) and CN (connect node) bases. The 8536 Switcher and CN bases perform the same function as our currently marketed G9 and G12 systems.

OIM (Operator Interface Module) Features Tested:

- OIM channel commands
- OIM disconnect commands
- OIM dump commands
- OIM log-off messages

Testing Focus: To work appropriately, these features must be disabled by our patch files if an error occurs. Therefore, our testing focus will be the disabling capabilities.

Responsibilities: After testing the features, our software development group will send copies of this report and product acceptance releases to program management. They then will write the functional specifications and send them to administration. Administration is responsible for manufacturing and marketing.

The flowchart on the following page graphically depicts this sequence.

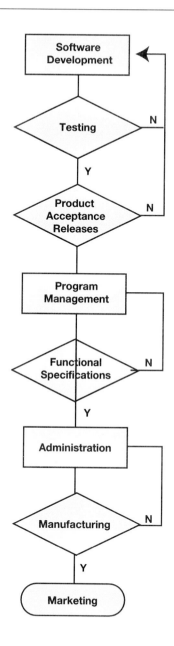

CHAPTER **5**

Memos

OBJECTIVES

Memos (correspondence written and read only within a company) are an important means by which employees communicate with each other. A memo is important not only because of its frequency of use and the wide range of subject matter presented in memo form but also because memos represent a component of your interpersonal communication skills within your work environment.

This chapter presents the following information regarding memos: (a) the differences among memos, letters, and e-mail; (b) criteria for writing successful memos; (c) the process for writing memos (with sample process log); and (d) examples.

THE DIFFERENCES AMONG MEMOS, LETTERS, AND E-MAIL

Before we discuss the criteria for memos, review the differences among memos, letters, and e-mail shown in Table 5.1.

TABLE 5.1 Memos versus Letters and E-mail			
CHARACTERISTIC	MEMOS	LETTERS	E-MAIL
Destination	Internal: correspondence written to colleagues within a company.	External: correspondence written outside the business.	Internal or external: correspondence written to personal friends and acquaintances as well as business associates.
Format	Identification lines include "Date," "To," From," and "Subject." The message follows.	Includes letterhead address, date, reader's address, salutation, text, complimentary close, and signatures.	Identification lines include "Subject," "From" (writer's name and e-mail address), "Sent" (date of transmission), and "To" (reader's name and e-mail address). The message follows.
Audience	Generally high-tech or low-tech, mostly business colleagues.	Generally low-tech and lay readers, such as vendors and clients.	Generally multiple readers with various levels of knowledge. Could include instructors, company supervisors, and subordinates as well as family and friends.
Topic	Generally high-tech to low-tech; abbreviations and acronyms often allowed.	Generally low-tech to lay; abbreviations and acronyms usually defined.	A wide range of diverse topics determined by audience.
Tone	Informal (peer audience).	More formal (audience of vendors and clients)	Usually informal (due to the "conversational" nature of electronic communication).

Attachments or enclosures	Hard-copy attachments can be stapled to the memo. Complimentary copies can be sent to other readers.	Additional information can be enclosed within the envelope. Complimentary copies can be sent to other readers.	Computer files, active Web links, and downloadable graphics can be attached. Complimentary copies can be sent to other readers.
Structure	Typically 8½″ × 11″ with 1″ margins, 80 characters per line, and about 55 lines per page.	Typically 8½″ × 11″ with 1″ inch margins, 80 characters per line, and about 55 lines per page.	Typically, one viewable screen with 60–70 characters per line and 12–14 lines per screen. Beyond these parameters, a reader must scroll.
Delivery Time	Determined by a company's in-house mail procedure. Memos could be delivered within 3 days (more or less).	Determined by the destination (within the city, state, or country). Letters could be delivered within 3 days but may take more than a week.	E-mail messages can be delivered within minutes, but delays are possible.
Security	If a company's mail delivery system is reliable, the memo will be placed in the reader's mailbox. Then, what the reader sees on the hard copy page will be exactly what the writer wrote. Security depends on the ethics of co-workers and whether the memo was sent within an envelope.	The U.S. Postal Service is very reliable. Once the reader opens the envelope, he or she sees exactly what the writer wrote. Privacy laws protect the letter's content.	E-mail systems, like all computer systems, malfunction from time to time. Sent e-mail might not arrive. What the reader sees on the screen might not be what the writer saw on the screen (due to problems with ASCII codes). The text will be the same, but the page layout could differ. E-mail can be tampered with and/or read by others with access to the system.

CRITERIA FOR WRITING SUCCESSFUL MEMOS

Although every memo will not look the same, Figure 5.1 shows an ideal organizational template that works well for most memos.

SUBJECT LINE

One hundred percent of your readers read the subject line (after all, we have to start reading any text at its beginning. The beginning of a memo is the subject line). The subject line, typed in all caps, is where you begin talking to your reader(s). Therefore, you want it to work for you. One-word subject lines don't communicate effectively, as in the following flawed subject line.

DATE:

TO:

FROM:

SUBJECT: Focus + Topic

Introduction: A lead-in, warm-up, overview stating *why* you're writing and *what* you're writing about

Discussion: Detailed development, made accessible through highlighting techniques, explaining *exactly what* you want to say

Conclusion: A summation stating *what's next, when* this will occur, and *why* the date is important

FIGURE 5.1 All-Purpose Memo Template

Subject: COMPTROLLERS

We've got a *topic* (a *what*), but we're missing a *focus* (a *what* about the *what*). An improved subject line would read as follows:

Subject: SALARY INCREASES FOR COMPTROLLERS

This subject line gives us the *topic*—comptrollers (the *what*)—plus a focus—salary increases (the *what* about the *what*)—all linked by a preposition.

Such a sequence works in all subject lines. For instance, in the following examples, note how the focus alters the meaning of the subject line's topic.

Subject: FOCUS	plus	TOPIC
Termination	of	Comptrollers
Hiring Procedures	for	Comptrollers
Vacation Schedules	for	Comptrollers
Training Seminars	for	Comptrollers

Although the topic stays the same, the focus changes and clarifies for the reader the actual subject matter of the memo.

INTRODUCTION

Once you've communicated your intent in the subject line, you want to get to the point in the introductory sentence(s). Readers are busy and don't want your memo to slow down their work. To avoid any delays for your audience, you want your first sentence or sentences to communicate immediately. A goal is to write one or two clear introductory sentences which tell your readers *what you want* and *why you're writing*. Remember, until you tell them, they don't know.

> In the third of our series of quality control meetings this quarter, I'd like to get together again to determine if improvements have been made.

This example invites the reader to a meeting, thereby communicating what the writer's intentions are. It also tells the reader that the meeting is one of a series of meetings, thus communicating why the meeting is being called.

> As a follow-up to our phone conversation yesterday (8/12/00), I have met with our VP regarding your suggestions. He'd like to meet with you to discuss the following ideas in more detail.

This introduction reminds the reader why this memo is being written—as a follow-up—and tells the reader what will happen next.

DISCUSSION

The discussion section allows you to develop your content specifically. You want to respond to the reporter's questions mentioned in Chapter 3 (*who, what, when, why, where, how*), but you also want to make your information accessible. Since very few readers read every line of your memo (tending instead to skip and skim), traditional blocks of data (paragraphing) aren't effective visually. The longer the paragraph, the more likely your audience is to avoid reading. Instead, try to make your text more reader-friendly by applying some of the highlighting techniques discussed in Chapters 8 and 9: (a) itemization, (b) white space, (c) boldface type, (d) headings, (e) columns, and (f) graphics.

Note the difference between the following examples: the first is reader-unfriendly, and the second is reader-friendly.

EXAMPLE 1—UNFRIENDLY TEXT

This year began with an increase, as we sold 4.5 million units in January compared to 3.7 for January 1999. In February we continued to improve with 4.6, compared with 3.6 for same time in 1999. March was not quite so good, as we sold 4.3 against the March 1999 figure of 3.9. April was about the same with 4.2, compared to 3.8 for April 1999.

EXAMPLE 2—READER-FRIENDLY TEXT

COMPARATIVE QUARTERLY SALES (IN MILLIONS)			
	1999	*2000*	INCREASE/DECREASE
Jan.	3.7	4.5	0.8+
Feb.	3.6	4.6	1.0+
Mar.	3.9	4.3	0.4+
Apr.	3.8	4.2	0.4+

The first block of data is unappealing to read and hard to follow because of its complexity. You can't see the relationship between figures and years clearly, nor can you see the variance from year to year easily. On the other hand, the table in the second example is accessible at a glance. It's clear and concise—perfect technical writing.

CONCLUSION

Conclude your memo with a *complimentary close* and/or a *directive close*. A complimentary close motivates your readers and leaves them happy, as in the following example.

> If our quarterly sales continue to improve at this rate, we will double our sales expectations by 2000. Congratulations!

A directive close tells your readers exactly what you want them to do next or what your plans are (and provides dated action).

> Next Wednesday (12/22/00), Mr. Jones will provide each of you a timetable of events and a summary of accomplishments.

Why is a conclusion important? Without it, the reader senses a lack of closure and does not know what to do next or why the actions requested are important.

Without a conclusion, the reader's response is going to be "OK, but now what?" or "OK, but so what?" To write an effective memo, sum up and provide closure.

AUDIENCE

Another criterion for effective memo writing is audience recognition. In memos, audience is both easier and more complex than in letters. Since letters go outside your company, your audience is usually low tech or lay, demanding that you define your terms more specifically. In memos, on the other hand, your in-house audience is easier to define (usually low tech or high tech). Thus, you often can use more acronyms and internal abbreviations.

However, whereas your audience for letters is usually singular—one reader—your audience for memos might be multiple. Studies tell us that memos average six readers. You might be writing simultaneously to your immediate supervisor (high tech), to his or her boss (low tech), to your colleagues (high tech), and even to a CEO (low tech).

This diverse readership presents a problem. How do you communicate to a large and varied audience? How do you avoid offending your high-tech readers with seemingly unnecessary data, which your low-tech readers need?

The problem is not easy to solve. Sometimes you can use parenthetical definitions. For instance, don't just write *CIA*, which to most people means Central Intelligence Agency. Provide a parenthetical definition—*CIA* (cash in advance)—if your usage differs from what most people will automatically assume. You are always better off saying too much.

STYLE

The appropriate style for memos is the same technical writing style discussed in earlier chapters, depending on conciseness, clarity, and accessibility. Use simple words, readable sentences, specific detail, and highlighting techniques.

In addition, strive for an informal, friendly tone. Memos are part of your interpersonal communication abilities. Just as you should be concerned with how you speak to others on the telephone or at the water cooler, you should assess your tone within a memo. Do you sound natural, normal, informal, and friendly? Does your memo sound like it is written by the kind of person you would want to work with or for? If not, you might be creating a negative image.

Decide which of the following examples has the more positive "feel." Which boss would you prefer working with?

> *EXAMPLE 1*
> We will have a meeting next Tuesday, Jan. 11, 2000. Exert every effort to attend this meeting. Plan to make intelligent comments regarding the new quarter projections.

> *EXAMPLE 2*
> Let's meet next Tuesday (Jan. 11, 2000). Even if you're late, I'd appreciate your attending. By doing so you can have an opportunity to make an impact on the new quarter projections. I'm looking forward to hearing your comments.

Isn't the latter more personal, informal, appealing, and motivational? This tone is achieved through audience involvement (*you* usage), contractions, and positive words, aspects of effective writing discussed in Chapter 4.

When you write a memo, you are writing to someone with whom you work every day. The Golden Rule is write as you would like to be written to.

GRAMMAR

Abide by all grammatical conventions when writing memos. Poor grammar or typographical errors destroy your credibility.

PROCESS

Now that you know what to include in a memo, follow the three-step process to create the correspondence.

- Prewriting
- Writing
- Rewriting

PREWRITING

No single method of prewriting is more effective than another, so let's try *clustering/mind mapping.*

The type of prewriting you prefer defends on your unique learning style. Clustering/mind mapping is visual and free-form. Figure 5.2 shows a typical mind-map pattern.

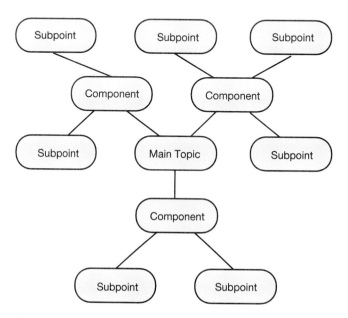

FIGURE 5.2 Clustering/Mind Mapping

The idea of clustering is to start with your main topic, whatever that might be, and then determine what that topic is composed of. For example, if your main topic is a company picnic, the next question is to ask what that picnic should entail. If you're planning such a picnic, you would want to consider food, entertainment, and a schedule for activities. These items represent the major components and are diagrammed in Figure 5.3.

Writing always requires that you develop your points so you are clear and detailed. Thus, the next step in clustering is to provide more detail through the addition of subpoints. Subpoints clustered around the food component might include the entree (BBQ beef), condiments (pickles, relish, potato salad), drinks (soda and lemonade), and desserts (apple pie). Subpoints clustered around the entertainment component might include a live band and games for both adults and children.

Figure 5.4 shows a mind map including subpoints.

Clustering/mind mapping is valuable because it allows you to sketch your ideas freely without falling prey to the rigid structure of an outline (which often deters some writers' creativity). Clustering/mind mapping also allows you to see graphically the relationship between subpoints and components of an idea. This helps you determine if you've omitted any key concerns or included any irrelevant ideas. If so, you can develop your ideas further by adding new subpoints or delete irrelevancies simply by scratching off one of your subpoint balloons. Thus, clustering/mind mapping is an excellent prewriting technique because it easily allows you to gather information and organize your thoughts.

WRITING

Once you've gathered your data and determined your objectives in prewriting, your next step is to draft your memo. In doing so, consider the following drafting techniques.

1. Review your prewriting. Before you begin writing, look back over your clustering to see if you've missed anything important. This review also provides one final reminder of what you want to say (the content of the memo) and why you are going to say it (your objectives).

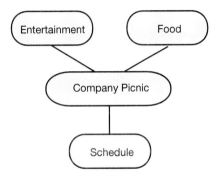

FIGURE 5.3 Major Components of a Company Picnic Mind Map

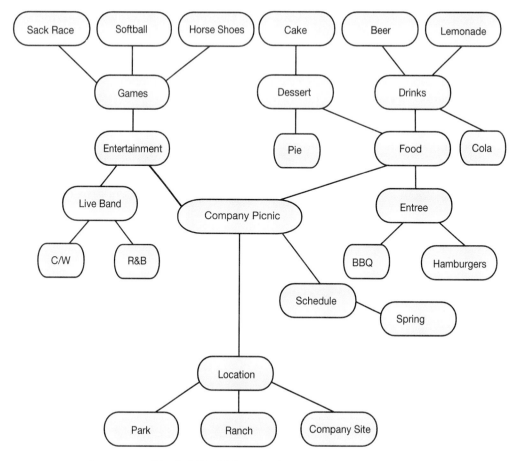

FIGURE 5.4 Company Picnic Mind Map, Including Subpoints

2. Determine your focus (topic sentence, thesis statement, objective of the memo, etc.). Prior to drafting the memo, write down in one or two sentences your objective for this correspondence. For example, regarding the company picnic, you might want to write, "My plan for the picnic includes food, entertainment, and a schedule for activities." Such a sentence provides you with a clear focus for your draft. It helps keep you from straying off course in your memo.

3. Clarify your audience. Again, before writing the draft, be sure that you know who you are writing to. Is your audience high tech, low tech, or lay? Is the audience composed of management, subordinates, or lateral colleagues? Is the audience multiple, consisting of many of the above levels? Make this determination and then write accordingly.

4. Review your memo criteria. Remember what a memo entails. In addition to listing date, to, from, and subject lines, try to use the template shown earlier (introduction, body, and conclusion).

5. *Organize your ideas.* If your supporting details are presented randomly, your audience will be confused. As a writer, you are obligated to develop your content in a manner which will allow your readers to follow your train of thought easily. Therefore, when you draft your memo, choose a method of organization that will help your readers understand your objectives. In Chapter 3, we discussed organizing by importance. You may want to use this method of organization in your memo.

Another method of organization is *chronology*. A chronological organization follows a time sequence, from first to last, from 8:00 A.M. to 5:00 P.M., from Monday through Friday, from first quarter to fourth quarter, from step 1 through step 27. This method of organization is useful if your memo focuses on any of the following:

Past Histories	Future Events
Incidents or accidents	Instructions
Meeting minutes	Deadlines
Performance appraisals	Agendas
Customer complaints	Itineraries

The memo shown in Figure 5.5 is organized chronologically.

6. *Write the draft.* Once you've decided on content (by reviewing your prewriting), determined your focus, clarified your audience, reviewed your criteria for effective memos, and determined how best to organize your thoughts, all that's left is to draft the text. The key to successful drafting is to avoid worry. When you worry about the end result, you get anxious and you can't write. To draft, you must throw words on the page freely. Don't delay, worry, or procrastinate. Just write it down! Use the sufficing technique. Get it on the page and let it suffice, for now. You can go back and revise later.

REWRITING

To revise your memo and make it as good as it can be, follow these revision techniques:

- *Add new detail for clarity.* Reread your draft. If you've omitted any information stemming from the reporter's questions (*who, what, when, why, where, how*), insert answers to these questions.

- *Delete dead words and phrases for conciseness.* Chapter 3, "Objectives in Technical Writing," will help you.

- *Simplify words and phrases.* For example, *in lieu of* might confuse someone. Why not simplify this to *instead of*? Don't "write down" to your audience. A good rule of thumb is to write as you speak, unless you speak poorly. Be more casual. Chapters 3 and 4 will help you achieve these goals.

- *Move information from top to bottom or bottom to top for emphasis.* Place your most important information first, or move information around to maintain an effective chronological order.

MEMORANDUM

DATE: May 1, 2000
TO: Planning Committee
FROM: Steve Janasz
SUBJECT: PLANNING AGENDA FOR JULY 4 COMPANY PICNIC

Congratulations! We had a series of outstanding meetings with excellent input from all team members. I appreciate your involvement.

To confirm our decisions reached yesterday at the final planning committee meeting, here's the agenda for everyone's July 4 Company Picnic responsibilities:

Entertainment

Mary, you'll need to focus on two topics: music and games. By *May 10*, select and book a country-western band. The team members' first choice is the band Rattlesnake, but they also would be happy with Texas River. By *May 15*, make arrangements with the local community college to lend us softball equipment (balls, bats, and gloves).

Food

Tasha, you're in charge of food. We've already chosen the menu (hamburgers, hot dogs, potato salad, baked beans, cole slaw, chocolate cake, and vanilla ice cream). By *May 20*, select a caterer. Remember, budget is important!

Location

Galen, since the committee members voted on softball for our company entertainment, you'll need to reserve a ball field. By *May 25*, contact the city Parks and Recreation Department and reserve field #2 at Pease Park. This field is the only one with covered eating facilities nearby.

Invitations

Jordan, you'll need to write and mail invitations by *June 10*. Use our desktop publishing software to produce the mailers. Be sure to include where we'll hold the picnic, when to arrive, when the festivities plan to end, costs per family, and the agenda for events.

Thank you for all your support on this project. Together we'll make this team effort a success. July 4 will be a company festival guaranteed to raise corporate morale.

FIGURE 5.5 Directive Memo Organized Chronologically

- *Reformat for access.* Use highlighting techniques (white space, bullets, numbers, headings, etc.) for reader-friendly appeal. (See Chapter 8 for information on document design.)
- *Enhance the tone and style of your memo.* If you want to sound tough, then revise your memo to sound that way. If you want to tone down your correspondence, now is the time to do so. If you want to personalize the memo (remembering that memos are part of your interpersonal communication skills within the office), then add pronouns and/or names to give the memo a more person-to-person tone. Chapter 4 discusses this important aspect of writing.
- *Correct for accuracy.* Proofread!
- *Avoid sexist language.* Women and men are equal. Sexist language implies that only men work and have positions of prominence.

Once you have revised your draft in the rewriting stage, review it once more according to the following Effective Memo Checklist.

EFFECTIVE MEMO CHECKLIST

✔ *Have you used the correct memo format,* including date, to, from, and subject lines?

✔ *Is your subject line correct?*
  - Is it typed in all caps?
  - Does it contain a topic and a focus?

✔ *Does your introduction tell why you are writing and what you are writing about?*

✔ *Does the body explain exactly what you want to say?*

✔ *Does the conclusion tell what's next,* providing either a complimentary or a directive close?

✔ *Is your page layout reader-friendly?*
  - Do you use highlighting techniques for accessibility, such as bullets, boldface, underlining, and white space?

  Refer to Chapter 8 for help.

✔ *Is your writing style concise?*
  - Do you limit the number of words per sentence, the number of syllables per word, and the number of lines per paragraph?

  Refer to Chapter 3 for help.

✔ *Is your writing clear?*
  - Do you answer reporter's questions?
  - Do you avoid vague words, such as *some, several, many,* or *few,* specifying instead?

  Refer to Chapter 3 for help.

✔ *Have you written appropriately to your audience?*
  • Have you defined high-tech terms where necessary for low-tech or lay readers?
  • Have you achieved the correct tone and personalized your memo, based on your audience and purpose?
  Refer to Chapter 4 for help.
✔ *Are errors eliminated?* Remember, a memo with grammatical or mechanical errors will destroy your credibility. You must prooofread to catch erros like the ones in this sentence—*prooofread* and *erros.*

PROCESS LOG

PREWRITING

Figure 5.6 shows one student's prewriting, written in response to a request that all students write a memo telling how to write a memo.

WRITING

Following is the student's rough draft (Figure 5.7).

REWRITING

Student peer evaluators suggested the revisions in Figure 5.8 (on p. 106). The finished copy is shown in Figure 5.9 (on p. 107).

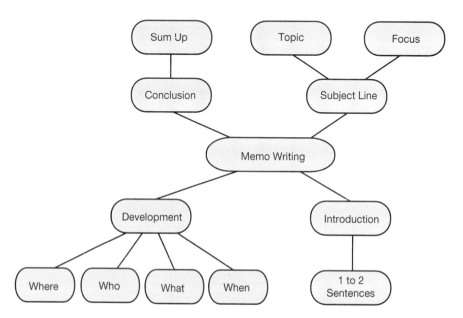

FIGURE 5.6 Student Mind Mapping

Subject: Memo Writting

Following are ways to write better memos.

Include a Subject Line, which provides a topic and a focus. Then have an Introduction to tell your readers what he needs to know. This Introduction could be about one or two sentences. Next, develope your ideas in a Discussion. Finally, sum it up in a Conclusion.

These suggestions should help you write better memos.

FIGURE 5.7 Rough Draft

SAMPLE MEMOS

Following are additional examples of both student and professionally written memos (complete with our evaluations of the memos' successes and needed improvements). The first, a student-written memo (Figure 5.10), is well organized and developed, but it has several stylistic flaws.

Let's discuss the successes first. We like the way the memo is formatted (indented body for accessibility). It has an introduction, discussion, and conclusion, following our template. The writer also has developed her ideas thoroughly: The introductory paragraph answers the reporters' questions *who* and *what*; the body discusses *why*; and the conclusion focuses on *when*. Finally, the tone of the memo is appropriate for management (especially in the last paragraph).

However, the memo is flawed stylistically. Look how long that first sentence is! This type of writing requires a high degree of patience on the reader's part. In addition, the writer has made two punctuation errors. No period is needed in the subject line. The colon after *because* in the body is incorrect. Either no punctuation is needed, or the sentence should be rewritten so a colon could be used. For example, if the sentence reads ". . . it would benefit our company for the following reasons," then you could insert a colon after *reasons*. (Chapter 18 covers this grammatical rule.)

The example in Figure 5.11 (on p. 109) is professionally written. It too has successes and failures. This memo succeeds in the following ways. As with the earlier student-written example, this memo uses the introduction-body-conclusion template. It ends with an excellent series of directives, specifying what is to occur next. The body is formatted for reader-friendly ease of access. The memo answers the reporter's questions *what*, *who*, *why*, and *when*.

The memo is flawed, however, in several ways. First, the subject line has a topic (operating procedure) but no focus. What about this procedure? Will the memo discuss the termination of the procedure, the writing of a new procedure, or changes in the procedure? A subject line such as "Revision of Operating Procedure 329 dated 5–9–00" would provide the necessary focus.

Add Date, To, From   (No Focus - what about it?)
Subject: Memo Writting   and needs to be typed
            SP   in all caps

Following are ways to write better memos.   Add why!
                                    What's their motivation?

            Reformat
Include a Subject Line, which provides a topic and a focus.

            Reformat   agreement?
Then have an Introduction to tell your readers what he

needs to know. This Introduction could be about one or

                    SP
two sentences. Next, develope your ideas in a Discussion.

Finally, sum it up in a Conclusion.   Reformat the text—make
    Your entire   it more accessible through
    memo needs   Highlighting
    More Development
These suggestions should help you write better memos.
                        Poor tone—motivate & tell
                        them when (Dated action)

FIGURE 5.8 Suggested Revisions

Next, note that this memo is written to a multiple audience (Distribution). Will each reader recognize that *PQR*, *OP*, and *EN* are abbreviations for *product quality requirement*, *operating procedure*, and *engineering notice*? You'd have to guess at an answer, and guessing gets you in trouble in technical writing. When in doubt, spell it out. To avoid any problems with audience recognition, add your abbreviation parenthetically. For example, write, "operating procedure (OP) 329." Doing so takes very little time but eliminates any audience doubt.

Finally, the memo uses sexist language in the sentence "They are then to conduct reviews with their supervisors to assure that each supervisor is aware of *his* responsibility." This must read "*his* or *her* responsibility."

DATE: October 19, 2000
TO: George Calvert
FROM: Candice Millard
SUBJECT: SUGGESTIONS FOR IMPROVED MEMO WRITING

George, I've just returned from a training seminar titled "Better Business Writing." Our department manager asked me to share tips I've learned about writing effective memos. Following are the improved memo-writing techniques I learned at this training seminar.

1. *Subject Line*   One hundred percent of readers read the subject line. Thus, to make this first line of communication effective, include a topic ("memo writing") and a focus ("suggestions for").

2. *Introduction*   Limit your introduction to one or two sentences. State why you're writing and what you're writing about.

3. *Discussion*   In the body, develop your points specifically (answering *who, what, where,* and *why*). Then itemize your points for easy access.

4. *Conclusion*   End by telling the reader what to do next and when to accomplish this task.

By using these techniques, you'll be able to communicate more successfully. If you have any questions, please drop by my office. I'd be happy to share more information with you.

FIGURE 5.9 Revised Memo

The two professionally written memos in Figures 5.12 and 5.13 (on pp. 110-111) succeed in all ways: organization, visual appeal, development, grammar, and tone.

MEMO

Date: January 19, 2000
To: Tom Lisk
From: Juliet Kincaid
Subject: Producing the Annoy-No-More Call Screening Machine.

Our product development staff has come up with a breakthrough new product, a phone answering machine that answers the phone, screens calls, and, by voice activation, routes the call to an appropriate message and gives a busy signal when the person doesn't want to be bothered.

We have researched this product fully and believe it would benefit our company because:

1. The product has a relatively simple design that occupies no more space than a regular phone and can be designed for wall or desk mounting.

2. The cost for manufacturing is low compared to the retail value. The unit costs approximately $100 to produce and retails for $999.99.

3. The unit can be interchanged easily with all phone networks including MBB and Shout.

We would like to begin mass production around April 1, 2000. Our Westport facility could handle the production with the addition of 10 technicians and engineers to aid in production. We hope you give this product consideration and endorse it fully. Thanks.

FIGURE 5.10 Student-Written Memo

MEMORANDUM

DATE: June 4, 2000
TO: Distribution
FROM: Michelle Fields
SUBJECT: OPERATING PROCEDURE 329 DATED 5-9-00

The reissue of this procedure was the result of extensive changes requested by Quality, Manufacturing, and Engineering.

The procedural changes that must be implemented according to Engineering Notice 195 are as follows:

- The asterisk (*) can no longer be used to identify guidance information or substitutable items in requirement lists. Substitutions must be called out by catalog numbers or descriptive names.

- Product quality requirements must be included in work directions either by stating the requirement or by referring to the appropriate PQR number.

- When oral instructions are given by supervisors, all engineers must be present.

Please review this revised OP and EN with your supervisors. They are then to conduct reviews with their supervisors to ensure that each supervisor is aware of his responsibility. Advise me by memo when all reviews are complete.

Distribution: H.P. Adams
              S.D. Smith
              P.R. Braun
              A.J. Rogers
              T.D. Peters
              L.E. Moss

FIGURE 5.11 Professionally Written Memo

 Memo

Data Coding Corp.
1818 Bond Ave.
Comstock, CO 44210

DATE: June 4, 2000
TO: Simon Maxwell
FROM: Larry Dodge
SUBJECT: INCREASED REJECTS DUE TO FAULTY ORDER CODES

Simon, last week our technicians recorded 8 to 10 percent increases in reject rates. Our target reject rate is 3 to 5 percent. I studied the problem to find out possible causes and to suggest solutions. Below are the results of this study.

Rejected Orders	Causes for Reject	Solution
10	Duplicate orders created when line workers failed to close the order data tape (ODT) file.	Have line workers report each ODT file closure.
28	Lost orders caused when a change order save (COS) did not execute.	Report COS problems to Information Services.
46	Orders misfiled under the wrong categories when a scale was incorrectly calibrated.	Recalibrate all scales monthly.

These solutions should place reject rates within our desired range. If additional problems occur, we'll probably need to form a TQM team to develop new procedures. Let me know what you decide.

FIGURE 5.12 Professionally Written Problem-Solution Memo

Diamond Delivery
101 Starburst Dr.
Antelope, GA 54210

MEMORANDUM

DATE: December 12, 2000
TO: Distribution
FROM: Luann Brunson
SUBJECT: REPLACEMENT OF MAINTENANCE RADIOS

On December 5, the manufacturing department supervisor informed the purchasing department that our company's maintenance radios were malfunctioning. Purchasing was asked to evaluate three radio options (the RPAD, XPO 1690, and MX16 radios). Based on my findings, I have issued a purchase order for 12 RPAD radios.

The following points summarize my findings.

1. *Performance*
   During a one-week test period, I found that the RPAD outperformed our current XPO's reception. The RPAD could send and receive within a range of 5 miles with minimal interference. The XPO's range was limited to 2 miles, and transmissions from distant parts of our building broke up due to electrical interference.

2. *Specifications*
Both the RPAD and the MX16 were easier to carry, because of their reduced weight and size, than our current XPO 1690s.

	RPAD	*XPO 1690*	*MX16*
Weight	1 lb.	2 lbs.	1 lb.
Size	5" x 2"	8" x 4"	6" x 1"

3. *Cost*
   The RPAD is our most cost-effective option because of quantity cost breaks and maintenance guarantees.

	RPAD	*XPO 1690*	*MX16*
Cost per unit	$70.00	$215.00	$100.00
Cost per doz.	$750.00	$2,580.00	$1,100.00
Guarantees	1 year	6 months	1 year

Purchase of the RPAD will give us improved performance and comfort. In addition, we can buy 12 RPAD radios for approximately the cost of 4 EXPOs. If I can provide you with additional information, please call. I'd be happy to meet with you at your convenience.

Distribution:	M. Ellis	M. Rhinehart	T. Schroeder
	P. Michelson	R. Travers	R. Xidis

FIGURE 5.13 Professionally Written Compare-and-Contrast Memo

CHAPTER HIGHLIGHTS

1. Memos are an important part of your interpersonal communication on the job.

2. Subject lines are read by 100 percent of your audience.

3. Use a topic and focus in your subject line.

4. In the introduction, state what you want and why you are writing.

5. In the discussion section, state the details.

6. Conclude by telling the reader what you plan to do next or what you expect him or her to do next. You might also want to date this action.

7. Consider the level of your audience when you write a memo.

ACTIVITIES

Writing Memos

1. Write a memo on one of the following topics. To do so, first prewrite (using mind mapping/clustering), then write a draft, and finally rewrite (revising your draft).

a. Your work environment is experiencing a problem (with scheduling, layoffs, turnover, production, quality, morale, etc.). Your boss has asked you to write a memo noting the problem and suggesting solutions.

b. An employee under your supervision sees a problem in his or her work environment and has written a memo suggesting a solution. You don't believe the suggested solution will work. Write a memo providing your response.

c. A major project is being introduced at work. Write a directive memo informing your work team of their individual responsibilities and schedules.

d. Your department needs a new piece of equipment to perform work. Write a memo requesting this equipment. Justify the need for the equipment and give the date when the equipment is needed.

e. You work in the purchasing department and must buy a new piece of equipment. You must first compare bids. You've done so and now must write a comparison/contrast memo explaining why you plan to purchase one piece of equipment versus another.

f. It's time for your quality circle team to meet again (or any committee you chair). Write a memo calling the meeting. Provide an agenda.

2. Practice mind mapping/clustering. Choose any of the topics presented in activity 1 and create a mind map/cluster.

Team Projects

1. Choose one of the topics listed in activity 1. In groups of three to five students, write the memo requested. Once the group memos are written, evaluate them. You can do this as a class by projecting transparencies, or you could exchange group memos and evaluate them in peer review groups. Decide which memos are successful and explain why. Decide which memos are flawed and explain why.

2. Gather memos from your work site (or the work site of your friends or relatives). In small groups of three to five, discuss whether these memos are successful or unsuccessful. To make your judgments, use the criteria for memos discussed in this chapter. Explain to the class why the good memos succeed and why the flawed memos fail. Then revise the flawed memos to improve them.

3. In groups of three to five students, revise the following flawed memo. Consider the criteria for good memos and rewriting techniques presented in this chapter. You can invent any information you consider necessary to improve the memo.

DATE: April 6, 2000
TO: Lenny Goodman
FROM: Wilkes Berry
SUBJECT: Engineering

To solve the recent problems with sewer design, the engineering shift will perform some studies on corrosion, water seepage, pipe decay, alignment, depth, soil composite, and so forth. We'll need to do this soon. If you have any questions, let me know.

Case Study

Apex Corporation is experiencing repeated failures with its Maxwell air compressors. The uncoated radiators tend to corrode, which causes pinhole leaks that reduce air pressure and volume. And, because the radiators were built before current energy-saving devices were mandated, the Maxwells use excess electricity. This forces compressors to make up for lost air. Finally, the Maxwells, which are now ten years old, have a high failure rate (24 percent downtime). This leads to increased maintenance costs and a lack of dependability.

Cheryl Huff, the department supervisor for maintenance engineering, needs to write a memo to the plant supervisor, Mark Levin, suggesting a solution. Her options are to retrofit the current Maxwells with modern energy-saving devices, to coat the Maxwells to prevent additional corrosion, and/or to purchase new replacement radiators. The cost for replacements would be approximately $15,000. Although that's high, the cost for continued maintenance and downtime is also high. Cheryl received this $15,000 bid contingent upon ordering before a November 15 deadline. After that date, the costs will be $16,750 because of a year-end price increase. Retrofitting and coating the radiators are stop-gap solutions that can be undertaken by in-house maintenance crews. Whichever choice Cheryl promotes must happen soon. Winter is fast approaching, and the air demand will increase.

Make the decision for Cheryl, based on the information provided (you can invent any additional information you choose). Then write the memo. Abide by the criteria provided in this chapter.

6 Letters

OBJECTIVES

Letters are external correspondence that you send from your company to a colleague working at another company, to a vendor, to a prospective client, to an agency, or to a friend who lives around the corner or across the continent. Letters leave your work site (as opposed to memos, which stay within the company).

Because letters are sent to readers in other locations, you must write them effectively. Your letters not only reflect your communication abilities but also are a reflection of your company. If your letter is sloppy, is marred with grammatical errors, or creates a negative tone, you will look bad and so will your company. Even worse, if your letter communicates incorrect information regarding prices, guarantees, due-date promises, or equipment and labor commitments, your company will be held legally responsible. When you write on your company's letterhead stationery and sign it, your letter constitutes a legally binding contract. Therefore, no matter what type of letter you're writing, take care to follow accepted letter formats, maintain the proper tone, and avoid errors.

This chapter helps you write the following kinds of letters: (a) inquiry, (b) cover (transmittal), (c) good news, (d) bad news, (e) complaint,

(f) adjustment, and (g) sales. So that you can write these letters effectively, we provide letter components and formats, examples, criteria for writing different types of letters, and a process log to follow.

LETTER COMPONENTS

ESSENTIAL COMPONENTS

Your letter should be typed or printed on 8½″ x 11″ unlined paper. Leave 1″ to 1½″ margins at the top and on both sides.

Your letter should contain the following components (see Figure 6.1).

1. Writer's Address

This section contains either your personal address or your company's address (you may be writing on printed, letterhead stationery).

If the heading consists of your address, you will include (a) your street address (do not abbreviate the words *street*, *avenue*, *road*, *drive*, etc.) and (b) the city, state, and zip code. Do not include your name. The state may be abbreviated with the appropriate two-letter abbreviation (see Chapter 18).

If the heading consists of your company's address, you will include (a) the company's name; (b) the street address; and (c) the city, state, and zip code.

2. Date

Document the month, day, and year when you write your letter. You can set up your date in one of two ways: May 31, 2000, or 31 May 2000.

Place the date one or two spaces below the writer's address (if the letter is long) or three to eight spaces below the writer's address (for shorter letters). For a short letter, leave more space so the spacing on the letter is balanced.

3. Inside Address

This is the address of the person or people to whom you are writing. The inside address contains these elements in the following order.

- Your reader's name (If you do not know the name of this person, begin the reader's address with a job title or the name of the department.)
- Your reader's title (optional if you include the name)
- The company name
- The company street address
- The company city, state, and zip code

Place this information two lines below the date.

4. Salutation

The traditional salutation, placed two spaces beneath the inside address, is your reader's last name, preceded by *Dear* and followed by a colon.

Dear Mr. Smith:

UNITED SPECTOGRAPH

(1) **Writer's address**
19015 Lakeview Avenue
Columbus, OH 43212

(2) **Date**
June 10, 2000

(3) **Inside address**
Christy Pieburn
Corporate Communications, Inc.
1245 Grant
Chicago, IL 60611

(4) **Salutation**
Dear Ms. Pieburn:

(5) **Letter body**
Here is the schedule for your presentation during our Basic Information Training for Supervisors (BITS) seminar. You will hold your sessions in our Training Center at 19015 Lakeview Avenue.

The schedule and important information regarding your two 8-hour seminars is as follows. Group 1, which will consist of 15 employees, will meet Monday, July 6, 2000, from 8:00 a.m. to 5:00 p.m. Group 2, which will consist of 18 employees, will meet Tuesday, July 7, 2000, from 8:00 a.m. to 5:00 p.m.

As you requested, the training room will have an overhead projector, screen, flip chart, and pad. During the morning sessions, coffee, juices, and rolls will be served. In the afternoon, seminar participants will be offered cold drinks and cookies. We look forward to working with you again. If I can answer any questions, please let me know.

(6) **Complimentary close**
Sincerely,

(7) **Signed name**
Martha Lee

(8) **Typed name**
Martha Lee

FIGURE 6.1 Essential Letter Components

You can also address your reader by his or her first name if you are on a first-name basis with this person.

Dear John:

If you are writing to a woman and are unfamiliar with her marital status, address the letter as follows:

Dear Ms. Jones:

However, if you know the woman's marital status, you can address the letter accordingly.

Dear Miss Jones:

or

Dear Mrs. Jones:

Occasionally, you will write to someone whom you do not know and who has a non-gender-specific first name, such as Pat, Kim, Kelly, Stacy, or Chris. In this case, avoid either Mr., Miss, Mrs., or Ms., and write as follows:

Dear Chris Evans:

Often, when writing to another company, you will not know your reader's name. In this instance, you have limited options. You could address the reader by his or her title (*Dear Vice President of Operations:*) or address the letter to the reader's department (*Accounting Department:*). If you address the letter to the department, *Dear* would be inappropriate and should be omitted.

Each of the above salutations is acceptable. However, several formerly common salutations are no longer useful in business communications. For example, as discussed in Chapter 4, sexist salutations such as *Dear Sir* and *Gentlemen* cannot be used (unless you know beyond all doubt that each of your readers is a man; however, we suggest that you eliminate the problem by avoiding the sexist salutation).

Other awkward salutations include *Good morning* (since mail is delivered twice daily at many corporations), *To Whom it May Concern* (which is trite and imprecise), and *Greetings* (which either has a seasonal ring to it or a military connotation). *Dear Sir/Madam* is also ineffective and should be avoided.

To avoid these pitfalls in salutations, you might want to call the company to which you're writing and ask for the name of your intended reader (and the correct spelling of the name).

5. Letter Body

Begin the body of the letter two spaces below the salutation. The body includes your introductory paragraph, discussion paragraph(s), and concluding paragraph. The body should be single spaced with double spacing between paragraphs. Whether you indent the beginning of each paragraph or leave them flush with the left margin is determined by the letter format you employ. We discuss this issue later in this chapter.

6. Complimentary Close

Place the complimentary close, followed by a comma, two spaces below the concluding paragraph. Although several different complimentary closes are acceptable, such as *Yours truly* and *Sincerely yours*, we suggest that you limit your close to *Sincerely*.

7. Signed Name

Sign your name legibly beneath the complimentary close.

8. Typed Name

Type your name four spaces below the complimentary close. If you wish, you may type your title one space beneath your typed name.

Sincerely,

Henry Marshall

Henry Marshall
Comptroller

OPTIONAL LETTER COMPONENTS

In addition to the letter essentials, you can include the following optional components.

Subject Line

A subject line, which is mandatory in memos (see Chapter 5), is also applicable in letters. You can type the subject line in all capital letters two spaces below the inside address and two spaces above the salutation.

Dr. Ron Schaefer
Linguistics Department
Southern Illinois University
Edwardsville, IL 66205

Subject: LINGUISTICS CONFERENCE REGISTRATION PAYMENT

Dear Dr. Schaefer:

You also could use a subject line instead of a salutation.

Linguistics Department
Southern Illinois University
Edwardsville, IL 66205

Subject: LINGUISTICS CONFERENCE REGISTRATION PAYMENT

A subject line not only helps readers understand the letter's intent but also (if you are uncertain of your reader's name) helps you avoid such flawed salutations as *To Whom It May Concern, Dear Sirs,* and *Ladies and Gentlemen.*

In the simplified letter, both the salutation and the complimentary close are omitted, and a subject line is included.

New-Page Notations

If your letter is longer than one page, you'll need to cite your name, the page number, and the date on all pages after page 1. Place this notation either flush with the left margin at the top of subsequent pages or across the top of subsequent pages. (You must have at least two lines of text on the next page to justify another page.)

LEFT, MARGIN, SUBSEQUENT PAGE NOTATION
Mabel Tinjaca
Page 2
May 31, 2000

ACROSS TOP OF SUBSEQUENT PAGES
Mabel Tinjaca   2   May 31, 2000

Writer's and Typist's Initials

If the letter has been typed by someone other than the writer, include both the writer's and the typist's initials two spaces below the typed signature. The writer's initials are capitalized, the typist's initials are typed in lowercase, and the two sets of initials are separated by a colon. If the typist and the writer are the same person, this notation is not necessary.

Sincerely,

W. T. Winnery

W. T. Winnery

WTW:mm

Enclosure Notation

If your letter prefaces enclosed information, such as an invoice, a report, or graphics, mention this enclosure in the letter and then type an enclosure notation two spaces below the typed signature (or two spaces below the writer and typist initials). The enclosure notation can be abbreviated *Enc.*; written out—*Enclosure*; show the number of enclosures, such as *Enclosures (2)*; or specify what has been enclosed—*Enclosure: January Invoice.*

Copy Notation

If you have made a carbon copy or a photocopy of your letter, show this in a copy notation. A complimentary copy is designated by a lowercase *cc*: a photocopy is designated by a lowercase *pc*. Type the copy notation two spaces below the typed signature or two spaces below either the writer's and typist's initials or the enclosure notation.

Sincerely,

Brian Altman

Brian Altman

Enclosure: August Status Report

pc

If you are sending copies of the letter to other readers, list these readers' names following the copy notation.

Sincerely,

Brian Altman

Brian Altman

Enclosure: August Status Report

pc: Marcia Rittmaster
Erica Nochlin

LETTER FORMATS

Type your letter using any of the formats shown in Figures 6.2 through 6.5.

CRITERIA FOR DIFFERENT TYPES OF LETTERS

LETTERS OF INQUIRY

If you want information about degree requirements, equipment costs, performance records, turnaround time, employee credentials, or any other matter of interest to you, you write a letter requesting that data. Letters of inquiry require that you be specific and precise. For example, if you write, "Please send me any information you have on the XYZ transformer," you're in trouble. You'll either receive any information the reader chooses to dispose of or none at all. Look at the following letter of request.

Dear Mr. Jernigan:

Please send us information about the following filter pools:

1. East Lime Pool
2. West Sulphate Pool
3. East Aggregate Pool

Thank you.

1119 South Bend
Chico, CA 95926
November 1, 2000

Dr. Robert Cottrell
7806 Northway
Austin, TX 78752

Dear Dr. Cottrell:

——————————— Introduction ———————————

——————————— Discussion ———————————

——————————— Conclusion ———————————

Sincerely,

Sue Timmons

Sue Timmons

ST:sg

cc: Jeremy Isaacs

FIGURE 6.2 Full Block Format

7501 Carriage
Coconut Grove, FL 33133
September 7, 2000

Ms. Jordan Cottrell
7926 Candle Lane
Pittsburgh, PA 15237

Subject: MARKETING OF ROBOTICS APPLICATIONS

Dear Ms. Cottrell:

———————————— Introduction ————————

————————————
———————————— Discussion ————————
————————————

———————————— Conclusion ————————

Sincerely,

Ruth Schneider

Ruth Schneider

RS:sl

FIGURE 6.3 Full Block Format with Subject Line

3628 East Vista Road
Chamblee, GA 30341
August 13, 2000

Mr. Kenny Frankel
7810 Warrensburg
Hackensack, NJ 07649

Dear Mr. Frankel:

———————————— Introduction ————————————

———————————— Discussion ————————————

———————————— Conclusion ————————————

Sincerely,

Sylvia Light

Sylvia Light

FIGURE 6.4 Modified Block Format

The disgruntled worker replied as follows:

Dear Mr. Scholl:

I would be happy to provide you with any information you would like. However, you need to tell me *what* information you require about the pools.

I look forward to your response.

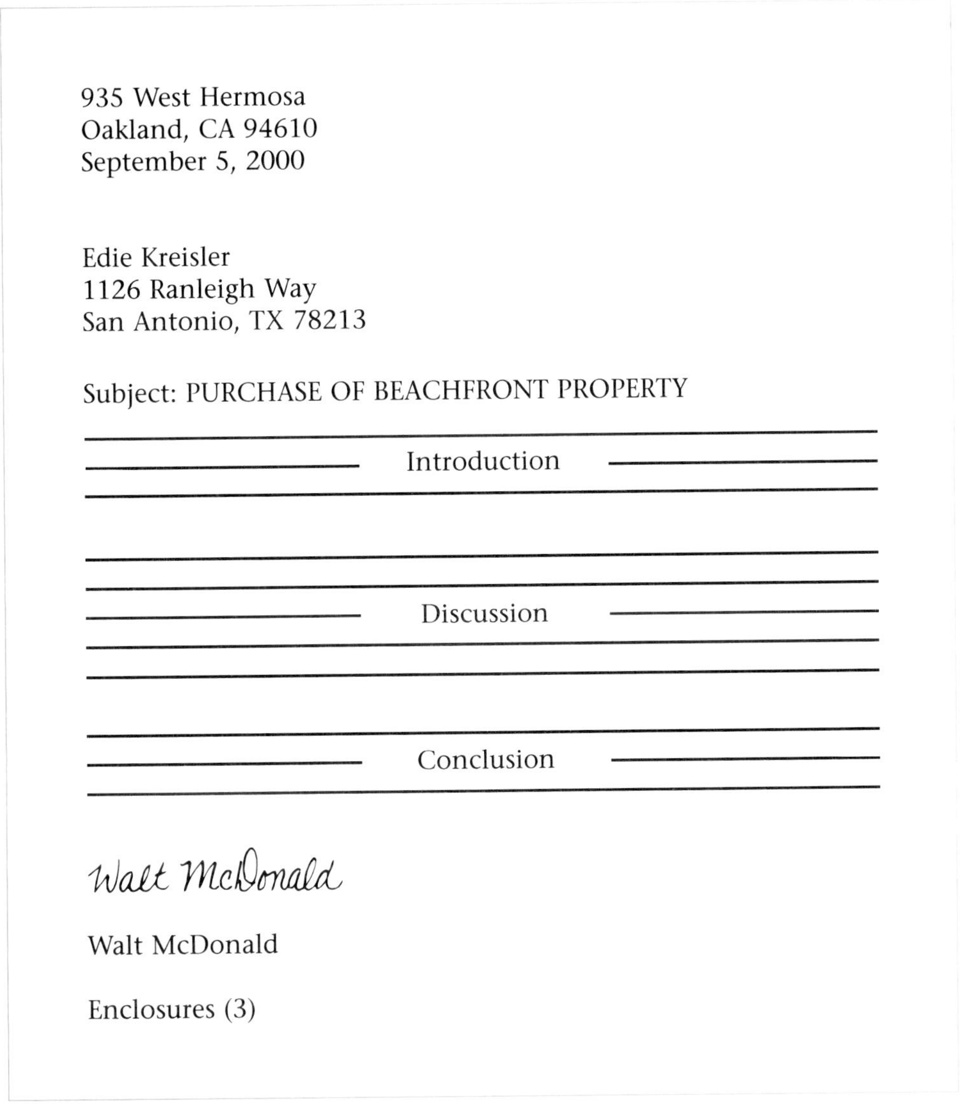

935 West Hermosa
Oakland, CA 94610
September 5, 2000

Edie Kreisler
1126 Ranleigh Way
San Antonio, TX 78213

Subject: PURCHASE OF BEACHFRONT PROPERTY

———————————————— Introduction ————————————————

———————————————— Discussion ————————————————

———————————————— Conclusion ————————————————

Walt McDonald

Walt McDonald

Enclosures (3)

FIGURE 6.5 Simplified Format

To receive the information you need, write an effective letter of inquiry. Successful letters of inquiry contain all the letter essentials, maintain an effective technical writing style, achieve audience involvement through pronoun usage, and avoid grammatical and mechanical errors. In addition, you must accomplish the following tasks.

Introduction

1. Clarify your intent in the introduction. Until you tell your readers why you are writing, they do not know. It's your responsibility to clarify your intent and explain your rationale for writing. Also tell your reader immediately what you are writing about (the subject matter of your inquiry). You can state your intent and subject matter in one to three sentences, as follows:

> My company is planning to purchase 30 new tractors by the end of the year. Your 2000 tractor with the onboard computer might be the answer to our problems. In addition to sending us a picture brochure of the CP95 tractor, could you also answer the following questions about your vehicle?

In this example, the first two sentences explain why you are writing. The last sentence explains what you want.

Discussion

2. *Specify your needs in the discussion.* To ensure that you get the response you want and need, ask precise questions or list specific topics of inquiry. You must quantify. For example, rather than vaguely asking about machinery specifications, you should ask more precisely about "specifications for the 12R403B Copier." Rather than asking, "Will the roofing material cover a large surface?" you need to quantify—"Will the roofing material cover 150′ × 180′?"

Conclusion

3. *Conclude precisely.* First, explain when you need a response (because until you tell the reader, he or she does not know). Don't write, "Please respond as soon as possible." Provide dated action—tell the reader exactly when you need your answers. Dated action doesn't mean you'll get the answers when you want them, but your chances are better. Second, to sell your readers on the importance of this date, explain why you need answers by the date given.

> Please send the above information by August 15. This will allow my shop supervisor to study the manuals and arrange for a test drive prior to our board of directors meeting on September 1.

Figure 6.6 will help you understand the requirements for effective letters of inquiry.

COVER LETTERS

In business, you are often required to send information to a client, vendor, or colleague. You might send multipage copies of reports, invoices, drawings, maps, letters, memos, specifications, instructions, questionnaires, or proposals. The reader may have requested this information. Maybe you're sending the data of your own accord. Whatever the situation, if you merely submit the data without a cover letter, your readers will be overwhelmed with information that they must wade through.

A cover letter accomplishes two goals. First, it lets you tell readers up front what they are receiving. Second, it helps you focus your readers' attention on key points within the enclosures. Thus, the cover letter is a reader-friendly gesture geared toward assisting your audience.

H O R I Z O O N

Enterprises
1020 Sunshine Road
Jamaica Plain, MA 02130

July 23, 2000

Mr. Leonard Liss
Director of Marketing
Johnson Motor Company
4718 West 91st Street
Tacoma, WA 82492

Dear Mr. Liss:

Thank you for your sales letter regarding the Model CP95 tractor with fuel-saving devices. My company plans to purchase 30 new tractors by the end of the year. Your tractors may be what we need.

Before we decide, however, we would like more information. Please send us a brochure of your CP95, along with answers to these questions.

1. What is the cost per tractor?
2. What are your financing options for quantity purchases?
3. Do you provide both a paper-bound instructional manual and an intranet manual with online help screens?
4. What are your tractor's comfort features?
5. When can you deliver 30 tractors?

Please send this information by August 15. Then my shop supervisor and I can study the material and arrange a test drive. I plan to present a proposal for purchase to our board of directors on September 1. I look forward to your response.

Sincerely,

Kathy Brewington

Kathy Brewington

FIGURE 6.6 Letter of Inquiry

As with the previously discussed letters, successful cover letters include the letter essentials, maintain an effective technical writing style, avoid grammatical and mechanical errors, and achieve audience recognition and involvement. In addition, a well-written cover letter contains an introduction, a discussion, and a conclusion.

Introduction

In the introductory paragraph, tell your reader *why* you are writing and *what* you are writing about. An appropriate introduction providing your reader a rationale for your writing might read like this: "To help you prepare for next week's audit, we have enclosed the following necessary forms."

What if the reader has asked you to send the documentation? Do you still need to explain why you're writing? The answer is yes. Although the reader requested the data, time has passed, other correspondence has been written, and your reader might have forgotten the initial request. Your job is to remind him or her. Write something like, "In response to your request, enclosed are the completed questionnaires from our engineers." These introductory sentences provide the reader information about *why* you are writing and *what* you are sending. Note that neither exceeds one sentence. Clarity of intent can be provided *concisely*.

Discussion

In the body of the letter, you want to accomplish two things. First, you either want to tell your reader *exactly what you've enclosed* or *exactly what of value is within the enclosures*. In both instances, you should provide an itemized list. For example, you can itemize enclosures as follows:

To help you prepare for next week's audit, we have enclosed the following necessary forms.

- Year-to-date sales figures
- Fiscal year sales figures
- Year-to-date expenditures
- Fiscal year expenditures

Similarly, you can itemize the important facts within the enclosures as follows:

In response to your request, enclosed are the completed questionnaires from our engineers.

Of special interest to you within the questionnaires are their answers to questions regarding the following:

- Easement dimensions .page 1
- Sewer construction materials . . . . .page 3
- Residential/commercial ratios . . . . .page 5

This example shows where important information can be found within the enclosure. Page numbers are a friendly gesture toward your audience. Instead of just dumping the data on your reader and saying, "Best of luck—hope you find what you want," you're helping the reader locate the important information. In so doing, you're achieving audience recognition and involvement.

However, including page numbers has a greater benefit than audience involvement. These page numbers also allow you to focus your reader's attention on what you want him or her to see. In other words, if you merely provide the enclosure without a cover letter, then you leave it up to readers to sift through the information and decide on their own what's important. In contrast, by providing an itemized list with page numbers, you direct the reader's attention. You can emphasize, for your benefit, the points in the enclosure which you consider to be important.

Conclusion

Your conclusion should tell your readers *what* you want to happen next, *when* you want this to happen, and *why* the data are important. Without such a conclusion, your reader will review the documentation and say, "OK, so what? What do you want me to do with the data?" or "OK, but what are your future plans?"

The *what* clarifies your intentions. The *when* specifies the date. The *why* either sells the importance of this date, possibly hinting at the urgency, or suggests your conscientiousness.

Here are examples of cover letters, complete with conclusions.

To help you prepare for next week's audit, we have enclosed the following necessary forms:

- Year-to-date sales figures
- Fiscal year sales figures
- Year-to-date expenditures
- Fiscal year expenditures

Please fill out these forms before our Wednesday arrival. Doing so will facilitate the audit, thereby allowing you and your colleagues to return to your regular work activities more rapidly.

In response to your request, enclosed are the completed questionnaires from our engineers.

Of special interest to you within the questionnaires are their answers to questions regarding the following:

- Easement dimensions .page 1
- Sewer construction materials . . . . .page 3
- Residential/commercial ratios . . . . .page 5

> We would be happy to meet with you at your convenience to discuss this issue in greater detail. After you have had a chance to review these enclosures, please call my office at 469-8500 to set up a meeting. We want to answer your questions regarding our construction planning.

The cover letter seen in Figure 6.7 contains letter essentials.

GOOD-NEWS LETTERS

You'll often have the opportunity to write good-news letters. For example, you might write a letter promoting an employee or offering an individual a job at your corporation. You might write to commend a colleague for a job well done. Maybe you'll write to tell a customer that his or her request for a refund was justified.

Introduction

As in other letters discussed in this chapter, the introduction should immediately explain *why* you're writing and tell *what* you're writing about. The point of these letters is good news. Furthermore, since your goal is to convey good news, you should begin with positive word usage, as in the following examples.

> EXAMPLE 1
> COMMENDING A COLLEAGUE (INTRODUCTION)
> Judy, you've proved to be indispensable again. Your work for the textbook committee has made all of our jobs easier.

> EXAMPLE 2
> PROMOTING AN EMPLOYEE (INTRODUCTION)
> Congratulations! We're proud to offer you early promotion, Jan.

Discussion

Once you've explained why you're writing and what you're writing about, the next step is to provide the detail—to explain *exactly what* has justified the commendation or the promotion.

> EXAMPLE 1
> COMMENDING A COLLEAGUE (INTRODUCTION AND DISCUSSION)
> Judy, you've proved to be indispensable again. Your work for the textbook committee has made all of our jobs easier.
>
> Thank you for performing the following services:
> - Meeting with all the sales reps to convey our departmental requirements
> - Reviewing those texts with computer-aided design components
> - Screening the textbook options and selecting the three most suited to our needs

A·M·E·R·I·C·A·N
....
HEALTHCARE
1401 Laurel Drive
Denton, TX 76201

November 11, 2000

Jan Pascal
Director of Outpatient Care
St. Michael's Hospital
Westlake Village, CA 91362

Dear Ms. Pascal:

Thank you for your recent request for information about our specialized outpatient care equipment. ***American Healthcare's*** stair lifts, bath lifts, and vertical wheelchair lifts can help your patients. To show how we can serve you, we've enclosed a brochure including the following information:

Maintenance, warranty, and guarantee information. 1–3
Technical specifications for our products, including
    sizes, weight limitations, colors, and installation
    instructions 4–6
Visuals and price lists for our products. 7–8
An ***American Healthcare*** order form 9
Our 24-hour hotline for immediate service 10

Early next month, I'll call to make an appointment at your convenience. Then we can discuss any questions you might have. Thank you for considering ***American Healthcare***, a company that has provided exceptional outpatient care for over 30 years.

Sincerely,

Toby Sommers

Toby Sommers

Enclosure

FIGURE 6.7 Cover Letter

Example 2
Promoting an employee (Introduction and discussion)
Congratulations! We're proud to offer you early promotion, Jan.

You've earned a grade raise to E30 for the following reasons:

1. *Productivity*. Your line personnel produced 2,000 units per month throughout this quarter.
2. *Efficiency*. You maintained a 95 percent manufacturing efficiency rating.
3. *Supervisory skills*. You received only four grievances; your annual performance appraisals showed that your subordinates appreciated your motivational management techniques.

Conclusion

Once you've clarified exactly what has resulted in the good news, your last paragraph should state *what* you plan next (or expect from your reader next), *when* this action will occur, and *why* the date is important. The following good-news letters are complete with introductions, discussions, and conclusions.

Judy, you've proved to be indispensable again. Your work for the textbook committee has made all of our jobs easier.

Thank you for performing the following services:

- Meeting with all the sales reps to convey our departmental requirements
- Reviewing those texts with computer-aided design components
- Screening the textbook options and selecting the three most suited to our needs

Due to your assistance, we have decided on the Wilkes text and plan to place our orders this April. That will allow us to stock the bookstore for the fall semester. Your work has made a difference, and we appreciate your efforts—another job well done!

Congratulations! We're proud to offer you early promotion, Jan.

You've earned a grade raise to E30 for the following reasons:

| 1. *Productivity* | Your line personnel produced 2,000 units per month throughout this quarter. |
| 2. *Efficiency* | You maintained a 95 percent manufacturing efficiency rating. |

| 3. *Supervisory Skills* | You received only four grievances; your annual performance appraisals showed that your subordinates appreciated your motivational management techniques. |

Because of your excellent work, you will receive your pay increase the first of next month. You deserve it. Good work, Jan.

BAD-NEWS LETTERS

Unfortunately, you occasionally will be required to write bad-news letters. These letters might reject a job applicant, deny an employee a raise, tell a vendor that his or her company's proposal has not been accepted, or reject a customer's request for a refund. Maybe you'll have to write to a corporation to report that its manufacturing is not meeting your company's specifications. Maybe you'll need to write a letter to a union documenting a grievance. You might even have to write a bad-news letter to fire an employee.

In any of these instances, tact is required. You can't berate a customer or client. You shouldn't reject a job applicant offensively. Even your grievance must be worded carefully to avoid future problems.

Since the point of a bad-news letter is bad news, you'll need to structure your correspondence carefully to avoid offending your reader.

Introduction

We've suggested throughout this chapter that your introduction should explain why you're writing and what you're writing about. However, such conciseness and clarity in a bad-news letter would be harsh and abrupt. Therefore, to avoid these lapses in diplomacy, begin your bad-news letter with a buffer. Start your letter with information that your reader can accept as valid but that will sway your reader to accept the bad news to come.

EXAMPLE 1
REJECTING A JOB APPLICANT (INTRODUCTION)
Thank you for your recent letter of application. As you can imagine, we received many letters from highly qualified applicants.

EXAMPLE 2
TERMINATING A CLIENT/VENDOR RELATIONSHIP (INTRODUCTION)
John, as you know, our business demands exact tolerances and precise workmanship. Because of these requirements and the reputation your company has for quality production, we were happy to pursue a long-term contract with you.

Discussion

Once you've provided the buffer, swaying your audience to your point of view, you can no longer delay the inevitable. The discussion paragraph states the bad news. However, to ensure that the reader accepts the bad news, preface your assertions with quantifiable proof.

EXAMPLE 1
REJECTING A JOB APPLICANT (INTRODUCTION AND DISCUSSION)
Thank you for your recent letter of application. As you can imagine, we received many letters from highly qualified applicants.

   Although we appreciate your interest in Acme, the advertisement specifically required that all applicants have an M.S. in computer science and at least five years of experience in telecommunications. We also suggested that a knowledge of fiber optics would be preferred. Your degree meets our criteria successfully. However, your years of experience fall below our requirements, and your résumé does not mention fiber optics expertise. Therefore, we must reject your application.

EXAMPLE 2
TERMINATING A CLIENT/VENDOR RELATIONSHIP (INTRODUCTION AND DISCUSSION)
John, as you know, our business demands exact tolerances and precise workmanship. Because of these requirements and the reputation your company has for quality production, we were happy to pursue a long-term contract with you.

However, your last two shipments contained flawed goods. In fact, we found these problems:

- 37 percent of your shipped components were off tolerance by .025 mm.
- Your O-rings suffered stress fractures when under 2,000 lb of pressure.

Because of these failures, we are returning the products.

Conclusion

If you end your bad-news letter with the bad news, then you leave your reader feeling defeated and without hope. You want to maintain a good customer-client, supervisor-subordinate, or employer-employee relationship. Therefore, you need to conclude your letter by giving your readers an opportunity for future success. Provide your readers options which will allow them to get back in your good graces, seek employment in the future, or reapply for the refund you've denied. Then, to leave your readers feeling as happy as possible, given the circumstances, end upbeat and positively.

   The following examples are complete with introductions, discussions, and conclusions.

Thank you for your recent letter of application. As you can imagine, we received many letters from highly qualified applicants.

Although we appreciate your interest in Acme, the advertisement specifically required that all applicants have an M.S. in computer science and at least five years of experience in telecommunications. We also suggested that a knowledge of fiber optics would be preferred. Your degree met our criteria. However, your years of experience fell below our requirements, and your résumé did not mention fiber optics expertise. Therefore, we must reject your application.

If you have fiber optics knowledge or have acquired additional job experiences which pertain to our work requirements, we would be happy to reconsider your application. In any case, we will keep your letter on file. When new positions open up, your letter will be reassessed. Good luck in your job search.

John, as you know, our business demands exact tolerances and precise workmanship. Because of these requirements and the reputation your company has for quality production, we were happy to pursue a long-term contract with you. However, your last two shipments contained flawed goods. In fact, we found these problems:

- 37 percent of your shipped components were off tolerance by .025 mm.
- Your O-rings suffered stress fractures when under 2,000 lbs. of pressure.

Because of these failures, we are returning the products.

If you can correct these problems and document to our satisfaction that the errors have been eliminated, we would be willing to reconsider our stance. We have enjoyed working with you, John, and look forward to the possibility of future contracts.

Although the preceding two letters convey bad news, they couch the negatives in positive terms. For example, although the first letter rejects a job applicant, the writer uses words and phrases such as *thank you*, *appreciate*, *happy*, and *good luck*. The second letter does the same: *quality*, *happy*, *enjoyed*, and *look forward to*. As the writer of the bad-news letter, you're in charge of the tone. Why not make it positive?

The example in Figure 6.8, which is from an insurance company cancelling an insurance policy, shows how a bad-news letter should be constructed.

State of Mind Insurance
11031 Bellbrook Drive
Stamford, CT 23091
(213) 333-8989

September 7, 2000

John Chavez
4249 Uvalde
Stamford, CT 23091

Dear Mr. Chavez:

Thank you for letting us provide you and your family with insurance for the last 10 years. We have appreciated your business.

However, according to our records, you have filed three claims in the past three years.

1. Damage to your basement due to a failed sump pump
2. Flooring damage due to a broken water seal in your second-floor bathroom
3. Fender repair coverage for a no-fault automobile accident

SMI's policy stipulates that no more than two claims may be filed by a client within a three-year period. Therefore, we must cancel your policy as of October 15, 2000.

If you have any questions about this cancellation, please call our 24-hour assistance line (913-482-0000). Our transition team can help you find new coverage. Thank you for your patronage and your understanding.

Sincerely,

Darryl Kennedy

Darryl Kennedy

FIGURE 6.8 Bad-News Letter

COMPLAINT LETTERS

You are purchasing director at an electronics firm. Although you ordinarily receive excellent products and support from a local manufacturing firm, two of your recent orders have been filled incorrectly and included defective merchandise. You don't want to have to hunt for a new supplier.

How can you improve your relations with the company and receive the service you want? Write a letter of complaint.

Introduction

In the introduction, *politely* state the problem. Although you might be angry over the service you've received, you want to suppress that anger. Blatantly negative comments don't lead to communication; they lead to combat. Because angry readers won't necessarily go out of their way to help you, your best approach is diplomacy.

To strengthen your assertions, in the introduction, include supporting documents, such as the following:

- Serial numbers
- Dates of purchase
- Invoice numbers
- Check numbers
- Names of salespeople involved in the purchase

Also state that copies of these documents are enclosed.

> *EXAMPLE*
> On August 15, you shipped to my company 36 XYZ digital oscilloscopes (copies of invoice #3492 are enclosed). Several of our customers have since complained that the o-scopes are malfunctioning.

Discussion

In the discussion paragraph(s), explain in detail the problems experienced. This could include dates, contact names, information about shipping, breakage information, and/or an itemized listing of defects. Be specific. Generalized information won't sway your readers to accept your point of view. In a complaint letter, you suffer the burden of proof. Help your audience understand the extent of the problem. After documenting your claims, state what you want done and why.

> *EXAMPLE*
> The following occurred after we delivered the o-scopes to three of our customers.
>
> 1. August 20—AAA Electronics stated that two of their five o-scopes were malfunctioning, giving incorrect readings.
> 2. August 27—Richards Electronics, Inc., said three of their o-scopes were incorrectly calibrated.
> 3. September 5—Five of ABC Computers' o-scopes would not interface with their computers.
>
> These ten o-scopes need to be repaired. You also should send service technicians to troubleshoot our clients' other o-scopes to avoid future problems.

Conclusion

End your letter positively. Remember, you want to ensure cooperation with the vendor, and you want to be courteous, reflecting your company's professionalism. And, in this paragraph, include your phone number and the time you can best be reached.

> *EXAMPLE*
> You can reach me at 469-8200 from 8:00 A.M. to 5:00 P.M., Monday through Friday. Thank you for your help. After working with your company for the past eight years, I have been impressed with the way you stand behind your products. I know you'll help us with these defective o-scopes.

Figure 6.9 is an example of a complaint letter.

ADJUSTMENT LETTERS

Responses to letters of complaint, also called adjustment letters, can take three different forms.

- 100 percent yes—you could agree 100 percent with the writer of the complaint letter.
- 100 percent no—you could disagree 100 percent with the writer of the complaint letter.
- Partial adjustment—you could agree with some of the writer's complaints but disagree with other aspects of the complaint.

100 Percent Yes Letter

This letter is like the good-news letter discussed earlier in the chapter. Since you've got good news for the reader, there's no reason to delay the message. In the introductory paragraph, state that you agree with your reader's complaint and will honor his or her recommendations for adjustment.

> *EXAMPLE*
> Thank you for calling our attention to your problem. We will be happy to replace your defective oscilloscopes.

In the discussion paragraph(s), explain what happened, why the problem occurred, and how the problem will be avoided in the future.

> *EXAMPLE*
> I have looked into your order and discovered the cause of the problem. The o-scopes were damaged during shipment. They were stacked too high, which led to breakage while handling. I have met with the carrier, who agrees that the packages were handled improperly. I assure you that none of these problems will occur again. Also, I am sending a service technician to check on your other o-scopes from this order.

1234 18th Street
Galveston, TX 77001
May 10, 2000

Mr. Holbert Lang
Customer Service Manager
Gulfstream Auto
1101 21st Street
Galveston, TX 77001

Dear Mr. Lang:

On February 12, 2000, I purchased two shock absorbers in your automotive department. Enclosed are copies of the receipt and the warranty for that purchase. One of those shocks has since proved defective.

I attempted to exchange the defective shock at your store on May 2,
2000. The mechanic on duty, Vernon Blanton, informed me that the
warranty was invalid because your service staff did not install the part. I
believe that your company should honor the warranty agreement and
replace the part for the following reasons:

1. The warranty states that the shock is covered for 48 months and
   48,000 miles.
2. The warranty does not state that installation by someone other
   than the dealership will result in warranty invalidation.
3. The defective shock absorber is causing potentially expensive
   damage to my tire and suspension system.

I can be reached between 1 p.m. and 6 p.m. on weekdays at 763–9280
or at 763–9821 anytime on weekends. I look forward to hearing from
you. Thank you for helping me with this misunderstanding.

Sincerely,

Carlos De La Torre

Carlos De La Torre

Enclosures (2)

FIGURE 6.9 Letter of Complaint

In your conclusion paragraph, resell to maintain customer satisfaction. End the letter upbeat.

EXAMPLE

Acme has worked with your company for many years, and I look forward to continuing this business relationship. I apologize for the inconvenience.
Please call me at 392-8270 if you have any other questions.

100 Percent No Letter

Unfortunately, sometimes you have to say no to a letter of complaint. If the customer is wrong, you have no alternative but to say so. However, you still want to maintain positive customer relationships. Although you will deny the customer's request for adjustment, you want to keep that customer. To do so, follow the pattern discussed earlier in this chapter for a bad-news letter.

In the introductory paragraph, begin with a buffer. This should include a positive statement and facts that all readers can accept, regardless of the situation.

EXAMPLE

Thank you for your letter regarding the XYZ oscilloscopes. We were sorry to hear about the malfunctions your customers are experiencing. Although such occurrences are rare, as you know, products occasionally are damaged in shipment.

In the discussion paragraph(s), precisely explain what happened, providing facts, figures, and incontestable detail. Then state the bad news.

EXAMPLE

After researching the matter thoroughly, we found that the problem was created by the shipping company. They stacked the boxes too high, which led to damage during handling. Since the damage was caused by the carriers and was not due to our negligence during manufacturing, we cannot replace the o-scopes.

In the conclusion, end on an upbeat note. To do so, use positive words and provide the reader an alternative.

EXAMPLE

To assist you in solving this problem, we have contacted Archie Cox, the shipping manager at ABC Freight (535–2020, ext. 345). He expects to hear from you soon. If we can help you in any other ways, please call.

Partial Adjustment

Life is complex. Sometimes you can agree with part of a writer's complaint but disagree with other parts. If you're not going to agree completely or disagree completely, then you'll want to write a partial adjustment letter.

In the introductory paragraph, state the good news first. As always, you want to be diplomatic, winning the reader to your point of view.

EXAMPLE
Thank you for your letter regarding the XYZ oscilloscopes. I was sorry to hear about the malfunctions your customers are experiencing. Since we pride ourselves on our products' quality, I am happy to replace the ten defective o-scopes.

In the discussion paragraph(s), precisely explain what happened, and then state the bad news.

EXAMPLE
The ten o-scopes, which were damaged during shipment due to packaging problems, will be replaced within six to eight weeks. I am unable to send a service technician to troubleshoot the remaining 26 o-scopes. That service is not part of our contract with you. (Please see section 4.1 of your service contract: "On-site service will not be provided in case of malfunction.")

In the conclusion, resell. Your reader will not be completely happy with you. You've provided some relief but not all that was requested. The last paragraph is your opportunity to win back a bit of the reader's good faith in your company. After all, you want to keep the customer for future contracts.

EXAMPLE
If any of the other o-scopes malfunction, please call us immediately. We stand by our products, as you know from having been one of our satisfied customers for over a decade. I'll be happy to help you in any way I can.

Figure 6.10 is an example of a 100 percent yes adjustment letter.

SALES LETTERS

You've just manufactured a new product (an electronic testing device, a fuel injection mechanism, a fiber optic cable, or a high-tech, state-of-the-art computer). Perhaps you've just created a new service (computer maintenance, automotive diagnosis, home repair, or telecommunications networking). Congratulations!

However, if your product just sits in your basement gathering dust or your service exists only in your imagination, what have you accomplished? To benefit from your labors, you must market your invention. Connect with your end users. Let the public know that you exist. The question is, how? If you have unlimited time, money, and personnel, you can try telemarketing or door-to-door cold calls. However, a more time-efficient and cost-effective way to market your product or service is to write a sales letter. You write a sales letter once (which saves money) and then mass mail it (which saves time).

To write your sales letter, you'll have to include the letter essentials discussed earlier in this chapter, maintain an effective technical writing style with a low fog

```
┌─────────────────────────────────────────────┐
│  GULFSTREAM AUTO                              │
│  1101 21st                                    │
│  Galveston, TX 77001                          │
│  (712) 451-1010                               │
└─────────────────────────────────────────────┘
```

May 31, 2000

Mr. Carlos De La Torre
1234 18th Street
Galveston, TX 77001

Dear Mr. De La Torre:

Thank you for your recent letter. I am pleased to inform you that Gulfstream will happily replace your defective shock absorber according to the warranty agreement.

The Trailhandler Performance XT shock absorbers that you purchased were discontinued in October 1999. Mr. Blanton, the mechanic to whom you spoke, incorrectly assumed that Gulfstream was no longer honoring the warranty on that product. Because we no longer carry that product, we either will replace it with a comparable model or refund the purchase price—you make the decision. Just ask to see Mrs. Brandsgard at the automotive desk on your next visit to our store. She is expecting you and will handle the exchange.

We appreciate your business, Mr. De La Torre. I'm glad you brought this problem to my attention. If I can help you in the future, please contact me.

Sincerely,

Holbert Lang

Holbert Lang

FIGURE 6.10 100 Percent Yes Adjustment Letter

index, and avoid grammatical and mechanical errors. You will also have to accomplish the following objectives.

Arouse Reader Interest

The introductory paragraph of your sales letter tells your readers *why* you are writing (you want to increase their happiness or reduce their anxieties). You also tell your readers *what* you are writing about (the product or service you are marketing). However, studies tell us that you have only about 30 seconds to grab your readers' attention, after which they will lose interest and toss out your letter. To lure readers into your correspondence, you must arouse their interest imaginatively in the first few sentences. Try any of the following lead-ins.

- *An anecdote*—a brief, dramatic story.

  It's late at night, the service stations are closed, and you've just had a blowout on Highway 35. Don't worry. Our new Tire-Right will solve your problems!

- *A question*—to make your readers read on for an answer.

  "Where will I get the money for my kid's college education?" "How am I going to afford to retire?" "Will my insurance cover all the medical bills?" You've asked yourself these questions. Our estate planning video has the answers.

- *A quotation*—to give your letter the credibility of authority.

  "Omit needless words," technical writers say. If you can't, let us help you. *Write Now*, our new office communications service, can help you write your in-house newsletters—clearly and concisely.

- *Data*—researched information, again ensuring your credibility.

  You're not alone. In fact, if you're at least 25 years old, you're in the majority. Today, 51 percent of our students are older than 25. So why not enroll now? Our college offers the nontraditional student many benefits.

Develop Your Assertions

In the discussion paragraph(s), specify exactly what you offer to benefit your audience or how you'll solve your readers' problems. You can do this in a traditional paragraph. In contrast, you might want to itemize your suggestions in a numerical or bulletized list. Whichever option you choose, the discussion should accomplish any of the following:

- *Provide data* to document your assertions.

  Eighty-five percent of the homeowners contend that . . .

  or

  Seven out of ten buyers said they would . . .

  or

Ten thousand retired metalworkers can't be wrong!

- Give *testimony* from satisfied customers.

The Job Corps of Blue Valley, Titan Co., Amex Inc., and the Southwestern Parish swear by our product.

- *Document your credentials*—years in business, certification of employees, number of items sold, and/or satisfied customers (this last point overlaps with our suggestions regarding data).

In business since 1889.
> or

Each of our mechanics is ASE certified.
> or

We've sold over 2 billion widgets.
> or

Over 3,000 corporate managers and supervisors have benefited from our training services.

Make Your Readers Act

If your conclusion says, "We hope to hear from you soon," you've made a mistake. Such a weak conclusion forces you to stare at the phone, twiddle your thumbs, and hope for a customer response. The concluding paragraph of a sales letter can't let the reader off the hook. Do something to make the reader act. Here are some suggestions.

- Give directions (with a map) to your business location.
- Provide a tear-out to send back for further information.
- Supply a self-addressed, stamped envelope for customer response.
- Offer a discount if the customer responds within a given period of time.
- Give your name or a customer-contact name and a phone number (toll free if possible).

Style

Your sales letter not only should convey the necessary information but also should present an appealing style. Consider the following requirements.

- Use verbs. Verb usage, or power writing, will give your letter punch.
- Format for reader-friendly ease of access. Highlight through white space, underlining, boldface, bullets, numbers, and so on.
- Achieve audience recognition and involvement through *you* orientation.
- Show ease of use. Sprinkle the words *easy* and/or *simple* throughout your correspondence.
- Imply urgency. Use the words *now, today, soon,* or *don't delay* to force the reader to immediate action.

IMAGE
GRAPHICS

2629 Westport   Oswego, ND 55290   (405) 555-2121

October 9, 2000

Dave Moore
Media Publishing
10901 College Blvd.
Mountainview, ND 55292

Dear Mr. Moore:

"You *don't* have the software I use?" "You *can't* print my PC job?"
If you've asked these questions, let **Image Graphics** help.
Image Graphics has the printing technology, accessibilty, and
knowledge that you're looking for.

⇨ **Image Graphics** is the most technologically advanced
   graphic design firm in the Midwest, according to *Printer's
   Wheel Weekly*.
⇨ We are the only design firm in your area with an in-house,
   24-hour service bureau.
⇨ Our service bureau can print any job to any media with no
   color or resolution limitations.
⇨ We support all brands of software.
⇨ Each member of our support staff has a Master's degree in
   graphic technology.

I'm glad I could introduce you to **Image Graphics**, Mr. Moore.
Look at the enclosed brochure and pricing sheet. Then call me
personally (ext. 2055) this week for a ***50 percent*** discount on
your first output order.

Sincerely,

Macy Hart

Macy Hart

Enclosure

FIGURE 6.11 Successful Sales Letter

PROCESS

Now that you know what letters of inquiry, cover letters, good-news letters, bad-news letters, and sales letters entail, the next question is, "How do I write these letters?" If you merely get out a blank piece of paper, or turn on your computer screen, and then fill it with words, assuming that they'll be good enough, you're making a mistake. No one can write a successful letter the first time through. In contrast, you need to approach writing more systematically. As we discussed in Chapter 2, to construct effective correspondence, approach writing as a step-by-step, sequential process.

You should do the following:

- Prewrite (to gather your data and to determine your objectives).
- Write a draft (to organize and format your text).
- Rewrite (to revise your correspondence, making it as perfect as it can be).

PREWRITING

To overcome the blank page syndrome or writer's block, spend some time *before* writing your letter gathering as much information as you can about your subject matter. After all, how can you expect to write an effective letter if you aren't sure what you want to say? In addition, determine your objectives. Ask yourself why you're writing the letter and what you want to achieve.

No single way of prewriting is more effective than another. One prewriting technique helpful to many writers is answering reporter's questions. By providing answers to *who*, *what*, *when*, *why*, *where*, and *how*, you can help yourself write more effective letters. Answers to these questions provide you content for your correspondence as well as focus in your writing objectives.

Following is a Reporter's Questions Checklist for a Letter of Inquiry.

REPORTER'S QUESTIONS CHECKLIST FOR A LETTER OF INQUIRY

✔ *Who* is your audience?

— High-tech peer

— Low-tech peer

— Lay reader

— Management

— Subordinate

— Multiple audiences

✔ *Why* are you writing?

✔ *What* is the general topic of your request?

✔ *What exactly* do you want to know (the specifics itemized in the discussion paragraph of your letter)?

✔ *What* do you want the reader to do next?

✔ *When* do you want the reader to act (dated action)?

✔ *Why* is this date important?

WRITING

Once you've gathered your data and determined your objectives, the next step in the process is to begin your rough draft. Doing so requires the following:

1. Study the letter criteria. By studying the specific criteria for the type of letter you'll write, you can remind yourself of what information should be included in each paragraph.

2. Review your prewriting. Now that you have reminded yourself of what each paragraph should include, reviewing the prewriting will help you determine whether, in fact, you have provided the correct details. Have you omitted any significant information? If so, now's the time to add missing content. Have you included unnecessary information? If so, now's the time to delete these irrelevancies.

3. Organize the data for your discussion paragraph. One organizational pattern especially effective for most letters is *importance*. When you organize by importance, you place the most important information first and less important ideas later. To do so, you can use the inverted journalist's pyramid shown in Figure 6.12.

This method works well for itemizing the discussion paragraph in your sales letters, letters of inquiry, and cover letters. You also can use importance to organize the body paragraph in your good-news letter.

However, since a bad-news letter leads up to the bad news rather than begins with it, importance would not be an effective organizational pattern. Instead, you'll want to organize your ideas inductively (indirectly)—specifics to general result. For instance, in a bad-news letter, you'll begin with the specific causes leading to bad news (recession, depression, changes in the market, unfavorable work habits, etc.). Then, after setting the stage with these specifics, you'll relate the general conclusion reached—the bad news (layoffs, refunds denied, promotions rejected, etc.). The bad-news letter examples in this chapter reveal this inductive organizational pattern.

4. Draft your correspondence. Finally, after you've organized your information, the last step is to write your rough draft. Once you know what you want to say, you should just say it—and let it suffice for the moment. Write a rapid rough draft, focusing on content and organization, not on grammar, mechanics, or style. Your primary goal in a rough draft is to get words on the paper or screen. The time for fine tuning comes later in rewriting. During rewriting, you'll be able to check your spelling, consider your punctuation, and perfect your tone. Drafting, in contrast, should be a stage in the writing process still free of those fears caused by grammar anxiety. After all, if you're worrying about the correctness of every word while you write it, you'll never complete a sentence.

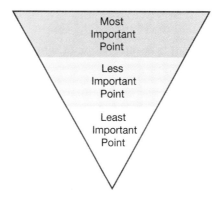

FIGURE 6.12 Inverted Journalist's Pyramid

REWRITING

After you've drafted your correspondence, the final step in the writing process and, in many people's opinion, the most important step, is to revise your rough copy. Rewriting is the step that most mediocre writers omit. Poor writers merely let the rough draft suffice as their finished version. They assume that their draft is sufficient. It isn't. Successful writers, in contrast, realize that an effective letter must be as perfect as it can be. To achieve this level of expertise, good writers revise their rough drafts—fine-tuning, honing, sculpting, molding, and polishing so what leaves their hands is something they can be proud of.

To revise your letters, consider the following revision criteria:

1. Add for clarity and correctness. Look back over your rough draft. In doing so, ask yourself what you might need to add for clarity. This might entail reviewing the reporter's questions. For example, has your letter clarified *who* will do the work (personnel), act as a contact person (liaison), manage the system (management), and so on? If not, then you need to provide further information by answering the reporter's question *who.* Have you correctly stated *when* you'll follow up, when you need the information, or when the bad news will occur? If not, add these dates. The same applies for all the reporter's questions. Add any missing information regarding *what, why, where,* or *how.*

In addition to adding answers to the reporter's questions, review your draft and add any missing letter essentials: letterhead address, date, inside address, salutation, complimentary close, typed and/or signed signature, and optional inclusions such as enclosure or copy notations.

2. Delete for conciseness. Deleting relates to style and content. Review your rough draft to delete any dead words and phrases which will raise your fog index (discussed in Chapter 3). Delete irrelevant content. This might include background sentences in the introductory paragraph which delay your main point. Writers often feel compelled to lead in to their thesis. This is a mistaken notion in technical writing. All the reader really wants to know is why you are writing and what you are writing about. Get to your point immediately; delete the unneeded background data.

3. *Simplify to aid easy understanding.* Chapter 3 stated that you should avoid old-fashioned words and phrases, such as *pursuant, accede, supersede,* and *in lieu of.* Rather than using such pompous words, simplify. Just write naturally. Why not say *after, accept, replace,* and *instead of?* This simplified style not only affects your fog index positively but also helps your readers understand your content, and that's the goal of good writing.

4. *Move information for emphasis.* This involves cutting and pasting— moving information within your letter. For example, let's say you've read your rough draft and realized that the itemized discussion paragraph is not organized according to importance. In fact, you have buried a key point deep in the body where readers might miss it. To focus your reader's attention on this idea, you should move it from the bottom of the list to the top. Doing so makes the idea more emphatic.

5. *Reformat for reader-friendly ease of access.* Review your letter's format. How does the letter look on the page? The letter's appearance affects your readers before they read one word. If the letter is open and appealing, then you're off to a good start. If the letter is blocky and inaccessible, then you've already done yourself a disservice.

Before you type your final copy, answer the following questions.

- Have you used enough (or too much) white space?
- Should you underline a key word or phrase?
- Are bullets appropriate in your discussion paragraph, or should you use numbers instead?
- Would boldfacing help you draw attention to a key concern?

6. *Enhance the letter's tone.* Letters aren't the inanimate objects they appear to be. They don't just dump data. Your letter is a reflection of your interpersonal communication skills and your company's attitudes. Reread your rough draft; is it revealing the personality or attitude that you want it to reveal? If the letter is stuffy, impersonal, high-handed, abrupt, offensive, or mean-spirited, then you must alter the tone.

To do so, add more pronouns, especially *you*-oriented ones; add verbs for punch; use contractions for a conversational tone; and/or add positive words and phrases to give your letter an upbeat tone. Using such personalization techniques (as discussed in Chapter 4) ensures customer and client cooperation.

7. *Correct errors.* Finally, before your letter leaves your office, correct any errors you've committed. The most important step of rewriting may be proof-reading. Check and double-check your grammar; check and double-check your mathematical computations and scientific notations. Your letter is a legal document. A mistake can cost your company money and you your job. Proofreading is boring, but reading the want ads to find a new job is worse. We presented several proofreading tips in Chapter 3. Save yourself future problems by proofing your letters.

PROCESS LOG

To give you a better idea of how this three-step process approach to letter writing works, we provide the following student-written process log. It includes the student's answers to the reporter's questions (prewriting), her rough draft (writing), her revisions (rewriting), and her finished copy.

PREWRITING

First, the student filled in the Reporter's Questions Checklist for a Letter of Inquiry (see Figure 6.13).

WRITING

After the student gathered her data, she wrote a rough draft (Figure 6.14).

REWRITING

Next, with the assistance of a peer review group consisting of two other students, the writer revised the letter (Figure 6.15).

Once the student made the suggested revisions, she submitted the finished copy (Figure 6.16).

CHAPTER HIGHLIGHTS

1. Letters are external correspondence.
2. When writing a letter, include these essential components:
   - Writer's address
   - Date
   - Inside address
   - Salutation
   - Body of the letter
   - Complimentary close
   - Your signed name
   - Your typed name
3. You could include optional letter components such as
   - A subject line
   - New-page notations
   - Writer's and typist's initials
   - Enclosure notation
   - Complimentary copy notation
4. In a letter of inquiry, you are asking for information.
5. In a cover letter, tell your reader what he or she is receiving in the enclosure, such as a questionnaire or a proposal.
6. Good-news letters provide your reader with positive information.

Who is your audience?

- ■ High-tech peer   _Vendor_
- ■ Low-tech peer
- ■ Lay reader
- ■ Management
- ■ Subordinate
- ■ Multiple audiences

Why are you writing? _In response to sales letter—to request information_

What is the general topic of your request? _Information about the_
personal shopping service

What exactly do you want to know (the specifics itemized in the discussion paragraph of your letter)?

- ■ _cost of services_
- ■ _number of employees needed_
- ■ _commission?_
- ■ _what department to work with_
- ■

What do you want the reader to do next? _Respond to questions_
_¹ e_

When do you want the reader to act (dated action)? _March 15_

Why is this date important? _for meeting with management_

FIGURE 6.13 Reporter's Questions Checklist for a Letter of Inquiry

Barney Allis Stores
5000 Main Street
Omaha, NE 32001

February 15, 2000

Fashion Pace
100 Eby Drive
New York, NY 01034

Dear Ms. Pace:

Thank you for the information about your personal shopping service. I am interested in Fashion Pace, and would like to receive the following information:

1. Sales figures
2. Customers serviced
3. Requirements and costs to start service

I would appreciate your response by March 1, 2000. Since I will be attending a meeting of the board on March 10, 2000 your early response will allow me to present your service at this time.

Sincerely,

Shirley Chandley

Shirley Chandley, Manager

FIGURE 6.14 Rough Draft of Letter of Inquiry

7. Bad-news letters require tactfulness and politeness on your part as the writer.

8. Complaint letters should politely state what you want and explain why your request is justified.

9. You can adjust (or answer) a letter of complaint in three ways:
   - 100 percent yes—you agree.
   - 100 percent no—you disagree.
   - Partial adjustment—you agree in part.

Barney Allis Stores
5000 Main Street
Omaha, NE 32001

February 15, 2000

Fashion Pace   *add name of contact*
100 Eby Drive
New York, NY 01034

Dear Ms. Pace:

Thank you for the information about your personal shopping
service. I am interested in Fashion Pace, and would like to receive
the following information:

Do you   1. Sales figures   *too vague - which ones?*
need the 2. Customers serviced *Awk! A list of satisfied customers*
periods? 3. Requirements and costs to start service *too vague*

        *Please... * Add more questions*
I would appreciate your response by March 1, 2000. Since I will
be attending a meeting of the board on March 10, 2000 your
early response will allow me to present your service at this time.

Sincerely,

Shirley Chandley, Manager
              move
          under name
                    * office space needed
                    * required training
                    * # of employees needed

FIGURE 6.15 Revision of Letter of Inquiry

10. Sales letters have to arouse interest, so explain how your product or service will benefit the reader.

11. The writing process will help you write any type of letter.

Barney Allis Stores
5000 Main Street
Omaha, NE 32001

February 15, 2000

Ms. Jackie Pace
Fashion Pace
100 Eby Drive
New York, NY 01034

Dear Ms. Pace:

Thank you for the information about your personal shopping service.

I am interested in Fashion Pace and would like to receive answers to the following questions:

1. What are your sales figures for the last two years?
2. Would you send me a satisfied customer list?
3. How much office space will your consultants need?
4. How many employees will you need to run your service?
5. What training is required for these employees?
6. What will you charge customers for your service?

Please send answers to these questions by March 1, 2000. Your early response will allow me to present your proposal at our March 10 sales meeting. I look forward to hearing from you.

Sincerely,

Shirley Chandley

Shirley Chandley
Manager

FIGURE 6.16 Letter of Inquiry Final Copy

ACTIVITIES

Writing Letters

1. Write a sales letter. To do so, imagine what product or service you could provide to a potential client. Select one from your area of expertise or academic major. Write your letter according to the criteria for sales letters and the writing process techniques discussed in this chapter.

2. Write a letter of inquiry. You might want to write to a college or university requesting information about a degree program, or to a manufacturer for information about a product or service. Whatever the subject matter, be specific in your request. Follow the criteria for writing letters of inquiry and the writing process techniques.

3. Write a cover letter. Perhaps your cover letter will preface a lab report you're working on in school, a report you're writing at work, or documentation you'll need to send to a client. Use the criteria for writing cover letters and the writing process techniques.

4. Write a good-news letter. Commend a coworker for a job well done. Congratulate a subordinate for his or her promotion. Better yet, write a thank-you letter to a teacher showing your appreciation for his or her professionalism and dedication. Follow the criteria for writing a good-news letter and the writing process techniques.

5. Write a bad-news letter. Inform a job candidate that he or she has not been offered the job, tell a customer that his or her request for a refund is denied, or explain to employees why they are to be laid off from their work. Use the criteria for writing bad-news letters and the writing process techniques.

6. Write a complaint letter. Perhaps you've purchased a product which has malfunctioned, received poor service from a salesperson, ordered one item but received another, failed to have a warranty honored, or received damaged or broken equipment in shipment. You want to write a letter of complaint to solve these problems. To do so, write your letter by following the suggestions provided in this chapter.

7. Write an adjustment letter. Envision that a client has written a complaint letter about a problem he or she has encountered with your product or service. Write a letter in response to their complaint. This could either be a 100 percent yes letter, a 100 percent no letter, or a partial adjustment letter. Follow the suggestions provided in this chapter when writing your letter.

Team Projects

1. Find business letters in your home and/or at work and bring them to class. In small groups of three to five students, categorize these letters as either sales letters, letters of inquiry, cover letters, good-news letters, bad-news letters, complaint letters, or adjustment letters. Then, in your groups, decide whether the letters are successful or unsuccessful, based on the cri-

teria provided in this chapter. If the letters are good, be prepared to explain why, either in writing or orally. If the letters fail, discuss how they could be improved.

2. In small groups of three to five students, select any of the flawed letters from activity 1 above and rewrite them. Use the revision techniques discussed in this chapter.

Case Studies

After reading the following case studies, write the appropriate letters required for each assignment.

1. Vivian Davis, who lives at 2939 Cactus in Santa Clara, CA 95054, has invented a new product, the VAST (a voice-activated speaker telephone). This telephone can be "dialed" without the use of one's hands, simply by saying names or numbers into the system's speaker. Vivian believes that such a machine could benefit the handicapped, the elderly, the infirm, homemakers, and businesspeople. She can sell this phone for $25 to electronic stores, hobby shops, general-purpose retailers, and so on, who then can market it for $50. Write a sales letter for Ms. Davis based on the preceding information.

2. Mark Shabbot works for Apex, Inc., at 1919 W. 23rd Street, Denver, CO 80204. Apex, a retailer of electronic equipment, wants to purchase 125 new oscilloscopes from a vendor, Omnico, located at 30467 Sheraton, Phoenix, AZ 85023. The oscilloscopes will be sold to a college in Denver (Northwest Hills Vocational-Technical College). However, before Apex purchases these oscilloscopes, Mark needs information regarding bulk rates, shipping schedules, maintenance agreements, equipment specifications, and machinery capabilities. Northwest Hills needs this equipment before the new term (August 15). Write a letter of inquiry for Mr. Shabbot based on the preceding information.

3. Sharon Baker works as a technical writer for Prismatic Consulting Engineering, 123 Park, Boston, MA 01755. In response to an RFP (request for proposal), she has written a proposal to the Oceanview City Council, 457 E. Cypress Street, Oceanview, MA 01759. The proposal suggests ways in which Oceanview can improve its flood control. Now Sharon needs to write a cover letter prefacing her proposal. In this cover letter, she wants to call her readers' attention to key concerns within the proposal: suggested costs, time frames, problems which could occur if the proposed suggestions are not implemented, ways in which the proposal will solve these problems, and Prismatic's credentials. Once the Oceanview City Council receives the proposal, Prismatic representatives will schedule follow-up discussions. Write Ms. Baker's cover letter based on the preceding information.

4. Bob Ward, a lineworker at HomeCare Health Equipment, 8025 Industrial Parkway, Ashley, NC 27709, deserves a letter of commendation for his excellent job record. He has not missed a day of work in five

years. In addition, his production line has achieved a 5 percent error reading (7 percent is considered acceptable). He has also trained new hires. Most importantly, he made six suggestions for improvements, three of which saved the company money. The company president, Peter Tsui, based at HomeCare's home office at 4791 Research Avenue, Wasa, MN 55900, wants to award Bob with a plaque at the annual awards dinner on September 7, 2000. Write Mr. Tsui's good-news letter based on the preceding information.

5. Stacy Helgoe works as a service technician for EEE Electronic Servicing, 11201 Blanco, Santa Fe, NM 88004. Yesterday, she went to Schoss-McGraw Associates, 1628 W. 18th Street, Taos, NM 88003, to service their computer systems. She billed them $75 for her time, but she did not bill them for parts since the machinery was supposedly under warranty. However, when she returned to EEE, Stacy's manager, Marilyn Hoover, informed Stacy that the machinery was not under warranty and that Schoss-McGraw would have to be billed an additional $45.87 for parts. Schoss-McGraw is an excellent client, so Marilyn wants Stacy to be especially tactful in requesting the additional money. Based on the preceding information, write Stacy's bad-news letter.

6. Diane Waisner (who lives at 1439 87th, Monroe, LA 67054) purchased a VCR on August 13 from Smiley's TV Town (8201 Magnolia, Monroe, LA 67056). The VCR came with a "ninety-day warranty against all defects" and a guarantee for "in-home free repairs and labor." On October 30, Diane's VCR showed a horizontal line across the screen when she replayed tapes. She called the store manager, Jill Miller, and explained the problem. Jill said the horizontal lines were caused by a dirty head and told Diane to bring the VCR in for cleaning. Jill also told Diane that she would be charged for this service since dirty VCR heads were basic wear and, therefore, not covered by the warranty. Diane was angered by this response from the store manager and decided to write a letter of complaint. Based on the information provided, write Diane's letter of complaint.

7. Gregory Peña (121 Mockingbird Lane, San Marcos, NV 87900) has written a letter of complaint to Donya Kahlili, the manager of CompuRam (4236 Silicon Dr., Reno, NV 87601). Mr. Peña purchased a computer from a CompuRam outlet in San Marcos. The San Marcos *Tattler* advertised that the computer "came loaded with all the software you'll need to write effective letters and perform basic accounting functions." (Mr. Peña has a copy of this advertisement.) When Mr. Peña booted up his computer, he expected to access word production software, multiple fonts, a graphics package, a grammar check, and a spreadsheet. All he got was a word processing package and a spreadsheet. Mr. Peña wants Ms. Kahlili to upgrade his software to include fonts, graphics, and a grammar check; wants a computer technician from CompuRam to load the software on his computer; and wants CompuRam to reimburse him $400 (the full price of the software) for his trouble.

Ms. Kahlili agrees that the advertisement is misleading and will provide Mr. Peña software including the fonts, graphics, and grammar check (complete with instructions for loading the software). However, she refuses Mr. Peña's other two requests.

Write Ms. Kahlili partial adjustment to Mr. Peña based on the information provided.

The Job Search

OBJECTIVES

A job search is demanding. You'll have to know how to look for a job, write a résumé and letter of application, interview for the job, and write a follow-up letter.

In this chapter, we help with your job search by giving you (a) ways to find job openings; (b) criteria for effective résumés; (c) criteria for effective letters of application; (d) techniques for interviewing effectively; (e) criteria for effective follow-up letters; and (f) sample résumés, letters of application, and follow-up letters.

HOW TO FIND JOB OPENINGS

When it's time for you to look for a job, how will you begin your search? You know you can't just wander up and down the street, knocking on doors randomly. That would be time-consuming, exhausting, and counterproductive. Instead, you must approach the job search more systematically.

1. Visit your college or university job placement center. This is an excellent place to begin a job search, for several reasons. First, your school's job placement service might have job counselors who will counsel you regarding your skills and job options. Second, your job placement center can give you helpful hints on preparing résumés, letters of application, and follow-up letters. Third, the center will post job openings within your community and possibly in other cities. Fourth, the center will be able to tell you when companies will visit campus for job recruiting. Finally, the service can keep on file your letters of recommendation or portfolio. The job placement center will send these out to interested companies upon your request.

2. Talk to your instructors. Whether in your major field or not, instructors can be excellent job sources. They will have contacts in business, industry, and/or education. They may know of job openings or people who may be helpful in your job search. Furthermore, since your instructors obviously know a great deal about you (having spent a semester or more working closely with you), they will know which types of jobs or work environments might best suit you.

3. Network with friends or past employers. Your friends and past employers know people. Those people know people. By visiting with acquaintances, you can network with numerous individuals. The more people you talk to about job openings, the more options you'll discover.

4. Check your professional affiliations and publications. If you belong to a professional organization, this could be a source of employment in three ways. First, your organization's sponsors or board members might be aware of job openings. Second, your organization might publish a listing of job openings. Finally, your organization's publications might list job opportunities.

5. Read the want ads. Check the classified sections of your local newspapers or in newspapers in cities you might like to live and work in. These want ads list job openings, requirements, and salary ranges.

6. Research the Internet. The Internet will help you research job opportunities. Most city newspaper Web sites have online classified pages, and a growing number of cities host Web sites dedicated to job searches. These sites list job openings and allow you to create online résumés. In addition, many companies list job openings on their Web sites and allow you to submit an application online. In 1997, an American Management Association survey found that 53 percent of major corporations found job candidates online. Some career placement experts believe that within a few years, "95 percent of jobs will be listed on the Internet" (Lorek).

CRITERIA FOR EFFECTIVE RÉSUMÉS

Once you've found a job which interests you, it's time to apply. Your job application will start when you send the prospective employer your résumé. Résumés are usually the first impression you make on a prospective employer. If your résumé is effective, you've opened the door to possible employment. You have given yourself the opportunity to sell your skills during an interview. If, in contrast, you write an ineffective résumé, you have closed the door to opportunity.

Your résumé should present an objective, easily accessible, detailed biographical sketch. However, don't try to include your entire history. Since the primary goal of your résumé, in conjunction with your letter of application, is to get you an interview, you can use your interview to explain in more detail information which does not appear on your resumé.

When writing your resumé, you have two optional approaches. You can write either a *reverse chronological resumé* or a *functional resumé*.

Write a reverse chronological resumé if you

- Are a traditional job applicant (recent high school or college graduate, age 18 to 25)
- Hope to enter the profession in which you have received college training or certification
- Have made steady progress in one profession (promotions and/or salary increases)
- Plan to stay in your present profession

Write a functional resumé if you

- Are a nontraditional job applicant (returning to the work force after a lengthy absence, older, not a recent high school or college graduate)
- Plan to enter a profession in which you have not received formal college training or certification
- Have changed jobs frequently
- Plan to enter a new profession

Whatever type of resumé you write, include the following:

- Identification
- Career objectives
- Employment
- Education
- Professional skills
- Military experience (if applicable)
- Professional affiliations
- Personal data
- References

IDENTIFICATION

Begin your resumé with the following:

- Your name, centered in all caps (full first name, middle initial, and last name). Your name can also be in boldface and printed in a larger type size (10 point, 12 point, etc.).
- Your address. Centered two spaces beneath your name, type your street address. Centered one space beneath your street address, type your city, state (use the correct two-letter abbreviation provided in Chapter 18), and zip code. If you're attending college, serving in the armed forces, or on your

own working, you might also want to include a permanent address. By including alternative addresses, you will enable your prospective employer to contact you more easily. What's the point of sending a résumé if you can't be reached?

- Your area code and phone number, centered one space beneath your city, state, and zip code. As with addresses, you can provide alternative phone numbers. However, limit yourself to two phone numbers, and don't provide a work phone. Having prospective employers call you at your present job is not wise. First, your current employer will not appreciate your receiving this sort of personal call; second, your future employer might believe that you often receive personal calls at work—and will continue to do so if he or she hires you. You can also include your e-mail and Web site addresses.

CAREER OBJECTIVES

A career objectives line is not mandatory. Some employers like you to state your career objectives; others do not. You will have to decide whether you want to include this section in your résumé.

The career objectives line is like a subject line in a memo or report. Just as the subject line clarifies your memo or report's intent, your career objective informs the reader of your résumé's focus.

Be sure your career objective is precise. Too often, career objectives are so generic that their vagueness does more harm than good, as in the following example:

FLAWED CAREER OBJECTIVE
Career Objective: Seeking employment in a business environment offering an opportunity for professional growth

This poorly constructed career objective provides no focus. What kind of business? What kind of opportunities for professional growth? Employers don't want to hire people who have only nebulous notions about their skills and objectives. Instead, look at the following clearly focused career objectives.

GOOD CAREER OBJECTIVE
Career Objective: An entry-level position as a computer programmer

BETTER CAREER OBJECTIVE
Career Objectives: To market financial planning programs and provide financial counseling to assure positive client relations

EMPLOYMENT

The employment section lists the jobs you've held. This information must be presented in reverse chronological order (your current job listed first, previous jobs listed next). This section must include the following:

- Your job title (if you have or had one)
- The name of the company you worked for
- The location of this company (either street address, city, and state or just the city and state)
- The time period during which you worked at this job
- Your job duties and responsibilities

EDUCATION

In addition to work experience, you must include your education. Document your educational experiences in reverse chronological order (most recent education first; previous schools, colleges, universities, military courses, and/or training seminars next). When listing your education, provide the following information.

- Degree. If you have not yet received your degree, you can write "Anticipated date of graduation June 2001" or "Degree expected in 2001."
- Area of specialization.
- School attended. Do not abbreviate. Although you might assume that everyone knows what *UT* means, your readers won't understand this abbreviation. Is UT the University of Texas, the University of Tennessee, the University of Tulsa, or the University of Toledo?
- Location. Include the city and state.
- Year of graduation or years attended.

As you can see, this information is just the facts and nothing else. Many people might have the same educational history as you. For instance, just imagine how many of your current classmates will graduate from your school, in the same year, with the same degree. Why are you more hirable than they are? The only way you can differentiate yourself from other job candidates with similar degrees is by highlighting your unique educational accomplishments. These might include any or all of the following:

- Grade point average
- Academic club memberships and leadership offices held
- Unique coursework
- Special class projects
- Academic honors, scholarships, and/or awards
- Fraternity/sorority leadership offices held
- Percentage of your college education costs you paid for while attending school
- Technical equipment you can operate

Please note two key concerns regarding your work experience and education. First, you should have no chronological gaps when all of your work and education are listed. You can't omit a year without a very good explanation. (A missing month or so is not a problem.) Second, the order in which you place your education or work is up to you. Our presentation of work before education was arbitrary. You should present your most significant section first. If, for example, your educational experience is what will get you that job, place it first. On the other hand, if your work experiences are superior to your education, then list your jobs first.

PROFESSIONAL SKILLS

If you are changing professions or reentering the workforce after a long absence, you will want to write a functional résumé. Therefore, rather than beginning with education or work experience, which won't necessarily help you get a job, focus your reader's attention on your unique skills. These could include

- Procedures you can perform
- Special accomplishments and awards
- On-the-job training you've received
- Training you've provided
- New techniques you've invented and/or implemented
- Numbers of and types of people you've managed
- Publications

These professional skills are important because they help show how you are different from all other applicants. In addition, they show that although you have not been trained in the job for which you are applying, you can still be a valuable employee.

MILITARY EXPERIENCE

If you served for several years in the military, you might want to describe this service in a separate section. You would state your

- Rank
- Service branch
- Location (city, state, country, ship, etc.)
- Years in service
- Discharge status
- Special clearances
- Achievements and professional skills
- Training seminars attended and/or education received

PROFESSIONAL AFFILIATIONS

If you belong to regional, national, or international clubs or have professional affiliations, you might want to mention these. Such memberships might include the Rotarians, Lions Club, Big Brothers, Campfire Girls, or Junior League. Maybe you belong to the Society for Technical Communication, the Institute of Electrical and Electronic Engineers, the National Office Machine Dealers Association, or the American Helicopter Society. Listing such associations emphasizes your social consciousness and your professional sincerity. However, if you maintain many such affiliations, your prospective employer might question your ability to handle your job and your numerous memberships.

PERSONAL DATA

Some job recruiters say that personal data rounds the person out. Others believe that personal data is just that, personal, and therefore irrelevant for job purposes. You be the judge. If you want to include personal data, here are some possibilities.

- Hobbies. Avoid generic ones, such as reading or movies. Similarly, avoid potentially negative ones, such as pool playing or heavy metal rock and roll.
- Languages fluently spoken. (Today's business environment is multilingual.)

Don't include any of the following information: birthdate, race, sex, religion, height, weight, religious affiliations, marital status, or pictures of yourself. Equal opportunity laws disallow employers from making decisions based on these factors.

REFERENCES

If you choose to include references (three or four colleagues, supervisors, teachers, or community individuals who will recommend you for employment), list their names, titles, addresses, and phone numbers. Instead of listing names, you might prefer to follow a reference heading by stating, "Supplied on request," "Available on request," or "Furnished on request."

References are not mandatory. As with all elements of a résumé, you should include only information which will benefit you in your job search. If you do not have anyone who will provide you an outstanding reference, omit this section. However, if you have great references, include these individuals—after asking their permission first. Excellent references, if checked before an interview, can give you an edge.

Another reason you might have for not listing several references by name, address, and phone number is that these references take up space on your résumé. If your job experiences and education are limited and you need additional information to fill up a page, include a list of references. However, your priority should be to emphasize your skills, the ways you will benefit a prospective employer. Don't waste precious space by listing references if you can provide more valuable information about yourself.

STYLE

The preceding information suggests *what* you should include in your résumé. Your next consideration is *how* this information should be presented. As mentioned throughout this textbook, format or page layout is essential for effective technical writing. The same holds true for your résumé.

1. Avoid sentences. Sentences create three problems in a résumé. First, if you use sentences, the majority of them will begin with the first-person pronoun *I*. You'll write, "I have...," "I graduated...," or "I worked...." Such sentences are repetitious and egocentric. Second, if you choose to use sentences you'll run the risk of committing grammatical errors: run-ons, dangling modifiers, agreement errors, and so forth. Third, sentences will take up room in your résumé, making it longer than necessary.

2. *Itemize your achievements.* Instead of sentences, highlight your résumé with easily accessible lists. Set apart your achievements by bulletizing your accomplishments, awards, unique skills, and so on.

3. *Begin your lists with verbs.* To convey a positive, assertive tone, use verbs when describing your achievements. Following is a list of verbs you might use.

accomplished	initiated
achieved	installed
analyzed	led
awarded	made
built	maintained
completed	managed
conducted	manufactured
coordinated	negotiated
created	ordered
customized	organized
designed	planned
developed	prepared
diagnosed	presented
directed	programmed
earned	reduced
established	resolved
expanded	reviewed
fabricated	sold
gained	supervised
implemented	trained
improved	

4. *Quantify.* Your résumé should not tell your readers how great you are; it should prove your worth. To do so, quantify, specify, precisely explain your achievements. For example, don't write, "Maintained positive customer relations with numerous clients." Instead write, "Maintained positive customer relations with 5,000 retail and 90 wholesale clients." Don't write, "Improved field representative efficiency through effective training." Instead write, "Improved field representative efficiency by writing corporate manuals for policies and procedures." Again, don't write, "Achieved production goals." Write, "Achieved 98 percent production, surpassing the company's desired goal of 90 percent."

5. Format your résumé for reader-friendly ease of access. You can make your résumé more accessible in many ways. For example, you can capitalize and bold-face headings. You can indent subheadings to create white space. You can underline subheadings and italicize major achievements. As already mentioned, you should use bullets to create lists. You also might want to use different font sizes to highlight headings, subheadings, and/or achievements. However, don't use every highlighting technique at your disposal. Select only a few; otherwise, your résumé will be overwhelmed with technique instead of substance.

6. Make it perfect. You cannot afford to make an error in your résumé. Remember, your résumé is the first impression you'll make on your prospective employer. Errors in your résumé will make a poor first impression.

ONLINE RÉSUMÉS

In addition to your paper-bound résumé, you might want to consider creating an online résumé for transmittal by e-mail or an HTML résumé for transmittal by the Internet. Online résumés provide several benefits not reaped by the traditional résumé printed on paper ("Benefits of Being Online"). Your online résumé will

- Allow thousands of international employers to access it instantly.
- Allow you to include portfolios of your work with lengthy documents, full-color graphics, animation, and sound.
- Allow you to update it quickly and easily.
- Allow you to show your expertise with HTML and Web-ready graphics.

Online résumés differ in some ways from the traditional paper résumé. For example, you will encounter no problems transmitting a paper résumé to your reader. You simply put the résumé in an envelope and mail it, and it will arrive at your reader's office looking exactly like it did when you sent it. However, this may not be the case when you e-mail a résumé. Let's say you've created your résumé using *Word*. You e-mail the file as an attachment. Your reader, however, is using a UNIX machine with *FrameMaker*. Will he or she be able to open your file easily, and will the résumé, once opened, look exactly like you the one you created? No. The reader may not be able to open the file, may have difficulty opening the file, or will see a résumé with skewed columns, formatting errors, and extra page breaks. These problems are caused by incompatible data processing systems (Robart).

Thus, you must construct your online résumé differently. If you want an online résumé to be *transmitted through e-mail*, you should create an ASCII version. If you want an online résumé to be *transmitted through the Internet*, you must use HTML.

ASCII RÉSUMÉS

You use ASCII for two reasons. First, all computers read standard ASCII; in contrast, all computers do not read extended ASCII (we discuss this problem in further detail in Chapter 10). Standard ASCII contains the following characters.

- Letters A through Z, both upper- and lowercase.
- Numbers 0 through 9
- Punctuation marks ! " " # $ & ' ( ) * + , - . / : ; <=> ? ` _ ~

If you are trying to send your online résumé to a computer that supports only standard ASCII but you have used extended ASCII, then the person receiving your online résumé will not be able to read all that you've written. For instance, computers supporting only standard ASCII cannot display em dashes or "smart" quotation marks, italics, colored fonts, dingbats, reverse text, underlining, and shadowing.

Second, you'll want to use standard ASCII to create your online résumé because many companies no longer hire people to read résumés. Instead, they use computers to screen résumés with a modern technique called electronic applicant tracking. The company's computer program scans résumés as raster (or bitmap) images. Then optical character recognition (OCR) software translates the bitmap images into standard ASCII. Next, the software uses artificial intelligence to read the ASCII text, scanning for keywords. If your résumé contains a sufficient number of these keywords, the résumé will then be given to someone in the human resources department for follow-up. If your online résumé is created using extended ASCII, OCR software will have trouble translating your text (McNair).

Use these techniques for creating your ASCII résumé for transmittal by e-mail (Skarzenski).

1. *Highlighting.* Avoid italics, underlining, colors, horizontal and vertical bars, and iconic bullets (instead, use asterisks for bullets). White space is still necessary, but don't use your tab key for spacing, since tabs will be interpreted differently in different computer environments (use your space bar instead). Finally, avoid organizing information in columns. Instead, use all-cap headings and place your text below the headings, spacing for visual appeal.

2. *Screen length.* Don't limit yourself to the traditional one-page résumé. You can't fit as many words on an ASCII page, so it's acceptable to submit an online résumé of two pages or more.

3. *Keywords.* If you're submitting your online résumé to a company that uses OCR, include precise keywords. For example, you don't want to write that you have knowledge of "various software products"—that's too vague. Be more precise: write that you can "create online help using *RoboHelp*" and that you can use "*PageMaker* and *Quark*." Similarly, don't say you are familiar with "computer technology." Instead, write that you have a working knowledge of "multimedia, hypertext, and Windows and Macintosh platforms" (McNair 14).

4. *Online protocol.* When you submit your résumé online, will you send it as an attachment or within the e-mail message? Not every e-mail system supports attachments. Thus, you will want to make sure your prospective employer can open an attachment before you send one.

If you submit an online résumé in response to a job opening advertised in a listserv, don't simply attach your ASCII to the original résumé request. If you do so, your résumé might be transmitted not only to the employer but also to the dozens of other job searchers and recipients of the original e-mail message. Instead, create a new e-mail document to ensure confidentiality.

HTML RÉSUMÉS

Another option is to create a Web site for your résumé, using HTML coding. We discuss hypertext and Web sites in great detail in Chapter 10, "Electronic Communication." When you develop an online résumé for the Internet, consider the criteria for effective Web pages we provide in Chapter 10.

- Tables for easy navigation between your Web site's screens
- Navigational buttons (either visual or textual links)
- Optimum contrast between text and background
- Headings and subheadings to ensure your readers won't get lost in cyberspace
- Limited line length, taking up approximately two-thirds of your viewable screen
- Highlighting techniques, such as graphics, white space, boldface headings, and font sizes appropriate for various heading/subheading levels

In addition, as with an ASCII résumé, you should include keywords in your HTML résumé. We discourage using the word *résumé* when you write a paperbound résumé because the fact that you are writing a résumé is obvious to the reader. However, the word *résumé* should be included on your Web résumé. Internet search engines usually categorize Web sites in one of two ways: either by your document title or by your URL (the Web site's address). Thus, to accommodate this fact, you should include the word *résumé* in your document title and in your URL as the directory or file name (Skarzenski 18).

Whichever of these options you pursue (ASCII or HTML résumé), help your reader by stating the software and version you've used to create the résumé. For example, don't just say you've used *Microsoft Word*; instead, specify which version of *Word* you used (Robart 14).

SAMPLE RÉSUMÉS

Figures 7.1 and 7.2 are examples of effective résumés.

CRITERIA FOR EFFECTIVE LETTERS OF APPLICATION

Your résumé, whether paper or online, will be prefaced by a letter of application. These two components of your job package serve different purposes.

The résumé is generic. You'll write one résumé and use it over and over again when applying for numerous jobs. In contrast, the letter of application is specific.

SHARON J. BARENBLATT
787 Rainbow Avenue
Boston, MA 12987
202-555-2121

OBJECTIVE
Employment as a computer maintenance technician in the telecommunications industry

EDUCATION
A.S., Computer Science. Boston Community College. Boston, MA. Anticipated date of graduation, 2001. 3.7 GPA while working 30 hours per week and going to college full time.
Areas of Expertise
- Troubleshooting computer systems
- Telecommunications
- Programming in Pascal
- Programming in BASIC
- Data entry

Frederick Douglass High School. Newcastle, MA. 1994.
Accomplishments
- 3.5 GPA
- Member, Honor Society
- Captain, Frederick Douglass High School tennis team

WORK EXPERIENCE
Salesperson/assistant department manager. Electronics Warehouse. Boston, MA. 1999 to present.
- Prepare nightly deposits, input daily receipts
- Open and close the store
- Troubleshoot computer systems
- Provide customer service
- Sell and support computer systems

Computer lab assistant. Boston Community College. 1999 to present.
- Assist students with computer application programs
- Administer and proctor tests
- Troubleshoot computer hardware and software problems

Salesperson, On-Line Computer Center. Boston, MA. 1997 to 1998.
- Demonstrated and sold computer systems

Market Researcher, PRM Research Company. Newcastle, MA. 1995 to 1996.
- Performed telephone surveys (20 hours per week while going to high school)

REFERENCES
Available on request

FIGURE 7.1 Reverse Chronological Résumé

MARY C. WOLTKAMP
1944 W. 112th Street
Salem, OR 64925
(513) 451–4978

OBJECTIVE
To manage a department or division, using administrative and supervisory skills

PROFESSIONAL SKILLS
- Operated sporting goods/sportswear mail-order house. Business began as home-based but experienced 125 percent growth and was purchased by a national retail sporting goods chain.
- Managed a retail design studio producing over $500,000 annually.
- Hired, trained, and supervised an administrative staff of 15 employees for a financial planning institution.
- Sold copiers through on-site demonstrations. Exceeded corporate sales goals by 10 percent annually.
- Provided purchaser training for office equipment, reducing labor costs by 25 percent.
- Acquired modern management skills through continuing education courses

WORK EXPERIENCE
Office manager, Simcoe Designs, Salem, OR. 1996 to present.
Sales representative, Hi-Tech Office Systems, Salem, OR. 1994 to 1996.
Office manager, Lueck Finances, Portland, OR. 1992 to 1994.
President, Good Sports, Inc., Portland, OR. 1991 to 1992.

EDUCATION
Continuing education courses, Salem Community College, 1996 to present.
- Effective Business Writing
- Managing Diversity
- Time Management
- Total Quality Management/Continuous Quality Improvement
Attended Portland State University, majoring in business, 1990 to 1993. Took courses in business accounting, business law, computer systems, and technical writing.

REFERENCES
Available on request

FIGURE 7.2 Functional Résumé

Each letter of application will be different, customized specifically for each job. Whereas the résumé will always stay the same, except for periodic updating, your letters of application will change with each new job search.

Criteria for an effective letter of application include the following.

LETTER ESSENTIALS

As discussed in Chapter 6, letters contain certain mandatory components: your address, the date, your reader's address, a salutation, the letter's body, a complimentary close, your signed name, your typed name, and an enclosure notation if applicable. If you are submitting an electronic résumé along with an online letter of application, you won't need these letter essentials. Instead, type the application letter as part of your online résumé or send the letter as an e-mail attachment (Robart 13).

INTRODUCTION

In your introductory paragraph, include the following:

- Tell where you discovered the job opening. You might write, "In response to your advertisement in the May 31, 2000, *Lubbock Avalanche Journal* . . ." or "Bob Ward, manager of human resources, informed me that. . . ."
- State which specific job you are applying for. Often, the classified section of your newspaper will advertise several jobs at one company. You must clarify which of those jobs you're interested in. For example, you could write, "Your advertisement for a computer maintenance technician is just what I have been looking for."
- Sum up your best credentials. "My B.S. in chemistry and five years of experience working in a hazardous materials lab qualify me for the position."

DISCUSSION

In the discussion paragraph(s), sell your skills. To do so, describe your work experience, your education, and/or your professional skills. This section of your letter of application, however, is not meant to be merely a replication of your résumé. The résumé is generic; the letter of application is specifically geared toward your reader's needs. Therefore, in the discussion, follow these guidelines.

- Focus on your assets uniquely applicable to the advertised position. Select only those skills from your résumé that relate to the advertisement and which will benefit the prospective employer.
- Don't explain how the job will make you happy: "I will benefit from this job because it will teach me valuable skills." Instead, using the pronouns *you* and *your*, show reader benefit: "My work with governmental agencies has provided me a wide variety of skills from which your company will benefit."
- Quantify your abilities. Don't just say you're great ("I am always looking for ways to improve my job performance"). Instead, prove your assertions with quantifiable facts: "I won the 1999 award for new ideas saving the company money."

CONCLUSION

Your final paragraph should be a call to action. You could say, "I hope to hear from you soon" or "I am looking forward to discussing my application with you in greater detail." You could tell your reader how to get in touch with you: "I can be reached at (913) 469–8500." If you are more daring, you could write, "I will contact you within the next two weeks to make an appointment at your convenience. At that time I would be happy to discuss my credentials more thoroughly."

In addition to these suggestions, you should mention that you've enclosed a résumé. You can do this either in the introduction, discussion, or conclusion. Select the place which lends itself best to doing so.

STYLE

Your letter of application should follow the guidelines for all successful technical writing.

- *Accuracy*. You cannot commit any errors (whether grammatical or autobiographical). Grammatical errors in your letter of application will suggest that you might make errors at work as well. Autobiographical errors constitute a lie. In either instance, you will have hurt your employment opportunities.

- *Conciseness*. A low fog index will help your readers focus their attention on your skills rather than on your elaborate sentence structure. Don't ramble on and on. Decide exactly what needs to be said, and then say it.

- *Audience recognition*. Show how you will benefit the employer. Don't tell the employer how his or her job will help you. Try to limit use of *I*; emphasize *you* and *your*.

- *Clarity*. Show clearly how your skills relate to the advertised position, and quantify your assertions.

SAMPLE LETTERS OF APPLICATION

Figures 7.3 and 7.4 are examples of effective letters of application.

TECHNIQUES FOR INTERVIEWING EFFECTIVELY

The goal of writing an effective résumé and letter of application is to get an interview. The résumé and letter of application open the door; only a successful interview will win you the job. Here's how to interview successfully.

1. Dress for success. Although this adage has become a cliché, it's still valid. The key to successful dressing is to wear clean, conservative clothing. No one expects you to spend money on high-fashion, stylish clothes, but everyone expects you to look neat and acceptable. Men and women's suits are still best.

2. Be on time. In fact, plan to arrive at your interview at least twenty to thirty minutes ahead of schedule. This will help you stay on time if you get lost or have difficulty finding a parking place. Also, being early will allow you to stop off

11944 West 112th Street
Denton, TX 77892
December 11, 2000

Mr. Oscar Holway
Valley Telephone Cooperative, Inc.
Raymonville, TX 78580

Dear Mr. Holway:

I am responding to your advertisement in the November 24, 2000, issue of *Telephony* for an accounting office supervisor. Because of my five years experience in the telecommunications industry and my B.S. in electrical engineering, I believe I have the skills you require.

Although I have enclosed a résumé, let me elaborate on my achievements. While working for Southwestern Telephone Company, I have managed their accounting activities for forecasting, expense, and nonregulated FCC accounts. As listed in one of your requirements, I have a working knowledge of FCC Part 32 accounts. In fact, my current project involves converting our Part 31 accounts to the Part 32 uniform system of accounting. Furthermore, my engineering education has enabled me to develop mechanized office functions. In one year, I implemented three systems which reduced Southwestern's expenses and saved the company $50,000.

I would like to meet with you and discuss employment possibilities at your company. Please call me at (913) 469–8500 so that we can set up an interview at your convenience. I appreciate your consideration.

Sincerely,

Leonard Liss

Leonard Liss

Enclosure

FIGURE 7.3 Letter of Application

4628 Little Avenue
Coral Gables, FL 33133
September 9, 2000

Ms. Karen Pechis
King Petroleum Company
Oakland Division
P.O. Box 189924
Oakland, CA 94610

Dear Ms. Pechis:

The September 4, 2000, *Oakland Voice* contained an advertisement from King Petroleum Company for an environmental protection specialist. After reading the required qualifications, I believe I can meet King's needs.

As my enclosed résumé indicates, I have an M.S. in groundwater chemistry. I currently work for a county environmental agency where I have gained six years' experience with the various applications you mentioned in your advertisement. I have performed landfill groundwater monitoring at regional and national parks and supervised a team of twelve solid waste disposal monitoring specialists. In addition, I coordinated spill cleanups off the Atlantic and Gulf coasts and reported hazardous waste disposal procedures for local, state, and federal environmental agencies. In each of these instances, I successfully achieved governmental compliance. In fact, my supervisors cited my work as exemplary and entrusted me with training our new personnel.

I welcome the challenge of working with King Petroleum as an environmental specialist. So that I can answer any of your questions, I would appreciate your calling me at (312) 555–2121. I look forward to interviewing with you.

Sincerely,

Glenn Pfeiffer

Glen Pfeiffer

Enclosure: résumé

FIGURE 7.4 Letter of Application

in the restroom for a final hair brushing or tie straightening. Finally, being 5 or 10 minutes early makes a good impression on your prospective employer, who will recognize your enthusiasm.

3. Don't chew gum, smoke, or drink coffee. The gum will distort your speech, the cigarette could be offensive, and the coffee might make you jittery.

4. Watch your body language. To make the best impression, don't slouch. Instead, sit straight in your chair, even leaning forward a little. Look your interviewer in the eye.

5. Control your speaking. Speak slowly, and don't ramble on and on. Once you've answered the questions satisfactorily, stop. Don't monopolize the interview.

6. Come prepared. This entails three things. First, you can bring to the interview a few supporting documents, including letters of recommendation, articles you've published or papers you've written, employer performance appraisals, transcripts, extra copies of your résumé, laboratory results, schematics, etc. Don't go overboard, however. Just two or three items will do. Second, research the company so you can ask a few informed questions about its organization. These will show the interviewer that you are sincerely interested in the company. Finally, know the types of questions you'll be asked and be ready with answers. Some typical questions include the following:

- What are your strengths?
- What are your weaknesses?
- Why do you want to work for this company?
- Why are you leaving your present employment?
- What did you like most about your last job?
- What did you like least about your last job?
- How will you benefit this company?
- What computer hardware are you familiar with, and what computer languages do you know?
- What machines can you use?
- What special techniques do you know, or what special skills do you have?
- How do you get along with colleagues and with management?
- Can you travel?
- Will you relocate?
- What starting salary would you expect?
- What do you want to be doing in five years, ten years?
- What about this job appealed to you?
- How would you handle this (hypothetical) situation?
- What was your biggest accomplishment in your last job or while in college?
- What questions do you have for us?

CRITERIA FOR EFFECTIVE FOLLOW-UP CORRESPONDENCE

Once you've interviewed, don't just sit back and wait, hoping that you'll be offered the job. Write a follow-up letter or e-mail. This follow-up accomplishes three primary things: it thanks your interviewers for their time, keeps your name fresh in their memories, and gives you an opportunity to introduce new reasons for hiring you.

A follow-up letter or e-mail message contains an introduction, discussion, and conclusion.

1. *Introduction.* Tell the reader(s) how much you appreciated meeting them. Be sure to state the date on which you met and the job for which you applied.
2. *Discussion.* In this paragraph, emphasize or add important information concerning your suitability for the job. Add details which you forgot to mention during the interview, clarify details which you covered insufficiently, and/or highlight your skills that match the job requirements. In any case, sell yourself one last time.
3. *Conclusion.* Thank the reader(s) for their consideration, and/or remind them how they can get in touch with you for further information. Don't, however, give them any deadlines for making a decision.

SAMPLE FOLLOW-UP LETTER

Figure 7.5 is an example of an effective follow-up letter.

CHECKLISTS

Seeking a job is not easy. However, it can be a manageable activity if you approach it as a series of separate but equal tasks. To get the job you want, you must search for an appropriate position, write an effective résumé and letter of application, interview successfully, and write a follow-up letter.

JOB SEARCH CHECKLIST

Job Openings

✔ Did you visit your college or university job placement center?

✔ Did you talk to your professors about job openings?

✔ Have you networked with friends or past employers?

✔ Have you checked with your professional affiliations or looked for job openings in trade journals?

✔ Did you read the want ads in the newspapers?

✔ Did you search the Internet for job openings?

4628 Little Avenue
Coral Gables, FL 33133
November 1, 2000

Ms. Karen Pechis
King Petroleum Company
Oakland Division
P.O. Box 189924
Oakland, CA 94610

Dear Ms. Pechis:

I enjoyed meeting you and your colleagues last Wednesday,
October 27, 2000, to discuss the opening for an environmental
protection specialist.

You stated in the interview that King Petroleum is planning to
expand into offshore exploration. With my training skills and
governmental compliance reporting abilities, I would welcome
the opportunity to become involved in this exciting expansion.

Again, thank you for your time and consideration.

Sincerely,

Glenn Pfeiffer

Glen Pfeiffer

FIGURE 7.5 Follow-up Letter

Résumé

✔ Are your name, address, and phone number correct?

✔ Is your job objective specific?

✔ Are the dates within your education, work experience, and military
experience sections accurate?

✔ Have you included your degree, school, city, and state within the education
section?

✓ Have you included your job title, company name, city, and state within the work experience section?

✓ Have you used verbs to introduce each of your professional skills?

✓ Have you quantified each of your achievements?

✓ Have you avoided using sentences and the word *I*?

✓ Does your résumé use highlighting techniques to make it reader-friendly?

✓ Have you proofread your résumé to find grammatical and mechanical errors?

✓ Have you decided whether you should write a reverse chronological résumé or a functional résumé?

✓ Have you decided to create an online résumé? If so, have you considered the ways in which it differs from the traditional paper résumé?

Letter of Application

✓ Have you included all of the letter essentials?

✓ Does your introductory paragraph state where you learned of the job, which job you are applying for, and your interest in the position?

✓ Does your letter's discussion unit pinpoint the ways in which you will benefit the company?

✓ Does your letter's concluding paragraph end cordially and explain what you will do next or what you hope your reader will do next?

✓ Is your letter free of all errors?

Interview

✓ Will you dress appropriately?

✓ Will you arrive ahead of time?

✓ Will you avoid gum, cigarettes, and coffee?

✓ Have you practiced answering potential questions?

✓ Have you researched the company so you can ask informed questions?

✓ Will you bring to the interview additional examples of your work or copies of your résumé?

Follow-Up Letter

✓ Have you included all the letter essentials?

✓ Does your introductory paragraph remind the readers when you interviewed and what position you interviewed for?

✓ Does the discussion unit highlight additional ways in which you might benefit the company?

✓ Does the concluding paragraph thank the readers for their time and consideration?

✓ Does your letter avoid all errors?

CHAPTER HIGHLIGHTS

1. Use many different resources to locate possible jobs, such as college placement centers, instructors, friends, professional affiliations, want ads, and the Internet.

2. Use either a reverse chronological résumé or a functional résumé.

3. Write a traditional paper-bound résumé and/or an online résumé in ASCII or HTML format.

4. Indicate a specific career objective on your résumé.

5. Write your letter of application so that it targets a specific job.

6. Prepare before your interview so you can anticipate possible questions.

7. Your follow-up letter or e-mail message will impress the interviewer and remind him or her of your strengths.

ACTIVITIES

1. Practice a job search. To do so, find examples of job openings in newspapers, professional journals, at your college or university's job placement service, and online. Bring these job possibilities to class for group discussions. From this job search, you and your peers will get a better understanding of what employers want in new hires.

2. Make an appointment to visit a personnel manager at a company in your field of interest. Interview him or her, using the interviewing techniques discussed in Chapter 13. Discover what traits this manager looks for in a new employee and what he or she considers to be important in a résumé and letter of application. Share your findings with your technical writing class through a brief oral presentation.

3. Write a résumé, either online or traditional paper-bound. To do so, follow the suggestions provided in this chapter. Once you've constructed this résumé, bring it to class for peer review. In small groups, discuss each résumé's successes and areas needing improvement. If you prefer, make transparencies of each résumé and have an entire class review them for suggested improvements.

4. Using an Internet search engine, find job openings in your area of interest. Which companies are hiring, and what skills do they want from prospective employees? Make a list of companies advertising on the Internet, and share with your class your findings regarding their preferred skills.

5. Find examples of HTML résumés. You can do this by typing in the word *résumé* on an Internet search engine. Then, using the criteria provided in this chapter and in Chapter 10, determine which résumés are successful and which need improvement. Select one résumé that can be improved and revise it.

6. Write a traditional paper-bound résumé. Then, using our tips, rewrite it in either ASCII or HTML format.

7. Write a letter of application. To do so, find a job advertisement in your newspaper's classified section, in your school's career planning and placement office, at your work site's personnel office, or in a trade journal. Then construct the letter according to the suggestions provided in this chapter. Next, in small groups or on an overhead, review your letter of application for suggested improvements.

8. Practice a job interview in small groups, designating one student as the job applicant and other students as the interview committee. Ask the applicant the sample interview questions provided in this chapter or any others you consider valid. This will give you and your peers a feel for the interviewing process.

9. Write a follow-up letter. Again, use the suggestions provided in this chapter.

8 Document Design

OBJECTIVES

In technical writing, are words your only concern? The answer is an emphatic "No!" What you say is important, but how the text looks on the page is equally important. If you give your readers excessively long paragraphs, pages full of wall-to-wall words, you've made a mistake. Ugly blocks of unreadable, unappealing text will get you and your company in trouble. Remember your purposes in technical communication.

1. The technical writing context. Why do people read correspondence? Although individuals read poetry, short stories, novels, and drama for enjoyment, few people read memos, letters, reports, or instructions for fun. They read these types of technical writing for information about a product or service. They read this correspondence while they talk on the telephone, while they commute to work, or while they walk to meetings.

Given these contexts, your readers want you to provide them information quickly, information they can understand at a glance. Reading word after word, paragraph after paragraph takes time and effort, which most readers cannot spare. Therefore, if your technical writing is visually unappealing, your audience might not even read your words. Readers will either give up before they've begun or be unable to remember what they've read. You cannot assume they will labor over your text to uncover its worth. Good technical writing allows readers rapid access to information.

2. Damages and dangers. If your intended readers fail to read your text because it is visually inaccessible, imagine the possible repercussions. They could damage equipment by not recognizing important information which you've buried in dense blocks of text. The readers could give up on the text and call your company's toll-free hotline for assistance. This wastes your readers' and your co-workers' time and energy. Worse, your readers might hurt themselves and sue your company for failing to highlight potential dangers in the user manual.

3. Corporate identity. Your document—whether it's a memo, letter, report, instruction, newsletter, or brochure—is a visual representation of your company, graphically expressing your company's identity. It might be the only way you meet your clients. If your text is unappealing, that's the corporate image your company conveys to the customer. If your text is not reader-friendly, that's how your company will appear to your client. Visually unappealing and inaccessible correspondence can negatively affect your company's sales and reputation. In today's competitive workplace, any leverage you can provide your company is a plus. Document design is one way to appeal to a client.

To clarify how important document design is, look at the inaccessible meeting minutes in Figure 8.1. These minutes are neither clear nor concise. You are given so much data in such an unappealing format that your first response upon seeing the correspondence probably is to say, "I'm not going to read that."

How can you make these minutes more inviting? How can you break up the wall-to-wall words and make key points jump off the page? To achieve effective document design, you'll need to provide your readers visual (a) organization, (b) order, (c) access, and (d) variety.

CHUNKING

The easiest way to improve your document's design is to break text into smaller chunks of information, a technique called *chunking*. When you use chunking to separate blocks of text, you help your readers understand the overall organization of your correspondence. They can see which topics go together and which are distinct (Keyes 640; Watzman ATA-49).

Chunking to organize your text is accomplished by using any of the following techniques.

- *Headings* (one to three words that summarize the content of a unit of information)
- *White space* (horizontal spacing between paragraphs, created by double or triple spacing)
- *Rules* (horizontal lines typed across the page to separate units of information)
- *Section dividers and tabs* (used in longer reports to create smaller units).

Notice how using some of these techniques improves the document design in Figure 8.2.

MINUTES

The meeting at the Carriage Club was attended by thirty members and guests. After the dinner, Roger Traver introduced the guest speaker, George Smith, university chancellor, and noted his accomplishments and experiences prior to education—U.S. Navy commander, Oak Ridge Laboratory researcher, and politician. Dr. Smith's talk, "Industry and Education Collaboration," was very interesting and included a history of special projects enjoyed by both academics and corporate heads. Dr. Smith suggested that we engineers could work with education to (1) provide training seminars, (2) help in urban development, and (3) provide intern opportunities. Recent industry-education collaborations include training seminars in computers, fiber optics, and human resource options. The chancellor's primary thrust was a request for $100,000 in financial aid for urban development. He said money had already been donated from three sources: a large realty firm, Capital Homes, had given $20,000; a philanthropic group, We Care, had donated a matching $20,000; Dr. Smith's university gave a matching $20,000. The remaining $40,000, Dr. Smith hoped, would come from industry donations. Finally, the chancellor noted that industry could help itself, as well as the community, by providing internships for university undergraduate majors. These internships could either be semester- or year-long arrangements, whereby students would work for minimum wage to learn more about the day-to-day aspects of their chosen fields. The chancellor said that these internships would not only increase the students' theoretical knowledge of engineering by giving them hands-on experience but also make them better future employees for the host engineering companies. Everyone would benefit. Dr. Smith noted that the students would receive a grade and credit for their work. After the speech, our VP introduced new business, calling for nominations for next year's officers; gave us the agenda for our next meeting; and adjourned the meeting.

FIGURE 8.1 Inaccessible Meeting Minutes

ORDER

Once a wall of unbroken words has been separated through chunking to help the reader understand the text's organization, the next thing a reader wants from your text is a sense of order. What's most important on the page? What's less important? What's least important? You can help your audience prioritize information by ordering—or *queuing*—ideas (Keyes 640–41; Watzman ATA-49). The primary way to accomplish this goal is through a hierarchy of headings set apart from each other through various techniques.

MINUTES

The meeting at the Carriage Club was attended by 30 members and guests. After the dinner, Roger Traver introduced the guest speaker, George Smith, university chancellor, and noted his accomplishments and experiences prior to education—U.S. Navy commander, Oak Ridge Laboratory researcher, and politician. Dr. Smith's talk, "Industry and Education Collaboration" was very interesting and included a history of special projects enjoyed by both academics and corporate heads. Dr. Smith suggested that we engineers could work with education to accomplish three goals.

Training Seminars
Recent industry-education collaborations include training seminars in computers, fiber optics, and human resource options.

Urban Development
The chancellor's primary thrust was a request for $100,000 in financial aid for urban development. He said money had already been donated from three sources: a large realty firm, Capital Homes, had given $20,000; a philanthropic group, We Care, had donated a matching $20,000; Dr. Smith's university gave a matching $20,000. The remaining $40,000, Dr. Smith hoped, would come from industry donations.

Internships
The chancellor noted that industry could help itself, as well as the community, by providing internships for university undergraduate majors. These internships could either be semester- or year-long arrangements, whereby students would work for minimum wage to learn more about the day-to-day aspects of their chosen fields. The chancellor said that these internships would not only increase the students' theoretical knowledge of engineering by giving them hands-on experience but also make them better future employees for the host engineering companies. Everyone would benefit. Dr. Smith noted that the students would receive a grade and credit for their work.

Conclusion
After the speech, our VP introduced new business, calling for nominations for next year's officers; gave us the agenda for our next meeting; and adjourned the meeting.

FIGURE 8.2 Document Design Using Chunking to Organize the Information

- *Typeface*. There are many different *typefaces* (or *fonts*), including Courier, Prestige, Helvetica, Arial, Bosanova, Future, LCD, Key, and Chicago. Whichever typeface you choose, it will either be a *serif* or *sans serif* typeface. Serif type has "feet" or decorative strokes at the edges of each letter. This typeface is commonly used in text because it is easy to read, allowing the reader's eyes to glide across the page (Benson 36-37).

SERIF ◀──────── decorative feet

Sans serif (**as seen in this parenthetical comment**) is a block typeface that omits the feet or decorative lines. This typeface is best used for headings.

SANS SERIF ◀──── no decorative feet

- *Type size*. Another way of queuing for your readers is through the size of your type. A primary, first-level heading should be larger than subsequent, less important headings: second level, third level, and so forth. For example, a first-level heading could be in 18-point type. The second-level heading would then be set in 16-point type, the third-level heading in 14-point type, and the fourth-level heading in 12-point type (Keyes 641).

Figure 8.3 shows examples of different typefaces and type sizes.

- *Density*. The *weight* of the type also prioritizes your text. Type *density* is created by boldfacing or double-striking words (Keyes 641; Benson 37).
- *Spacing*. Another queuing technique to help your readers order their thoughts is the *amount of horizontal space* used after each heading. The first-level heading should have more space following it than the second-level heading, and so forth.
- *Position*. Your headings can be centered, aligned with the left margin, indented, or outdented (*hung heads*). No one approach is more valuable or more correct than another. The key is consistency. If you center your first-level heading, for example, and then place subsequent heads at the left margin, this should be your model for all chapters or sections of that report (Keyes 641; Watzman VC-86).

Figure 8.4 shows an *outdented* first-level heading with *indented* subsequent headings. Figure 8.5 shows a *centered* heading with subsequent headings aligned with the left margin.

Figure 8.6 reformats the meeting minutes seen in Figure 8.2 and uses queuing to order the hierarchy of ideas. The outdented first-level heading is set in an 12-point bold sans-serif typeface, all caps. The second-level heading is set in a 10-point bold serif typeface and is separated from the preceding text by horizontal white space. The third-level heading is set in a 10-point bold serif typeface and is separated from the preceding text by double spacing. It is also set on the same line as the following text.

Hierarchical heading levels allow the readers to visualize the order of information, to see clearly how the writer has prioritized text.

Sans-Serif Typefaces

Avant Garde 12 point

Avant Garde 14 point

Avant Garde 18 point

Futura 12 point

Futura 14 point

Futura 18 point

Helvetica 12 point

Helvetica 14 point

Helvetica 18 point

Serif Typefaces

Courier 12 point

Courier 14 point

Courier 18 point

Book Antiqua 12 point

Book Antiqua 14 point

Book Antiqua 18 point

Bookman 12 point

Bookman 14 point

Bookman 18 point

FIGURE 8.3 Examples of Typefaces and Type Sizes

ACCESS

Chunking helps the reader see which ideas go together, and a hierarchy of headings helps the reader understand the relative importance of each unit of information. Nonetheless, the document design in Figure 8.6 needs improvement. The reader still must read every word carefully to see the key points within each chunk of text. Readers are not that generous with their time. As writer, you should make your reader's task easier.

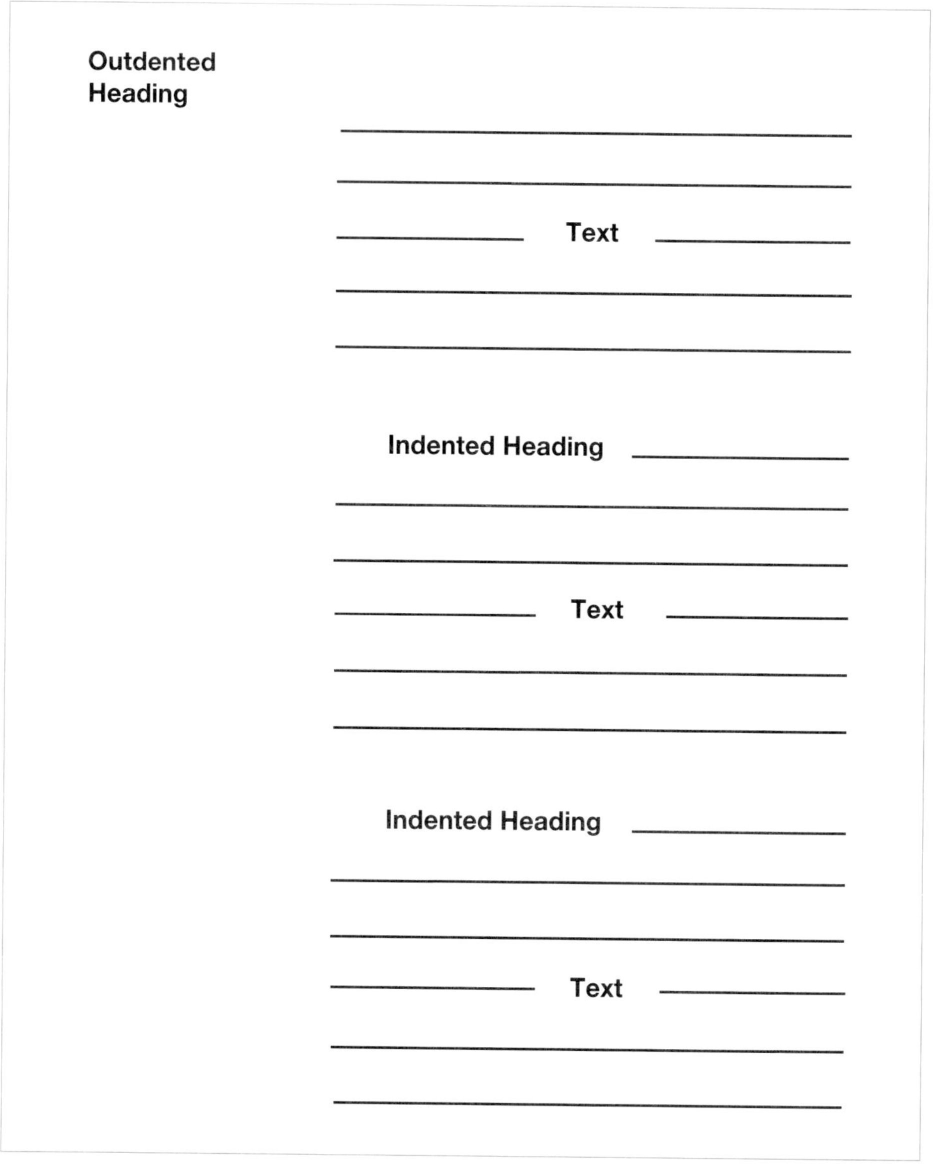

Outdented Heading

_____

_____

_____ Text _____

_____

_____

Indented Heading _____

_____

_____

_____ Text _____

_____

_____

Indented Heading _____

_____

_____

_____ Text _____

_____

_____

FIGURE 8.4 Outdented and Indented Headings

A third way to assist your audience is by helping them *access* information rapidly—at a glance. You can use any of the following highlighting techniques to help the readers filter out extraneous or tangential information and focus on key ideas (Keyes 641; Watzman ATA-49).

- *White space.* In addition to horizontal space, created by double or triple spacing, you also can create *vertical space* by indenting. This vertical white

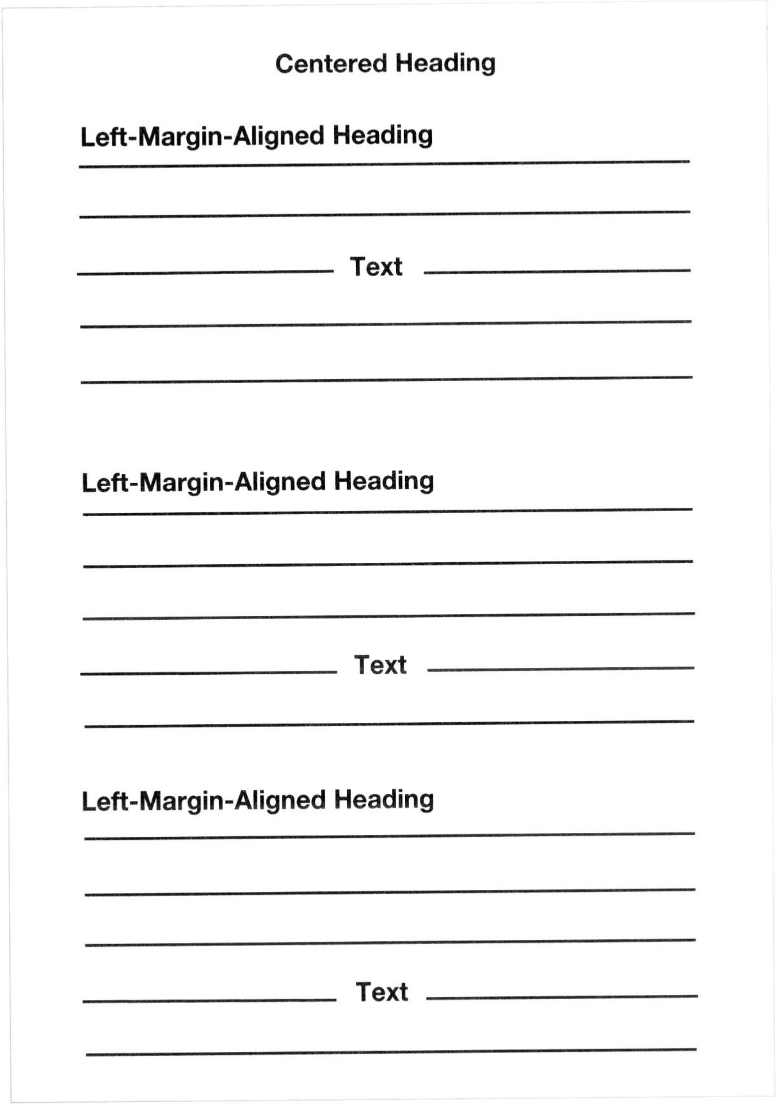

Centered Heading

Left-Margin-Aligned Heading

———————————————————————————————

——————————————————— Text ———————————————————

———————————————————————————————

Left-Margin-Aligned Heading

———————————————————————————————

——————————————————— Text ———————————————————

———————————————————————————————

Left-Margin-Aligned Heading

———————————————————————————————

——————————————————— Text ———————————————————

FIGURE 8.5 Centered and Left-Margin-Aligned Headings

space breaks up the monotony of wall-to-wall words and gives your readers breathing room. White space invites your readers into the text and helps the audience focus on the indented points you want to emphasize.

- *Bullets*. Bullets, used to emphasize items within an indented list, are created by using asterisks (*), hyphens (-), a lowercase "o", degree signs (°),typographic symbols (■, ❑, ●, or ♦), or iconic dingbats (☞, ✆, or ✓).
- *Numbering*. *Enumeration* creates itemized lists that can show sequence or importance and allow for easy reference.

MINUTES The meeting at the Carriage Club was attended by 30 members and guests. After the dinner, Roger Traver introduced the guest speaker, George Smith, university chancellor, and noted his accomplishments and experiences prior to education—U.S. Navy commander, Oak Ridge Laboratory researcher, and politician. Dr. Smith's talk, "Industry and Education Collaboration," was very interesting and included a history of special projects enjoyed by both academics and corporate heads. Dr. Smith suggested that we engineers could work with education to accomplish three goals.

Urban Development

The chancellor's primary thrust was a request for $100,000 in financial aid for urban development. He said money had already been donated from three sources: a large realty firm, Capital Homes, had given $20,000; a philanthropic group, We Care, had donated a matching $20,000; Dr. Smith's university gave a matching $20,000. The remaining $40,000, Dr. Smith hoped, would come from industry donations.

Internships

The chancellor noted that industry could help itself, as well as the community, by providing internships for university undergraduate majors. These internships could either be semester- or year-long arrangements, whereby students would work for minimum wage to learn more about the day-to-day aspects of their chosen fields. The chancellor said that these internships would not only increase the students' theoretical knowledge of engineering by giving them hands-on experience but also make them better future employees for the host engineering companies. Everyone would benefit. Dr. Smith noted that the students would receive a grade and credit for their work.

Training Seminars. Recent industry–education collaborations include training seminars in computers, fiber optics, and human resource options.

Conclusion

After the speech, our VP introduced new business, calling for nominations for next year's officers, gave us the agenda for our next meeting; and adjourned the meeting.

FIGURE 8.6 Document Design Using a Hierarchy of Heading Levels to Order the Information by Importance

- *Boldface*. Created on a computer or by double striking on a typewriter, boldface text emphasizes a key word or phrase.
- *All caps*. The technique of capitalizing text is an excellent way to highlight a WARNING, DANGER, CAUTION, or NOTE. However, capitalizing other types of information is not suggested, since reading lowercase words is easier for your audience. Reading all caps hinders the audience, since such text creates a block of letters in which individual letters aren't easily distinguished from each other.
- *Underlining*. Underlining should be used cautiously. If you underline too frequently, none of your information will be emphatic. One underlined word or phrase will call attention to itself and achieve reader access. Several underlined words or phrases will overwhelm your readers.
- *Italics*. Italics and underlining are used similarly as highlighting techniques. Notice how all subheadings in this bulleted list are italicized.
- *Windowing*. If you indent your key points and enclose them in a box, you're windowing. Here's an example.

> **NOTE**: Be sure to *hand-tighten* the nuts at this point. Once you've completed the installation, go back and securely tighten all nuts.

Windowing draws your reader's attention to an idea, thus making it more emphatic. You can also italicize, boldface, and use all caps within the windowed box, as we have in the preceding example.

- *Fills*. You can further highlight windowed text through fills (lines, patterns, waves, bricks, gradients, and shadings in a text box).
- *Color*. Another way to make key words and phrases leap off the page is to color them. *Danger* would be red, for example, *Warning* orange, and *Caution* yellow. You can also use color to help a reader access the first-level heading, a header, or a footer. (*Headers* contain information placed along the top margin of text; *footers* contain information placed along the bottom margin of text.)

All colors are not equal in visual value. Generally, darker colors provide the most contrast with surrounding text. Therefore, they help your readers access information more than lighter colors.

- *Inverse type*. You can also help readers access information by using inverse type—printing white on black, versus the usual black on white.

Here's a very important consideration! When it comes to using highlighting techniques, *more is not better*. A few highlighting techniques help your readers filter out background data and focus on key points. Too many highlighting techniques are a distraction and clutter the document design. Be careful not to overdo a good thing. Figure 8.7 gives examples of several highlighting methods. And, notice how Figure 8.8 uses highlighting techniques to help the readers access information in the meeting minutes.

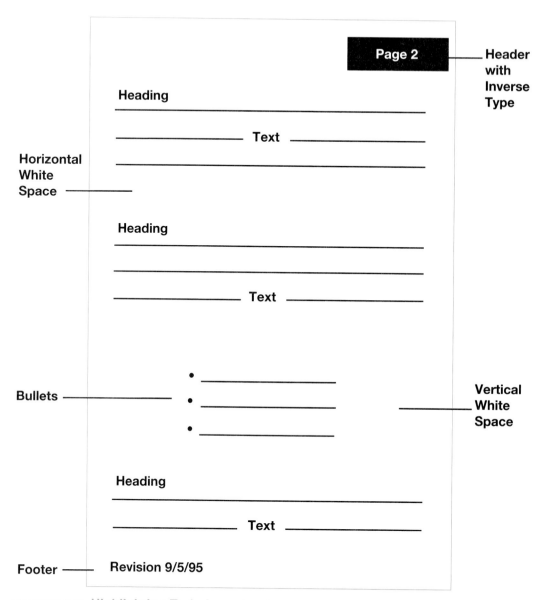

FIGURE 8.7 Highlighting Techniques for Access

VARIETY

Each of the document designs in Figures 8.1, 8.2, 8.6, and 8.8 uses one column and is printed vertically on a traditional 8½ × 11 page (a type of printing called *portrait*; see Figure 8.9). This is not your only option. Your reader might profit from more variety (Watzman ATA-48). For example, you might want to use a smaller or larger paper; to vary the weight of your paper, using 10-pound, 12-pound, or heavier card stock; or even to print your text on colored paper.

MINUTES The meeting at the Carriage Club was attended by 30 members and guests. After the dinner, Roger Traver introduced the guest speaker, George Smith, university chancellor, and noted his accomplishments and experiences prior to education:

- U.S. Navy commander
- Oak Ridge Laboratory researcher
- Politician

Dr. Smith's talk, "Industry and Education Collaboration," was very interesting and included a history of special projects enjoyed by both academics and corporate heads. Dr. Smith suggested that we engineers could work with education to accomplish three goals.

Urban Development

The chancellor's primary thrust was a request for $100,000 in financial aid for urban development. He said money had already been donated from three sources:

1. A large realty firm, Capital Homes, had given $20,000.
2. A philanthropic group, We Care, had donated a matching $20,000.
3. Dr. Smith's university also gave $20,000.

The remaining $40,000 would come from industry.

Internships

The chancellor noted that industry could help itself, as well as the community, by providing two types of internships for university undergraduate majors:

1. Semester-long internships
2. Year-long internships

Students would work for minimum wage to learn more about the day-to-day aspects of their chosen fields. The chancellor said that these internships would not only increase the students' theoretical knowledge of engineering by giving them hands-on experience but also make them better future employees for the host engineering companies. Everyone would benefit. Dr. Smith noted that the students would receive a grade and credit for their work.

Training Seminars. Recent industry-education collaborations include training seminars in computers, fiber optics, and human resource options.

Conclusion

After the speech, our VP introduced new business, calling for nominations for next year's officers; gave us the agenda for our next meeting; and adjourned the meeting.

FIGURE 8.8 Document Design Using Highlighting Techniques to Help Readers Access Key Ideas

FIGURE 8.9 Portrait with One Column Full Justified

More important, you can vary the document design as follows:

- *Print horizontally.* Rather than print your text vertically—8½ × 11 portrait—you could print *horizontally,* as an 11 × 8½ landscape.
- *Use more columns.* Provide your reader two to five columns of text.
- *Vary gutter width.* Columns of text are separated by vertical white space called the *gutter.*
- *Use ragged-right margins.* Most text is fully justified (both right and left margins are aligned). Once this was considered professional, giving the text a clean look. Now, however, studies confirm that right-margin-justified text is harder for the audience to read. It's too rigid. In contrast, *ragged-right* type (the right margin is not justified) is easier to read and more pleasing to the eye. You can use this method to vary page layout (Everson 397).

Figure 8.10 shows how you can use columns, landscape orientation, and ragged-right margins to vary your document design.

Although you can vary your document design by printing horizontally and by using multiple columns, the audience is still confronted by words, words, and more words. The majority of readers don't want to wade through text. Luckily, words are not your only means of communication. You can reach a larger audience with different learning styles by varying your method of communication.

Graphics are an excellent alternative. Many people are more comfortable grasping information visually than verbally. Although it's a cliché, a picture is often worth a thousand words.

FIGURE 8.10 Landscape with Two Columns and Ragged-Right Margins

We discuss graphics (figures and tables) extensively in Chapter 9. However, to clarify our point about the value of variety, see Figure 8.11, which adds a graphic to the meeting minutes.

SUMMARY

Creating effective document design is time-consuming and creatively demanding. It's easier just to write the old-fashioned way—word after word, paragraph after paragraph, blocks of inaccessible text. Although it's easier, it's also less effective.

Your job isn't just to dump data and hope your readers can figure it out. Instead, a successful technical writer works to involve the audience. Your goal is to help the readers understand the organization of your text, recognize its order, and access information at a glance. You also want to use varied types of communication, including graphics, for readers who absorb information better visually, for example, rather than logically.

What's the payoff? Studies tell us that effective document design saves money and time.

One international customs department, after revising the design of its lost-baggage forms, reduced its error rate by over 50 percent. When a utilities company changed the look of its billing statements, customers asked fewer questions, saving the company approximately $250,000 per year. In the early 1980s, the IRS redesigned Form 1040A. The new form helped taxpayers compute their income averaging more accurately. The U.S. Department of Commerce, Office of Consumer Affairs, reported that when several companies improved their documents' visual appeal, the companies increased business and reduced customer complaints (Schriver 250–51).

MINUTES The meeting at the Carriage Club was attended by 30 members and guests. After the dinner, Roger Traver introduced the guest speaker, George Smith, university chancellor, and noted his accomplishments and experiences prior to education:
- U.S. Navy commander
- Oak Ridge Laboratory researcher
- Politician

Dr. Smith's talk, "Industry and Education Collaboration," was very interesting and included a history of special projects enjoyed by both academics and corporate heads. Dr. Smith suggested that we engineers could work with education to accomplish three goals.

Urban Development

The chancellor's primary thrust was a request for $100,000 in financial aid for urban development. He said money had already been donated from three sources, but business and industry can still help significantly.

The following pie chart clarifies what money has been encumbered and how industry donations are still needed.

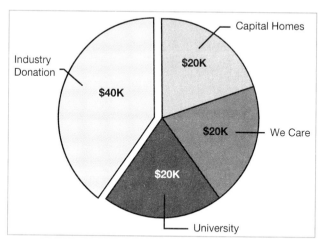

Figure 1 Donations

Internships

The chancellor noted that industry could help itself, as well as the community, by providing two types of internships for university undergraduate majors:

FIGURE 8.11 Document Design Using a Pie Chart to Vary the Communication

1. Semester-long internships
2. Year-long internships

Students would work for minimum wage to learn more about the day-to-day aspects of their chosen fields. The chancellor said that these internships would not only increase the students' theoretical knowledge of engineering by giving them hands-on experience but also make them better future employees for the host engineering companies. <u>Everyone would benefit.</u> Dr. Smith noted that the students would receive a grade and credit for their work.

Training Seminars. Recent industry-education collaborations include training seminars in computers, fiber optics, and human resource options.

Conclusion

After the speech, our VP introduced new business, calling for nominations for next year's officers, gave us the agenda for our next meeting, and adjourned the meeting.

FIGURE 8.11 Document Design Using a Pie Chart to Vary the Communication (Cont.)

Document design isn't a costly frill. Effective document design is good for your company's business.

CHAPTER HIGHLIGHTS

1. Breaking your text into smaller chunks of information will help you create a more readable document.

2. When you create an order among the items in a document through use of typeface and type size, your audience can more easily prioritize the information.

3. Your reader should be able to glance at the document and easily pick out the key ideas. Highlighting techniques will help accomplish this goal.

4. Vary the appearance of your document by using columns, varying gutter widths, and printing in portrait or landscape orientation.

ACTIVITIES

Team Projects

1. Bring samples of technical writing to class. These could include letters, brochures, newsletters, instructions, reports, or advertisements from magazines or newspapers.

In groups of three to five class members, assess the document design of each sample. Determine which samples have successful document designs and which samples have poor document designs. Base your decisions on the criteria provided in this chapter: organization, order, access, and variety.

Either orally or in writing, share your findings with other teams in the class.

2. In teams of three to five class members, take one of the less successful samples from team project 1 and reformat it to improve its document design. Focus on improving the sample's organization, order, access, and variety.

Once your team has completed reformatting the text, make photocopies or transparencies of your work. Then review the various team projects, determining which team created the best document design. Select a winner, and explain why this text now has a successful document design.

3. In teams of three to five class members, reformat the following memo to improve its document design.

DATE: November 30, 2000
TO: Jan Hunt
FROM: Tom Langford
SUBJECT: CLEANING PROCEDURES FOR MANUFACTURING
            WALK-IN OVENS #98731, #98732, AND #98733

The above-mentioned ovens need extensive cleaning. To do so, vacuum and wipe all doors, walls, roofs, and floors. All vents/dampers need to be removed, and a tack cloth must be used to remove all loose dust and dirt. Also, all filters need to be replaced.

I am requesting this because loose particles of dust/dirt are blown onto wet parts when placed in the air-circulating ovens to dry. This causes extensive rework. Please perform this procedure twice per week to ensure clean production.

---

4. In teams of three to five class members, reformat the following summary to improve its document design.

SUMMARY

The City of Waluska wants to provide its community a safe and reliable water treatment facility. The goal is to protect Waluska's environmental resources and to ensure community values.

To achieve these goals, the city has issued a request for proposal to update the Loon Lake Water Treatment Plant (LLWTP). The city recognizes that meeting its community's water treatment needs requires overcoming numerous challenges. These challenges include managing changing regulations and protection standards, developing financially responsible treatment services, planning land use for community expansion, and upholding community values.

For all of the above reasons, Hardtack and Sons (H & S) engineering is your best choice. We understand the project scope and recognize your community's needs.

We have worked successfully with your community for a decade, creating feasibility studies for Loon Lake toxic control, developing odor-abatement procedures for your streams and creeks, and assessing your water treatment plant's ability to meet regulatory standards.

H & S personnel are not just engineering experts. We are members of your community. Our dynamic project team has a close working relationship with your community's regulatory agencies. Our Partner in Charge, Julie Schopper, has experience with similar projects worldwide, demonstrated leadership, and the ability to communicate effectively with clients.

H & S offers the City of Waluska an integrated program that addresses all your community's needs. We believe that H & S is your best choice to ensure that your community receives a water treatment plant ready to meet the challenges of the twenty-first century.

5. On the Internet, access 10 corporate Web sites. Study them to determine how these electronic documents are designed. Make a list of the techniques they use for visual appeal. Which Web sites are successful, and why? Which Web sites are unsuccessful, and why? How would you redesign these less successful Web sites to achieve better document design?

OBJECTIVES

Although your writing may have no grammatical or mechanical errors and you may present valuable information, you won't communicate effectively if your information is inaccessible. Consider the following paragraph:

In January 2000, the actual rainfall was 1.50", but the average for that month was 2.00". In February 2000, the actual rainfall was 1.50", but the annual average had been 2.50". In March 2000, the actual rainfall was 1.00", but the yearly average was 2.50". In April 2000, the actual rainfall was 1.00", but annual averages were 2.50". The May 2000 actual rainfall was 0.50", whereas the annual average had been 1.50". No rainfall was recorded in June 2000. Annually, the average had been 0.50". In July 2000, only 0.25" rain fell. Usually, July had 0.50" rain. In August 2000, again no rain fell, whereas the annual August rainfall measured 0.25". In September and October 2000, the actual rainfall (0.50") matched the annual average. Similarly, the November actual rainfall matched the annual average of 1.50". Finally, in December 2000, 2.00" rain fell, compared to the annual average of 1.50".

If you read the preceding paragraph in its entirety, you are an unusually dedicated reader. Such wall-to-wall words mixed with statistics do not create easily readable writing.

The goal of effective technical writing is to communicate information easily. The example paragraph fails to meet this goal. No reader can digest the data easily or see clearly the comparative changes from one month's precipitation to the next.

To present large blocks of data and/or reveal comparisons, you can supplement, if not replace, your text with graphics. In technical writing, visual aids accomplish several goals. Graphics (whether hand drawn, photographed, or computer generated) will help you achieve conciseness, clarity, and cosmetic appeal.

CONCISENESS

Visual aids allow you to provide large amounts of information in a small space. Words used to convey data (such as in the example paragraph) double, triple, or even quadruple the space needed to report information. By using graphics, you can also delete many dead words and phrases.

CLARITY

Visual aids can clarify complex information. Graphics help readers see the following:

- *Trends* (increasing or decreasing sales figures, for instance). These are most evident in line graphs.
- *Comparisons between like components* (as in actual monthly versus average rainfalls). These can be seen in grouped bar charts.
- *Percentages*. Pie charts help readers discern these.
- *Facts and figures*. A table states statistics more clearly than a wordy paragraph.

COSMETIC APPEAL

Visual aids help you break up the monotony of wall-to-wall words. If you only give unbroken text, your reader will tire, lose interest, and overlook key concerns. Graphics help you sustain your reader's interest. Let's face it; readers like to look at pictures.

The two types of graphics important for technical writing are tables and figures. This chapter helps you correctly use both.

COLOR

All graphics look best in color, don't they? Not necessarily. Without a doubt, a graphic depicted in vivid colors will attract your reader's attention. However, the colors might not aid communication. For example, colored graphics could have these drawbacks (Reynolds and Marchetta 5–7):

1. The colors might be distracting (glaring orange, red, and yellow combinations on a bar chart would do more harm than good).

2. Colors that look good today might go out of style in time (like those avocado green and autumn gold appliances in your family's kitchen).

3. Colored graphics increase production costs.

4. Colored graphics consume more disk space and computer memory than black-and-white graphics.

5. The colors you use might not look the same to all readers.

Let's expand on this last point. Just because you see the colors one way does not mean your readers will see them the same. We're not talking about people with vision problems. Instead, we're talking about what happens to your color graphics when someone reproduces them as black-and-white copies. We're also talking about computer monitor variations.

We discuss computer monitors in great detail in Chapter 10 ("Electronic Communication"). The color on a computer monitor depends on its resolution (the number of pixels displayed) and the monitor's RGB values (how much red, green, and blue light is displayed). Because all monitors do not display these same values, what you see on your monitor will not necessarily be the same as what your reader sees. To solve this problem, test your graphics on several monitors. Also, limit your choices to primary colors instead of the infinite array of other color possibilities. More important, use patterns to distinguish your information so that the color becomes secondary to the design.

THREE-DIMENSIONAL GRAPHICS

Many people are attracted to three-dimensional graphics. After all, they have obvious appeal. Three-dimensional graphics are more interesting and vivid than flat, one-dimensional graphics. However, 3-D graphics have drawbacks. A 3-D graphic is visually appealing, but it does not convey information quantifiably. A word of caution: use 3-D graphics sparingly. Better yet, use the 3-D graphic to create an impression; then, include a table to quantify your data.

CRITERIA FOR EFFECTIVE GRAPHICS

Figure 9.1 is an example of a cosmetically appealing, clear, and concise graphic. At a glance, the reader can pinpoint the comparative prices per barrel of crude oil between 1996 and 2000. Thus, the line graph is clear and concise. In addition, the writer has included an interesting artistic touch. The oil gushing out of the tower shades just the parts of the graph that emphasize the dollar amounts. Envision this graph without the shading. Only the line would exist. The shading provides the right touch of artistry to enhance the information communicated.

The graph shown in Figure 9.1, although successful, does not include all the traits common to effective visual aids. Whether hand drawn or computer generated, successful tables and figures

1. Are integrated with the text (i.e., the graphic complements the text; the text explains the graphic).

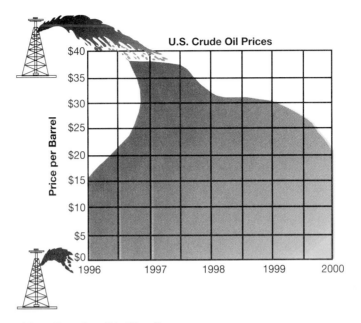

FIGURE 9.1 Line Graph with Shading

2. Are appropriately located (preferably immediately following the text referring to the graphic and not a page or pages later).

3. Add to the material explained in the text (without being redundant).

4. Communicate important information that could not be conveyed easily in a paragraph or longer text.

5. Do not contain details that detract from rather than enhance the information.

6. Are an effective size (not too small or too large).

7. Are neatly printed to be readable.

8. Are correctly labeled (with legends, headings, and titles).

9. Follow the style of other figures or tables in the text.

10. Are well conceived and carefully executed.

TYPES OF GRAPHICS

TABLES

Let's tabulate the information about rainfall in 2000 presented earlier. Because effective technical writing integrates text and graphic, you'll want to provide an introductory sentence prefacing the table, as follows:

Table 9.1 reveals the actual amount of rainfall for each month in 2000 versus the average documented rainfall for those same months.

TABLE 9.1	2000 Monthly Rainfall versus Average Rainfall (All Figures in Inches)	
MONTH	2000 RAINFALL	AVERAGE RAINFALL
January	1.50	2.00
February	1.50	2.50
March	1.00	2.50
April	1.00	2.50
May	0.50	1.50
June	0.00	0.50
July	0.25	0.50
August	0.00	0.25
September	0.50	0.50
October	0.50	0.50
November	1.50	1.50
December	2.00	1.50

This table has advantages for both the writer and the reader. First, the headings eliminate needless repetition of words, thereby making the text more readable. Second, the audience can see easily the comparison between the actual amount of rainfall and the monthly averages. Thus, the table highlights the content's significant differences. Third, the table allows for easy future reference. Tables could be created for each year. Then the reader could compare quickly the changes in annual precipitation. Finally, if this information is included in a report, the writer will reference the table in the Table of Contents' List of Illustrations. This creates ease of access for the reader.

Criteria for Effective Tables

To construct tables correctly, do the following:

1. Number tables in order of presentation (i.e., Table 1, Table 2, Table 3, etc.).

2. Title every table. In your writing, refer to the table by its number, not its title. Simply say, "Table 1 shows. . . ," "As seen in Table 1," or "The information in Table 1 reveals. . . ."

3. Present the table as soon as possible after you've mentioned it in your text. Preferably, place the table on the same page as the appropriate text, not on a subsequent, unrelated page or in an appendix.

4. Don't present the table until you've mentioned it.

5. Use an introductory sentence or two to lead into the table.

6. After you've presented the table, explain its significance. You might write, "Thus, the average rainfall in both March and April exceeded the actual rainfall by 1.50 inches, reminding us of how dry the spring has been."

7. Write headings for each column. Choose terms that summarize the information in the columns. For example, you could write "% of Error," "Length in Ft.," or "Amount in $."

8. Since the size of columns is determined by the width of the data and/or headings, you may want to abbreviate terms (as shown in item 7). If you use abbreviations, however, be sure your audience understands your terminology.

9. Center tables between right and left margins. Don't crowd them on the page.

10. Separate columns with ample white space, vertical lines, or dashes.

11. Show that you've omitted information by printing two or three periods or a hyphen or dash in an empty column.

12. Be consistent when using numbers. Use either decimals or numerators and denominators for fractions. You could write 3 1/4 and 3 3/4 or 3.25 and 3.75. If you use decimal points for some numbers but other numbers are whole, include zeroes. For example, write 9.00 for 9.

13. If you do not conclude a table on one page, on the second page write *Continued* in parentheses after the number of the table and the table's title.

Table 9.2 is an excellent example of a correctly prepared table.

TABLE 9.2	Student Headcount Enrollment by Age Group and Student Status, Fall 2000				
AGE GROUP	NEW STUDENTS	CONTINUING STUDENTS	READMITTED	OTHER	TOTAL
15–17	453	33	2	2	490
18–20	1,404	1,125	132	—	2,661
21–23	339	819	269	—	1,427
24–26	263	596	213	—	1,072
27–29	250	436	134	—	820
30–39	524	1,168	372	—	2,064
40–49	271	510	186	—	967
50–59	76	121	54	—	251
60+	19	48	16	—	83
Unknown	109	92	27	2	230
Total	3,708	4,948	1,405	4	10,065

FIGURES

Another way to enhance your technical writing is to use figures. Whereas tables eliminate needless repetition of words, figures highlight and supplement important points in your writing. Like tables, figures help you communicate with your reader.

Types of figures include the following:

- Bar charts
  - —Grouped bar charts
  - —3-D (tower) bar charts
  - —Pictographs
  - —Gantt charts
- Pie charts
- Line charts
  - —Broken line charts
  - —Curved line charts
- Flowcharts
- Organizational charts
- Schematics
- Line drawings
  - —Exploded views
  - —Cutaway views
  - —Super comic book look
- Photographs
- Icons
- Internet graphics

All of these types of figures (except photographs) can be computer generated using an assortment of computer programs. The program you use depends on your preference and hardware.

Criteria for Effective Figures

To construct figures correctly, do the following:

1. Number figures in order of presentation (i.e., Figure 1, Figure 2, Figure 3, etc.).
2. Title each figure. When you refer to the figure, use its number rather than its title: for example, "Figure 1 shows the relation between the average price for houses and the actual sales prices."
3. Preface each figure with an introductory sentence.
4. Don't use a figure until you've mentioned it in the text.
5. Present the figure as soon as possible after mentioning it instead of several paragraphs or pages later.
6. After you've presented the figure, explain its significance. Don't let the figure speak for itself. Remind the reader of the important facts you want to highlight.

Virtual reality of a proposed building's interior stairwell.
Courtesy of the Office of Facility Planning, Johnson County Community College.

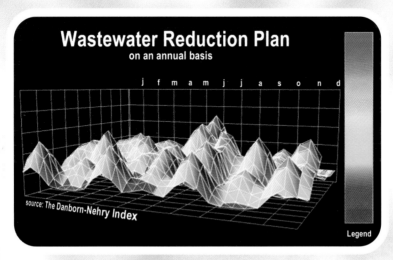

Topographical depiction of wastewater reduction.
Courtesy of Brandon Henry.

Architectural hand-drawn rendering.
Courtesy of George Butler Associates, Inc.

Comparison of planets with cutaway view.
Courtesy of Brandon Henry.

7. Label the figure's components. For example, if you're using a bar or line chart, label the *x*- and *y*-axes clearly. If you're using line drawings, pie charts, or photographs, use clear *call-outs* (names or numbers that indicate particular parts) to label each component.

8. When necessary, provide a legend or key at the bottom of the figure to explain information. For example, a key in a bar or line chart will explain what each differently colored line or bar means. In line drawings and photographs, you can use numbered call-outs in place of names. If you do so, you'll need a legend at the bottom of the figure explaining what each number means.

9. If you abbreviate any labels, define these in a footnote. Place an asterisk (*) or a superscript number ([1], [2], [3], . .) after the term and then at the bottom of the figure where you explain your terminology.

10. If you've drawn information from another source, note this at the bottom of the figure.

11. Frame the figure. Center it between the left and right margins and/or window it in a box.

12. Size figures appropriately. Don't make them too small or too large.

13. Try the super comic book look (figures drawn in cartoon-like characters to highlight parts of the graphic and to interest readers).

Bar Charts

Bar charts show either vertical bars (as in Figure 9.2) or horizontal bars (as in Figure 9.3). These bars are scaled to reveal quantities and comparative values. You can shade, color, and/or crosshatch the bars to emphasize the contrasts. If you do so, include a key explaining what each symbolizes, as in Figure 9.4. *Pictographs* (as in Figure 9.5) use picture symbols instead of bars to show quantities. To create effective pictographs, do the following:

1. The picture should be representative of the topic discussed.

2. Each symbol equals a unit of measurement. The size of the units depends on your value selection as noted in the key or on the *x*- and *y*-axes.

3. Use more symbols of the same size to indicate a higher quantity; do not use larger symbols.

Gantt Charts

Gantt charts, or *schedule charts* (as in Figure 9.6), use bars to show chronological activities. For example, your goal might be to show a client phases of a project. This could include planned start dates, planned reporting milestones, planned completion dates, actual progress made toward completing the project, and work remaining. Gantt charts are an excellent way to represent these activities visually. They are often included in proposals to project schedules or in reports to show work completed. To create successful Gantt charts, do the following:

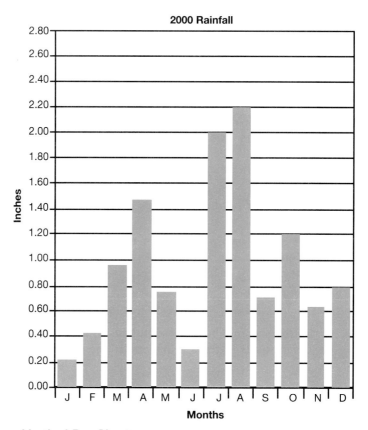

FIGURE 9.2 Vertical Bar Chart

1. Label your *x*- and *y*-axes. For example, if the *y*-axis represents the various activities scheduled, then the *x*-axis represents time (either days, weeks, months, or years).

2. Provide grid lines (either horizontal or vertical) to help your readers pinpoint the time accurately.

3. Label your bars with exact dates for start or completion.

4. Quantify the percentages of work accomplished and work remaining.

5. Provide a legend or key to differentiate between planned activities and actual progress.

Pie Charts

Use pie charts (as in Figure 9.7) to illustrate portions of a whole. The pie chart represents information as pie-shaped parts of a circle. The entire circle equals 100 percent, or 360 degrees. The pie pieces (the wedges) show the various divisions of the whole.

To create effective pie charts, do the following:

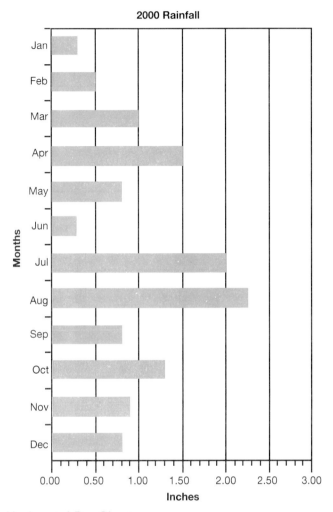

FIGURE 9.3 Horizontal Bar Chart

1. Be sure that the complete circle equals 100 percent, or 360 degrees.
2. Begin spacing wedges at the 12 o'clock position.
3. Use shading, color, and/or crosshatching to emphasize wedge distributions.
4. Use horizontal writing to label wedges.
5. If you don't have enough room for a label within each wedge, provide a key defining what each shade, color, and/or crosshatching symbolizes.
6. Provide percentages within wedges when possible.
7. Do not use too many wedges—this would crowd the chart and confuse readers.
8. Make sure that different sizes of wedges are fairly large and dramatic.

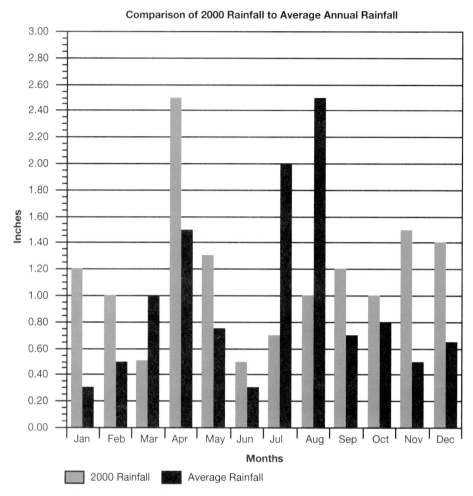

FIGURE 9.4 Grouped Bar Chart

Line Charts

Line charts reveal relationships between sets of figures. To make a line chart, plot sets of numbers and connect the sets with lines. These lines create a picture showing the upward and downward movement of quantities. Line charts of more than one line (see Figure 9.8) are useful in showing comparisons between two sets of values. However, avoid creating line charts with too many lines, which will confuse your readers.

Flowcharts

You can show chronological sequence of activities using a flowchart. Flowcharts are especially useful for writing technical instructions (see Chapter 12). When using a flowchart, remember that ovals represent starts and stops, rectangles represent steps, and diamonds equal decisions (see Figure 9.9).

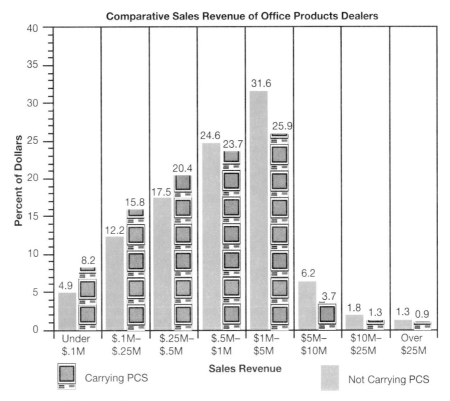

FIGURE 9.5 Pictograph

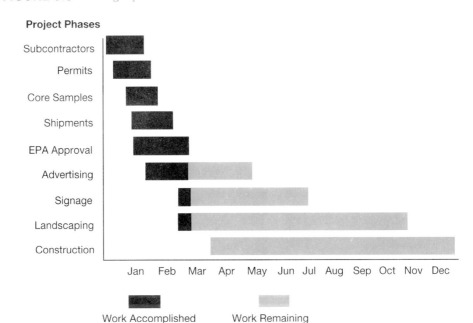

FIGURE 9.6 Gantt Chart

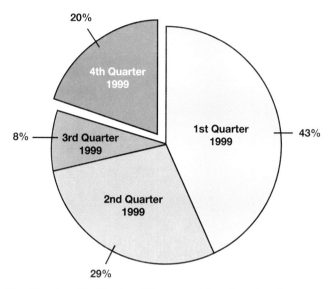

FIGURE 9.7 Pie Chart with Wedge Removed for Emphasis

Organizational Charts

These charts (as in Figure 9.10) show the chain of command in an organization. You can use boxes around the information or use white space to distinguish among levels in the chart. An organizational chart helps your readers see where individuals work within a business and their relation to other workers.

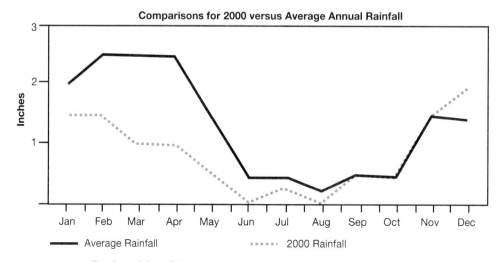

FIGURE 9.8 Broken Line Chart

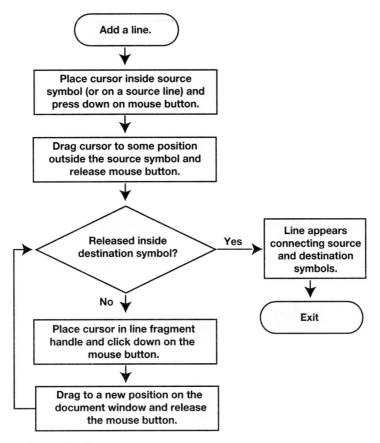

FIGURE 9.9 Flowchart

Schematics

Schematics are useful for presenting abstract information in technical fields such as electronics and engineering. A schematic diagrams the relationships among the parts of something such as an electrical circuit. The diagram uses symbols and abbreviations familiar to highly technical readers.

The schematic in Figure 9.11 shows various electronic parts (resistors, diodes, condensors) in a radio.

Line Drawings

Use line drawings to show the important parts of a mechanism or to enhance your text cosmetically. To create line drawings, do the following:

1. Maintain correct proportions in relation to each part of the object drawn.

2. If a sequence of drawings illustrates steps in a process, place the drawings in left-to-right or top-to-bottom order.

3. Using call-outs to name parts, label the components of the object drawn (see Figure 9.12).

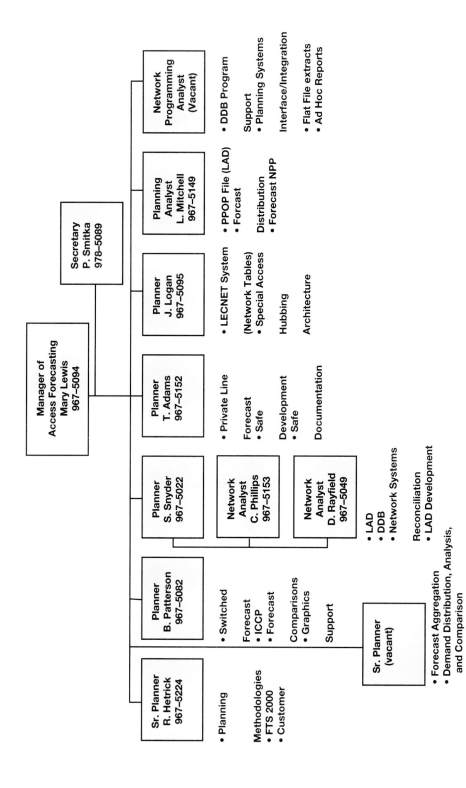

FIGURE 9.10 Organizational Chart

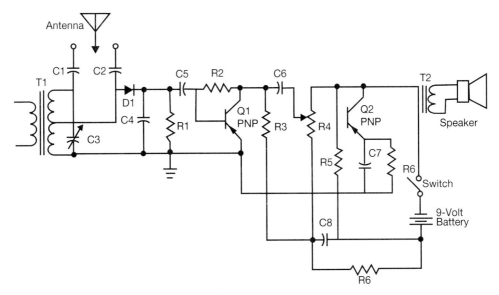

FIGURE 9.11 Schematic of a Radio

4. If there are numerous components, use a letter or number to refer to each part. Then reference this letter or number in a key (see Figure 9.13).

5. Use exploded views (Figures 9.12 and 9.13) or cutaways (Figure 9.14) to highlight a particular part of the drawing.

Photographs

A photograph can illustrate your text effectively. Like a line drawing, a photograph can show the components of a mechanism. If you use a photo for this purpose, you'll need to label (name), number, or letter parts and provide a key. Photographs are excellent visual aids because they emphasize all parts equally. Their primary advantage is that they show something as it truly is.

Photographs have one disadvantage, however. They are difficult to reproduce. Whereas line drawings photocopy well, photographs do not.

Icons

Approximately 20 percent of America's population is functionally illiterate. In today's global economy, consumers speak diverse languages. Given these two facts, how can technical writers communicate to people who can't read and to people who speak different languages? Icons offer one solution. Icons (as in Figure 9.15) are visual representations of a capability, a danger, a direction, an acceptable behavior, or an unacceptable behavior.

For example, the computer industry uses icons (e.g., and open manila folders to represent a computer file). In manuals, a jagged lightning stroke iconically represents the danger of electrocution. On streets, an arrow represents the direction we should travel; on computers, the arrow shows us which direction to scroll.

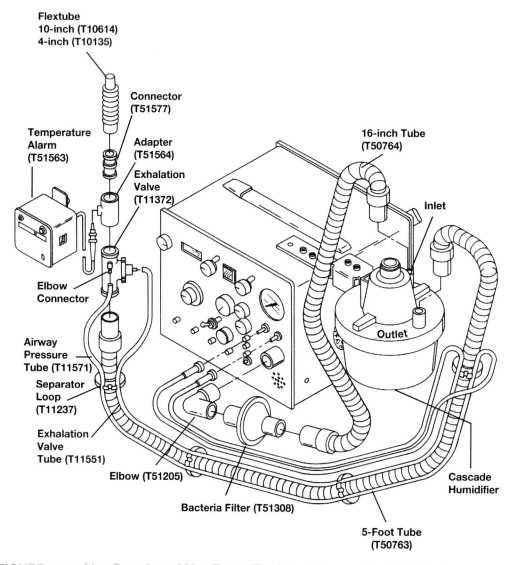

FIGURE 9.12 Line Drawing of Ventilator (Exploded View with Call-Outs)

Universally depicted stick figures of men and women greet us on restroom doors to show us which rooms we can enter and which rooms we must avoid.

When used correctly, icons can save space, communicate rapidly, and help readers with language problems understand the writer's intent.

To create effective icons, follow these suggestions:

1. *Keep it simple.* You should try to communicate a single idea. Icons are not appropriate for long discourse.

2. *Create a realistic image.* This could be accomplished by representing the idea as a photograph, drawing, caricature, outline, or silhouette.

Exhalation Valve Parts List		
Item	Part Number	Description
1	000723	Nut
2	003248	Cap
3	T50924	Diaphragm
4	Reference	Valve Body
5	Reference	Elbow Connector
–	T11372	Exhalation Valve

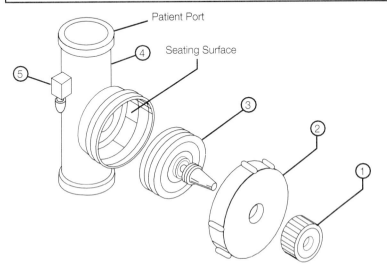

FIGURE 9.13 Line Drawing of Exhalation Valve (Exploded View with Key)

3. *Make the image recognizable.* A top view of a telephone or computer terminal is confusing. A side view of a playing card is completely unrecognizable. Select the view of the object which best communicates your intent.

4. *Avoid cultural and gender stereotyping.* For example, if you're drawing a hand, you should avoid showing any skin color, and you should stylize the hand so it is neither clearly male nor female.

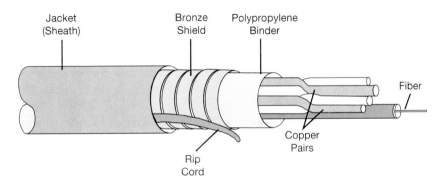

FIGURE 9.14 Line Drawing of Cable (Cutaway View)

FIGURE 9.15 An Icon

5. *Strive for universality.* Stick figures of men and women are recognizable worldwide. In contrast, letters—such as *P* for parking—will mean very little in China, Africa, or Europe. Even colors can cause trouble. In North America, red represents danger, but red is a joyous color in China. Yellow calls for caution in North America, but this color equals happiness and prosperity in the Arab culture (Horton, "Universal" 682–93).

INTERNET GRAPHICS

More and more, you will be writing online as the Internet, intranets, and extranets become prominent in technical writing. (We discuss the Internet in detail in Chapter 10.) You can create graphics for your Web site in three ways:

1. Download existing online graphics for use in your site. The Internet contains thousands of Web sites, which provide online clip art. These graphics include photographs, line drawings, cartoons, icons, animated images, arrows, buttons, horizontal lines, balls, letters, bullets, and hazard signs. In fact, you can download any image from any Web site. Many of these images are freeware, which you can download without cost and without infringing on copyright laws.

To download these images, just place your cursor arrow on the graphic you want, then right click on the mouse. A pop-up menu will appear. Scroll down to "Save image as." Once you have done this, a new menu will appear. You can save your image in the file of your choice, either on the hard drive or on your disk. The images from the Internet will already be *gif* (graphics interchange format) or *jpg* (joint photographic experts group) files. Thus, you will not have to convert them for use in your Web site.

2. Modify and customize existing online graphics for use in your site. If you plan to use an existing online graphic as your company's logo, for example, you will need to modify and/or customize the graphic. You'll want to do this for at least two reasons: to avoid infringing on copyright laws and to make the graphic uniquely yours.

To modify and customize graphics, you can download them in two ways. First, you can print the screen by pressing the "Print Screen" key (usually found on the upper right of your keyboard). This captures the entire screen image in a clipboard. Then you can open a graphics program and paste the captured image. Second, you can save the image in a file (as discussed above) and then open the graphic in a graphics package. Most graphics programs will allow you to customize a graphic. Popular programs include *Paint*, *Paint Shop Pro*, *PhotoShop*, *Corel Draw*, *Adobe Illustrator*, *Freehand*, and *Lview Pro*. In these graphics programs, you can manipulate the images: change colors, add text, reverse the images, crop, resize, redimension, rotate, retouch, delete or erase parts of the images, overlay multiple images, join multiple images, and so forth. After you make substantial changes, the new image becomes your property.

You could also take any existing graphic from hard-copy text (magazines, journals, books, newsletters, brochures, manuals, reports, etc.), scan the image, crop and retouch it, save it, then reopen this saved file in one of the above graphics programs for further manipulation.

Some graphics programs, such as *Paint*, save an image only as a *bmp*, a bitmap image. Once the image has been altered, you'll need to convert your bitmap file to a *gif* or *jpg* for use in your Web site. Doing so is important since the Internet will not read bmp images.

3. Create new graphics for use in your site. A final option is to create your own graphic. If you are artistic, draw your graphic in a program (see the list above), save the image as a *gif* or *jpg*, and then load the image into your Web site. This option might be more challenging and time-consuming. However, creating your own graphic gives you more control over the finished product, provides a graphic precisely suited to your company's needs, and helps to avoid infringement of copyright laws.

CHAPTER HIGHLIGHTS

1. Using graphics can allow you to create a more concise document.
2. Graphics often help you present information more clearly.
3. Graphics add variety to your text, breaking up wall-to-wall words.
4. Color and 3-D graphics can be effective. However, these two design elements also can cause problems. Your color choices might not be reproduced exactly as you planned, and a 3-D graphic could be misleading rather than informative.
5. Tables are effective for presenting numbers, dates, and columns of figures.
6. You can often communicate more easily with your audience when you use figures, such as bar charts, pie charts, line charts, flowcharts, and organizational charts, to highlight and supplement important parts of your text.

ACTIVITIES

Creating Figures and Tables

1. Present the following information in a pie chart, a bar chart, and a table.

In 1999, the Interstate Telephone Company bought and installed 100,000 relays. They used these for long-range testing programs that assessed failure rates. They purchased 40,000 Nestor 221s; 20,000 VanCourt 1200s; 20,000 Macro R40s; 10,000 Camrose Series 8s; and 10,000 Hardy SP6s.

2. Using the information presented in activity 1 and the following revised data, show the comparison between 1999 and 2000 purchases through two pie charts, a grouped bar chart, and a table.

In 2000, after assessing the success and failure of the relays, the Interstate Telephone Company made new purchases of 200,000 relays. It bought 90,000 VanCourt 1200s; 50,000 Macro R40s; 30,000 Camrose Series 8s; and 30,000 Hardy SP6s. No Nestors were purchased.

3. Create a table for the following information.

When the voltage out is 13 V, the frequency out is 926 Hz. When the voltage out is 12.5 V, the frequency out is 1.14 K. When the voltage out is 12 V, the frequency out is 1.4 K. When the voltage out is 11 V, the frequency out is 1.8 K. When the voltage out is 10 V, the frequency out is 2.3 K. When the voltage out is 9 V, the frequency out is 2.8 K. When the voltage out is 8 V, the frequency out is 3 K. When the voltage out is 7 V, the frequency out is 0 Hz.

4. Create a line chart. To do so, select any topic you like. The subject matter, however, must include varying values. For example, present a line chart of your grades in one class, your salary increases (or decreases) at work, the week's temperature ranges, your weight gain or loss throughout the year, the miles you've run during the week or month, amounts of money you've spent on junk food during the week, and so forth.

5. Create a flowchart. To do so, select a topic about which you can write an instruction—for example, the steps for changing oil, winterizing your house, planting trees, seeding your yard, refinishing furniture, making a cake, changing a tire, interviewing for a job, wallpapering a room, building a deck, installing a fence. Flowchart the sequential steps for any of these procedures.

6. Create three bar charts, using the information from activity 1. Make one of the bar charts in color, one as a 3-D bar chart, and the other as a one-dimensional bar chart. Print them and give them to others in class. Determine which of the three is most effective and why.

Case Study

Your company is submitting a proposal to a client. The proposal is about the creation of a company newsletter geared to the client's employees. Part of this proposal will include a tentative project schedule. You plan on informing the client that you will begin the project May 1 and conclude September 7.

Project activities include the following: (1) meetings between the newsletter staff and the client's marketing department, scheduled for May 1 through May 5, to determine objectives; (2) employee interviews regarding company recreational events and employee celebrations (birthdays, births, awards, etc.), scheduled for

May 8 through May 19; (3) meetings with the accounting department, scheduled for May 22 through May 31, to ascertain stock information, annual corporate earnings, and employee benefits; (4) research regarding newsletter layout options (paper size, weight, color, and number of pages suggested per edition), scheduled for June 5 through June 16; (5) cost estimates for newsletter production based on the above findings, scheduled for June 19 through June 30; (6) follow-up meetings with client management to approve the cost estimates and tentative page layout, scheduled for July 3 through July 5; (7) writing the first draft of a newsletter, scheduled for July 10 through July 21; (8) submission of the draft to client management for approval, scheduled for July 24; (9) follow-up interviews of employees and the accounting department to update information for the final newsletter draft, scheduled for August 1 through 12; (10) newsletter production, scheduled for August 15 through 26; and (11) project completion—newsletters delivered to each client employee—scheduled for September 7.

Create a Gantt chart conveying the above information.

Team Projects

1. In small groups, create new icons for any or all of the following common computer functions: open, retrieve, close, save, password, print, undo, search, cut, and paste. Don't use any icons you've seen elsewhere; invent new ones. To do so, abide by the criteria provided in this chapter. Once the icons have been created, share them with other groups and select the best icons. Be prepared to justify your selections either in a group oral presentation or in a group-written report.

2. Find examples of good and/or bad tables and figures in a popular magazine, a technical journal, and the local newspaper. Bring these to class and, in small groups, discuss their successes and failures. Be prepared to justify your opinions either in a group oral presentation or in a group-written report.

3. Select any of the unsuccessful graphics from team project 2. In small groups, decide how the flawed graphics could be improved and redraw or alter them by computer accordingly. Once these revisions are made, share your results with the class and select the best revisions.

CHAPTER 10

E-mail, Online Help, and Web Sites

For the first time in centuries, the written word has undergone a quantum metamorphosis, leaping from the printed page into cyberspace. This change has affected technical writing significantly. Correspondence, once limited to letters and memos, is now often online as e-mail. Corporate brochures and newsletters, once paper bound, now are online. Product and service manuals, once paper bound, now are online. Résumés are online. Research is online. Technical writing is increasingly electronic. To keep up with this communication revolution, you must go online also. By teaching you how to write successful e-mail, write online help screens, and create an effective Web site, this chapter will help you learn how to work in cyberspace.

THE INTERNET—THE "INFORMATION SUPERHIGHWAY"

Your e-mail, online help screens, and Web sites will be transmitted via the Internet, an intranet, or an extranet. But what do these words mean? Though the word "Internet" has been bandied about for years, finally on October 24, 1995, we received an official definition. The Federal Networking Council (FNC) unanimously resolved to define the

word, in consultation with leadership from the Internet and intellectual property rights (IPR) communities. Here's their definition ("FNC Resolution"):

> The Federal Networking Council (FNC) agrees that the following language
> reflects our definition of the term "Internet."
>
> > "Internet" refers to the global information system that—
> > 1. Is logically linked together by a globally unique address . . . based on
> >   the Internet Protocol (IP) or its subsequent extensions.
> > 2. Is able to support communications using the Transmission Control
> >   Protocol/Internet Protocol (TCP/IP) suite or its subsequent exten-
> >   sions . . . and other IP-compatible protocols.
> > 3. Provides, uses or makes accessible, either publicly or privately,
> >   high-level services layered on the communications and related
> >   infrastructures.

The FNC's definition, though official, is challenging. Let's try to simplify
matters. The Internet is a network that electronically connects over a million
computers internationally. And it is growing in importance daily. The number of
Internet data packets grew from 152 million in June 1988 to 60,587 million
packets in July 1994. Host growth escalated from 235 in May 1982 to over 3
million in 1994 (Pike et al., 14–15). As of January 1998, the Internet included
approximately 30 million users in over 100 countries ("Internet"). Only three
months later, an article in the *Kansas City Star* reported that "traffic on the
Internet is doubling every 100 days" and stated that "as many as 62 million
Americans are now using the world-wide network" ("Internet Traffic" B1).

Furthermore, the Internet is a decentralized medium, with very few external
controls. To use the Internet, you can choose which Internet service or provider
you wish, and you can access any Internet site you desire. Thus, as the synonym
"information superhighway" suggests, the Internet allows the user to access
data from various locations almost without restriction. The Internet lets you
travel from your computer to information sites worldwide. You can speak to
and hear from educators, engineers, lawyers, doctors, businesspeople, hobby-
ists, or anyone linked to this international communication web, day or night.

How do you find the information you need on the Internet? If you were on a
highway looking for a specific site, you might consult a road atlas. To help you
find information on the Internet, you can use something resembling a road atlas—
the World Wide Web (WWW or Web), along with Web browsers such as *Netscape
Navigator* and Microsoft's *Internet Explorer*. The World Wide Web supports documents (Web sites) formatted in a special computer language called HTML
(HyperText Markup Language). (See Table 10.1 for Web abbreviations and
acronyms.) HTML allows the webmaster (the creator of the Web site) to include
graphics, audio, video, and animation as well as hypertext links to other sites and
other Web screens ("World Wide Web" 1).

TABLE 10.1	Common Web Site/Internet Abbreviations and Acronyms
<A HREF>	Anchor Hyper-reference
<BG>	Background
CGI	Common Gateway Interface
.com	commercial business—domain name
.edu	educational institution—domain name
ftp	file transfer protocol
.gif	graphics interchange format
.gov	governmental—domain name
HTML	HyperText Markup Language
HTTP	hypertext transfer protocol
<IMG SRC>	image source
.jpeg, .jpg	joint photographic experts group
<LI>	listed item
.mil	military—domain name
MOO	MUD, object oriented
MUD	multi-user dialogue
.net	network—domain name
<OL>	ordered list
.org	organization—domain name
SGML	Standard Generalized Markup Language
TCP/IP	Transmission Control Protocol/Internet Protocol
.tif	tagged image file
<UL>	unordered list
URL	uniform resource locator

THE INTRANET—A COMPANY'S INTERNAL WEB

Whereas the Internet opens the entire world of information to us, an intranet is more limited. It is a contained network. An intranet might include several, linked local area networks (LANs), and it might include connections to the Internet. Like the Internet, an intranet uses TCP/IP, HTTP and other Internet protocols. The difference, however, is that an intranet belongs to a company. Its purpose is "to share company information and computing resources among employees" ("Intranet"). Individuals outside the company cannot access the intranet unless they have corporate authorization; a "firewall" prevents access to an intranet's private network ("Firewall" 1).

How does a company benefit by using an intranet? Rather than printing hundreds of copies of a document for employees who might be officed in buildings located around a city or around the world, a company can make this information accessible to its employees through an intranet. Companies can place on their corporate intranets

- Web-based discussion forums. These allow employees to post, read, respond to, and then archive corporate e-mail messages.
- Web-based multimedia and kiosks for instructional or informational purposes.
- Online polls. Employees can respond to corporate polls. Then the results can be tabulated, and employees around the company can view the results.
- Company forms. These can be filled out and submitted electronically.
- Policy and procedure manuals. Whereas paper documents take up space, get dusty, and get lost, online intranet manuals save space and are available instantly.
- Employee phone directories. Company employees come and go. Online, additions and deletions can be updated easily.
- Organizational charts. Company hierarchies change rapidly. Online, they can be updated and conveyed to employees immediately.

The benefits of an intranet are numerous. An intranet is paper free. Companies can lower mailing, binding, and printing costs. Information can be updated easily. Data can be archived without taking up space. Procedures, polls, directories, forums, and forms are at an employee's fingertips rather than buried under paper or on the top shelves of inaccessible bookshelves. Most important, every corporate employee throughout a company—whether in the office or off-site—can access corporate information online at the touch of a key. Speed, efficiency, and cost savings—these are the benefits of a corporate intranet ("Intranet Applications").

THE EXTRANET—A WEB WITHIN A WEB

We've seen that the Internet gives us worldwide access to information online, and we've explained that an intranet gives employees within a company access to corporate information. What's an extranet? An extranet "is a collaborative network" that uses Internet protocols to "link businesses with their suppliers, customers, or other businesses" that have common goals. An extranet thus allows several companies to share information but also to keep this information within the confines of their collaborative unit. Extranets can include Web-based training programs of mutual benefit to the participating companies, product catalogs, private newsgroups relating to a specific industry, and management approaches for companies working on a shared project ("Extranet").

ONLINE TECHNICAL WRITING APPLICATIONS: E-MAIL, ONLINE HELP, AND WEB SITES

E-MAIL

The use of e-mail has exploded in recent years. Over 25 million people used e-mail, sending approximately 15 billion e-mail messages in 1997 ("E-mail"). Others estimate that by the year 2000, electronic messages will total 60 billion worldwide (Freeman and Bacon 71).

Why has there been such explosive growth? E-mail is quick, cheap, and convenient. E-mail messages can arrive within minutes at your reader's computer. Many companies, such as Juno, Yahoo, Excite, and Hotmail, provide free e-mail use via the Internet. Most important, e-mail is convenient. You can type your message on your computer and click on "Send"—and that's it. No stamps, no envelopes, no driving to the post office or walking to the mailbox.

In many ways, e-mail is even more convenient than the telephone or a fax machine. E-mail, like a fax, can be sent to a person who is unavailable at the moment. Rather than playing "telephone tag," never reaching your intended audience, you can e-mail or fax that person. The message will be there when the person finally returns to the office. A significant difference, however, between e-mail and a fax is accessibility. You can send and receive e-mail messages from your own office. In contrast, many people have to travel to fax locations elsewhere in their buildings—or even to off-site vendors—to send a fax (Laquey 42). E-mail is popular because it is easy and immediate, a mere keystroke away versus a lengthy jog to a distant fax machine.

E-MAIL PROBLEMS

Despite e-mail's benefits, this form of correspondence also has some problems.

1. Poor or incomplete documentation. If you write or receive a hard-copy memo or letter, this correspondence can be filed for future reference. In contrast, e-mail resides only in your computer unless a hard copy is printed. Unfortunately, because e-mail has an instantaneous and chatty nature, people tend to delete their sent and received messages without considering the long-term ramifications. Without a printed copy filed for future reference, your company could have problems unless there is a backup archive of old e-mail. If a company needs to defend a decision it has made related to hirings, firings, purchases, returns, and so on, and that company does not archive its e-mail, what documents will it use to support its case? Having no files is a frightening possibility (Freeman and Bacon 71).

2. Computer limitations. As we all know, hardware and software are not universally compatible. The computer commands you use for your e-mail will not necessarily mean the same thing on someone else's computer. If you have used a certain command to create textual enhancements (underlining, italics, boldface, color, and so on), this command might be interpreted differently on your recipient's computer. Even a typed word might not transmit as you'd assume. Take the simple word "it's," for example. People have received e-mail in which this word is displayed as "it96s." Why? At a minimum, all computers support 128 ASCII (American Standard Code for Information Interchange) characters. ASCII is "the standard way that printable . . . characters are represented in the USA" (Shipman 1). Other computer systems, however, have additional ASCII codes, called extended ASCII. If the writer's computer has extended ASCII and the reader's computer doesn't, there is a problem. In the example regarding "it's," some computers, with only standard ASCII, don't know how to represent the apostrophe on screen. It was created using extended ASCII. Thus, users get "it96s," the computer's representation of the writer's apostrophe.

3. *Lack of privacy*. One of the wonderful aspects of e-mail is that everyone can talk to each other very quickly. In contrast to "snail mail," which takes days or weeks to deliver, e-mail almost allows "real-time" conversations. However, traditional mail is confidential. Not only is a letter sealed in an envelope, but also privacy laws protect your mail. In contrast, e-mail isn't necessarily confidential. A company's systems "operator is allowed to intercept messages during transmission. The operator is allowed to disclose user messages to anyone [he or she] chooses" (Rose 172). Even deleting e-mail doesn't ensure confidentiality. Deleted e-mail can be stored in back-up systems or recycle bins (Kim 52). A company's systems administrator can access anyone's e-mail after transmission.

More important, employees do not own their e-mail. A company does, and/or the recipient of the e-mail message does. The 1986 Electronic Communications Privacy Act asserts that an employee's e-mail content is owned by the company, sent e-mail may be saved internally, and archived e-mail messages are officially company documents. Once an e-mail message is sent through the Internet, copies of employee e-mail across Internet systems are not necessarily protected by any privacy laws. Finally, an e-mail message's content, after receipt by another company, may be reused without the writer's knowledge or consent.

4. *Misunderstandings and/or erroneous messages* (Freeman and Bacon 71). The ultimate problem with e-mail is caused by you, the e-mail writer. E-mail has an immediacy about it that makes communication intimate. Because you can "talk" back and forth with your e-mail reader, e-mail tends to become much friendlier than memos or letters—even chatty. We've read many e-mail messages complete with a conversational "gotta," "gonna," and "betcha" included in the text. That's OK . . . sometimes. If you're carrying on a casual conversation with a peer, there's nothing wrong with e-mail colloquialisms. However, if your e-mail message is sent to a boss, client, or vendor and will be printed for future reference, that conversational tone can get you in trouble. This failure to recognize audience is highlighted in e-mail since e-mail addresses don't also list the reader's title. You never know if he or she is a salesperson, purchasing agent, or vice president, unless you're writing to an acquaintance.

Similarly, when we talk to others, either face-to-face or on the telephone, we occasionally employ verbal sarcasm. The intonations of our voices and our physical gestures let our audience know that sarcasm is intended. When you use the correct voice inflection and appropriate gesture, your phone audience or face-to-face listener will recognize your sarcastic intent.

The casualness implicit in e-mail correspondence also often leads to a conversational tone in which sarcasm and verbal humor prevail. E-mail, however, leaves no room for gesture or inflection. Will your reader recognize your sarcasm? Let's say that your e-mail counterpart has written you a memo asking whether your company should select XYZ Corporation as a hardware vendor. You think the idea is terrible, so you sarcastically write back, "Great idea."

What's the reader supposed to think? Will he or she recognize the sarcasm? If the reader doesn't and then signs a contract with the vendor, you've made a terrible mistake by letting e-mail's inherent casualness affect your correspondence (Krol 94).

Because of the previously discussed problems, e-mail is not the best method for communicating in all instances. Sometimes, a more traditional memo or letter is more effective. Table 10.2 indicates the specific areas in which one method of communication might be preferred over another.

TECHNIQUES FOR WRITING EFFECTIVE E-MAIL

To avoid the previously discussed problems and to write successful e-mail correspondence, we suggest the following techniques (Figure 10.1 shows an e-mail message.)

1. Use the correct e-mail address. An e-mail address looks like this: name@arpnet.uh.hou.tx.us. You can imagine how easy it is to omit any part of this address or to punctuate it incorrectly. To connect with your intended recipient, confirm his or her e-mail address.

2. Provide an effective subject line. Memos and e-mail have similarities and differences (see Table 5.1). As with memos, an effective subject line will consist of a "what" (the topic of your e-mail) and a "what about the what" (the focus of your topic). Here's a successful subject line:

SUBJECT: MEETING DATES FOR TECH PREP CONFERENCE

The topic of this e-mail is a "Tech Prep Conference." If you had received only these words in the subject line, your question would have been, "What about the conference?" "Meeting Dates" provides the focus. Because the e-mail screen is so small, giving you little room to converse, an effective subject line is an important tool. You can concisely communicate a great deal in a limited space.

3. Limit your e-mail message to one screen (if possible). As noted already, readers don't want to scroll through several screens to read your message.

TABLE 10.2	Sending Printed Documents versus E-Mail	
PURPOSE	PRINTED DOCUMENTS (MEMO OR LETTER)	E-MAIL
A long message	Yes	No
A short message	Yes	Yes
A formal message	Yes	No
An informal message	Yes	Yes
For a permanent record	Yes	No
To one reader	Yes	Yes
To many readers	Yes	Yes
For immediate discussion	No	Yes
To send attachments	Yes	Yes
For confidentiality	Yes	No

Source: Hartman and Nantz 62–63.

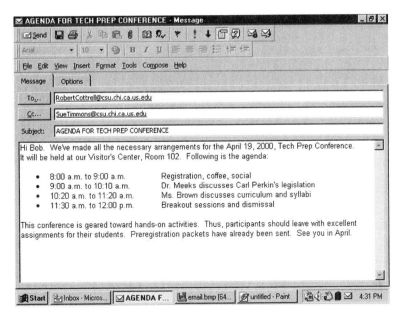

FIGURE 10.1 An E-Mail Message

Scrolling is time consuming and visually difficult. And, when one screen contains your entire message, the readers benefit from WYSIWYG (What You See Is What You Get). When confined to one screen, your message will have more impact, visually and in terms of your content.

To achieve e-mail limited to one screen, you will want to consider our suggestions regarding conciseness (discussed in Chapter 3). You also can limit the length of your e-mail by using conventional "cyberspeak" abbreviations, such as the following (Legeros 11):

BTW	by the way
FYI	for your information
IOW	in other words
LOL	laughing out loud
TIA	thanks in advance
WRT	with respect to

Although conciseness is always important in effective technical writing, because of the restricted space on an e-mail screen, conciseness becomes an even greater asset (Freeman and Bacon 72).

4. Organize your e-mail. Successful writing usually contains an introductory paragraph, a discussion paragraph, and a conclusion. Your e-mail also should contain these components. Use the introduction to tell the reader why you are writing and what you are writing about. In the discussion, clarify your points thoroughly. Use the conclusion to tell the reader what's next, possibly explaining when a follow-up is required and why that date is important.

5. Use highlighting techniques, but use them sparingly. Many e-mail packages will not let you **boldface**, *italicize*, or <u>underline</u>. Even if you could, you should avoid doing so. The commands required for these operations might not be compatible with your end user's computer. In that case, the text could be distorted (Krol 95). Still, you want your text to be accessible and emphatic. Try these options. You can indent the e-mail discussion unit (using the tab key) and then set off the itemized listings in this discussion either with an asterisk (*), hypens (-), or numbers. And use headings as a lead into new paragraphs. The added white space will help readers access your text.

The use of uppercase letters will also make a key word or phrase emphatic, but don't use all caps throughout the text. An entire e-mail message composed of capitalized letters is difficult to read. If you need to italicize or underline (to designate a book or article title, for example), use these standard techniques:

underline	=	underlined text	(an underline character before and after the appropriate text)
italics	=	italicized text	(an asterisk before and after the appropriate text)

6. Proofread your e-mail. E-mail can be written and transmitted rapidly. That's both good and bad. It's valuable to have electronic communication abilities that let you correspond with others around the world, in seconds. Rapid writing also can lead to mistakes, unfortunately. Whereas most word processing packages have spell checks and grammar checks, many e-mail systems do not. You can only avoid errors in your e-mail the old-fashioned way—read it and reread it. Don't let important correspondence that might be filed for future reference come back to haunt you. Errors will undermine your credibility.

7. Don't e-mail confidential information. E-mail, as already noted in this chapter, is not necessarily safe from roving eyes. If the content of your message, because of its confidential nature, could hurt you, someone else, or your company, stop and reconsider. For privacy, be careful about what you send via e-mail; better yet, use traditional posted mail, a slower but more confidential option (Hartman and Nantz 63).

8. Make hard copies for future reference. Like memos, letters, and reports, e-mail can communicate important business transactions. Your e-mail might document significant decisions. Though many companies would like to become paper-free, you still must make hard copies of correspondence for future reference. Don't be misled by the conversational nature of e-mail. Your company might need your message next week, next month, or next year. Save the e-mail printout.

9. Practice "netiquette." Electronic communication is not only influencing our methods for writing correspondence but also creating a new culture—cyberspace. This new culture requires a new ethic called *netiquette*—appropriate online behavior. Because e-mail has a tendency to become very casual and conversation-al, you need to be careful about the tone of your message. When you write your e-mail messages, strive to achieve the following (Fisher et al. 33–36):

- *Be courteous.* How different is this from any other communication? Not different at all. In every business instance, you should be courteous. This is especially important in e-mail, which can be copied to others and saved for future reference. Don't let the instantaneous quality of e-mail negate your need to be calm, cool, deliberate, and professional.

- *Don't write abusive, angry e-mail messages.* This is the converse of the above point. Courtesy and anger can't coexist. Yet e-mail, because of its quick turnaround abilities, can lead to negative correspondence called *flaming*—sending angry mail. Such correspondence is never acceptable. Figure 10.2 is an example of an angry e-mail that should not have been sent.

- *Learn how to express emotions visually.* E-mail readers can't tell if you're joking, aggravated, or sincere. To solve this communication problem, e-mail writers occasionally use *smileys* or *emoticons* to show attitude. Here are some common examples:

:-)	A happy face to show you are joking
;-)	A winking face to show you are being snide or sarcastic
:-(	A sad face to show disagreement or unhappiness
:-o	A startled face to show shock or amazement
:-D	A laughing face
%-)	A confused face
:-X	A face with lips sealed to show the information is being treated as confidential

  Earlier in this chapter, we provided the example of a company suggesting a new hardware vendor, followed by another employee's sarcastic comeback. Here's how the sarcastic e-mail response would have looked when accompanied by a smiley: Great idea. :-) The horizontal smiling face following the text clues the reader into your attitude. Of course, such smileys should be used only for casual e-mail, certainly not in a business document regarding an important corporate decision.

- *Read and reread your message before you send it.* If you haven't created the appropriate tone, either for your audience or for your desired corporate image, cancel the message and start over.

ONLINE HELP

Online help systems are growing in importance as a technical writing tool because of several factors (Pratt 33):

- The increased use of computers in business, industry, education, and the home.
- The reduced dependence on hard-copy manuals by consumers.
- The need for readily available online assistance.

TO: Office Manager
FROM: Unhappy Employee
SUBJECT: EXCESSIVE PRINTER PAPER

I'VE TALKED WITH SEVERAL PEOPLE AND THIS SEEMS TO BE A
PROBLEM IN YOUR DEPARTMENT. SOMEONE PRINTS INFOR-
MATION AND WON'T PICK IT UP AT THE PRINTER. THEN THE
NEXT PERSON HAS TO SORT THROUGH THE PRINTED MATERI-
AL TO FIND WHAT HE OR SHE WANTS. SOME OF THE PRINT-
OUTS ARE NEVER USED; THEY JUST SIT THERE FOR DAYS, GET-
TING IN OTHER PEOPLE'S WAY. PEOPLE SHOULD JUST PICK UP
THEIR PRINTING AND GET IT OUT OF EVERYONE ELSE'S WAY.
THAT'S ONLY COMMON COURTESY. BUT THE PEOPLE IN
YOUR DEPARTMENT AREN'T EVEN REMOTELY CONSIDERATE
OF OTHERS. IF YOU MANAGEMENT PEOPLE WOULD JUST DO
YOUR JOBS, NONE OF THIS WOULD HAPPEN. THE PROBLEM
WOULD BE ALLEVIATED, AND WE'D ALL BE HELPED. THAT'S
MY OPINION, THOUGH I DON'T GUESS YOU CARE. YOU GET
PAID THE BIG BUCKS. WHY NOT EARN SOME OF IT!

FIGURE 10.2 Flaming E-Mail

- Proof that people learn more effectively from online tutorials than from printed manuals.

What's an online help system? Online help systems employ computer software to help users complete a task. Help menus on your computer are excellent examples of online help systems. As the computer user, you pull down a help menu, search a help list, and click on the topic of your choice, revealed as a *hot button* (discussed under "Web Sites"). When you select your topic, you could get a *pop-up* (a small window superimposed on your text), or your computer might *jump* to another full-sized screen layered over your text. In either instance, the pop-up or jump gives more information about your topic (Henselmann 475; Pratt 34–36):

- Overviews—explanations of why a procedure is required and what outcomes are expected.
- Processes—discussions of how something works.
- Definitions—an online glossary of terms.
- Procedures—step-by-step instructions for completing a task.
- Examples—feedback verifying the completion of a task and/or graphic depictions of a completed task. These could include a screen capture or a description with call-outs.
- Cross references—hypertext links to additional information.
- Tutorials—opportunities to practice online.

Online help systems allow the technical writer to create interactive training tools and/or informational booths within a document. These online systems can be created using a wide variety of authoring tools. Some popular ones include *RoboHELP*, *HelpBreeze*, *ForeHelp*, *Help Magician*, *Visual Help*, and *WYSI–Help* (Zubak 11). The value of online help is immense. Because online help offers "just in time" learning or "as needed" information, readers can progress at their own pace while learning a program or performing a task.

TECHNIQUES FOR WRITING EFFECTIVE ONLINE HELP

To create effective online help screens, consider these suggestions:

1. Organize your information for easy navigation. Poorly organized screens lead to readers who are "lost in cyberspace." Either they can't find the information they need, or they have accessed so many hypertext links that they are five or six screens deep into text.

You can avoid such problems and help your readers access information in various ways:

- Allow users to record a history of the screens they've accessed through a bookmark or a "Held Topics" pull-down menu (Stevens 411).
- Provide an online "Contents" menu, allowing users to access other, cross-referenced help screens within the system (Stevens 411).
- Provide a "Back" button and/or a "home" button to allow the readers to return to a previous screen (Goldenbaum and Calvert).

A good test is the "three clicks rule." Readers should not have to access more than three screens to find the answer they need. Similarly, readers should not have to backtrack more than three screens to return to their place in the original text (Timpone).

2. Recognize your audience. Online help must be user-oriented. After all, the *only* goal of online help is to help the user complete a task. If the reader is not helped, what has the technical writer accomplished? Thus, a successful online help system must be "designed at the same level of detail as the user's knowledge and experience" (Wagner). This means that technical writers have to determine a user's level of knowledge.

If the system is transmitted by way of an intranet or extranet, you might be tempted to write at a high- or low-tech level. Your readers, you assume, will work within a defined industry and possess a certain level of knowledge. That, of course, is probably a false assumption. Even within a specific industry or within a specific company, you will have co-workers with widely diverse backgrounds: accountants, engineers, data processors, salespeople, human resource employees, management, technicians, and so on.

Don't assume. Find out what information your readers need. You can accomplish this goal through usability testing (discussed in Chapter 12), focus groups, brainstorming sessions, surveys (discussed in Chapter 16), and your company's hotline help desk logs (Timpone). Then, don't scrimp on the information you provide. Don't just provide the basic or the obvious. Provide more detailed information and numerous pop-ups or jumps. Pop-up definitions are an especially effective tool for helping a diverse audience (see Chapter 4 for help with definitions). Remember, with online help screens, readers who don't need the information can skip it. Those readers who do need the additional information will appreciate your efforts. They will more successfully complete the tasks, and your help desk will receive fewer calls.

3. Achieve a positive, personalized tone. Users want to be encouraged, especially if they are trying to accomplish a difficult task. Thus, your help screens should be constructive, not critical. Your text should be "written in the affirmative," a concept supported by HCI (human-computer interaction) concerns (Wagner). HCI-driven online help systems "coach" the users rather than "command" them ("Human Computer Interaction" 2). The messages also should be personalized, including pronouns to involve the reader.

Excellent examples of this positive, personalized tone can be seen in Microsoft's Office Assistants, found in Microsoft *Word*, *Excel*, and *Outlook*. This online help system allows you to choose your help wizard from several options. For example, you could choose an Einstein caricature called the Genius, who says, "Hello. Can I assist you with your work in electronic space?" Another wizard, Mother Nature, "provides gentle help and guidance." A third option, the Dot, introduces itself by stating, "When you need help of any kind, just give me a click."

4. Design your document. How will your help screens look? Document design is important because it helps your readers access the information they need. To achieve an effective document design, consider these points:

- Use color sparingly. Color causes several problems in online documents. Bright colors and too many colors strain your reader's eyes. Furthermore, a color that looks good on a high-resolution monitor might be difficult to read on a monitor with poor resolution. Your primary goal is contrast. To help your audience read your text, you want to maximize the contrast between the text color and the background color (Goldenbaum and Calvert). Black text on a white background offers the optimum contrast. Many help screens, such as those provided by *Microsoft Office* and our sample, use black text on a pale yellow background. This combination provides contrast and differentiates the help screens from the running text. *WordPerfect*'s help system also uses black text on a pale yellow background. Once you select a topic for assistance, however, the pop-up is shown with black text on a white background; headings are blue, and hot buttons are green. In all instances, contrast is achieved.

- Be consistent. Pick a color scheme and stick with it. Your headings should be consistent, along with your word usage, tone, placement of help screen jumps and pop-ups, graphics, wizards, and icons (Henselmann 475). Readers expect to find things in the same place each time they look. If your help screens are inconsistent, readers will be confused.

- Use a 10-point sans serif font. A 12-pont type size is standard for most printed documents, but 10-point type will save you valuable space online. Serif fonts are the standard for most technical writing (See our discussion of type-

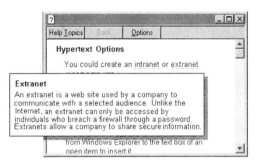

FIGURE 10.3 Help Screen

face and type size in Chapter 8.) A serif font, with small, horizontal "feet" at the bottom of each letter, helps guide the eye while reading printed text. However, on lower-resolution monitors, serif text is more difficult to read.

- Use white space. Don't clutter your help screens. "Avoid excessive emphasis techniques" (Stevens 410). Minimize your reader's overload by adding ample horizontal and vertical white space. Online, less is best.

5. Be concise. "It takes 20 to 30 percent longer to read on screen than in print, so you must minimize text" (Timpone). Furthermore, online space is limited. For example, a help screen pop-up will rarely exceed eight typed lines, with five to six words per line. That's short! Chapter 3 provides you many techniques for achieving conciseness. In addition to limiting word and sentence length, a help screen should avoid horizontal and vertical scrolling (Stevens 410). Each screen should include one "self-contained" message (Timpone).

6. Be clear. Your audience reads the help screen only to learn how to perform a task. Thus, your only job is to meet the reader's needs—clearly. To accomplish this goal, you need to be specific. (See Chapter 3 for suggestions.) In addition, clarity online could include the following:

- Tutorials to guide the reader through a task. Microsoft gives its readers a "How?" button to meet this need. *WordPerfect*'s help system offers "How Do I?" links to show step-by-step procedures.
- Graphics that depict the end result. Microsoft gives its readers a "Show me" button followed by screen captures. *WordPerfect*'s help system provides an "Examples" link with pictures of what you can do and how to do it.
- Cross references. *WordPerfect* provides an "Additional Help" link, complete with a glossary, an index, customer support, a "Contents button," and a speech recognition system (available on CD-ROM). Microsoft programs also provide a help index and a table of contents.
- Pop-up definitions.

7. Correct your grammar. As in all technical writing, incorrect grammar online leads to two negative results: a lack of clarity and a lack of professionalism.

Don't embarrass your company and/or confuse your reader with grammatical errors. Proofread.

WEB SITES

Like memos or letters, e-mail usually conveys short messages. Web sites, in contrast, can convey any amount of information. Web sites are created by companies, organizations, schools, and government agencies, not to mention individuals. That's why a Web site's URL (uniform resource locator—the Web site's address) reads *.com* (commercial), *.org* (organization), *.edu* (education), *.gov* (government), and so on. The content within these sites ranges from sales and marketing information about ceiling fans, to seating options at a baseball game, to biographical data about a country western singer, to the best fishing holes in the Ozarks, to course requirements for an engineering degree at your local college, to an individual's portfolio of poetry and photography, to the nutritional contents of a major corporation's fried chicken.

The Democratic National Party has a Web site. So do McDonalds, Ford Motor Company, the city of Las Vegas, the University of Texas, and Greenpeace. There are over 40 Web sites dedicated to Elvis Presley. Whatever topic you are interested in researching, you can find it online in someone's Web site. In December 1996, the Internet search engine Yahoo listed 161,068 company Web sites in its "Business and Economy" link. By March 1999, this number had almost tripled to 431,034 companies with Web sites listed in Yahoo's "Business and Economy" page. And that number increases every day.

Web sites range in size from one online screen—called a home page—to numerous screens linked by hypertext hot buttons. These hypertext buttons can be highlighted words or phrases that are boldfaced, underlined, or colored to distinguish them from other text. Or the hypertext links can be GUIs (graphical user interfaces), icons that are "hot." When you point your cursor at the highlighted hot button, the pointer turns into a small hand. Then you click on the hot button to jump to another screen.

WEB SITES VERSUS PAPER TEXT

Web sites are an entirely new mode of communication. Thus, Web sites and traditional paper text differ greatly. What are the differences between paper text and online communication? In the Internet's infancy, Web sites were not very different from paper documents. Early, poorly constructed Web sites did little more than recreate hard-copy brochures and newsletters. What you saw on paper was what you got online. That was a bad idea, a poor use of the Internet. Why? Successful Web sites have their own unique characteristics.

Margins

Most hard-copy text is $8\frac{1}{2}'' \times 11''$ with a $1''$ margin (top, sides, and bottom). When you get to the bottom of the page, you turn to a second page. Such is not the case with Web sites. Online, horizontal margins vary from monitor to monitor (21",

19″, 17″, 15″, etc.). On a 17″ monitor, for example, text runs approximately 12″, left margin to right margin. On a 15″ monitor, text runs approximately 10½″. This causes the first problem. On a Web page, your audience will read more words per line if you use the entire screen. That's difficult. Readers have trouble maintaining their focus (visually and mentally) when expected to read long lines of horizontal text. Vertically, the problem increases. Regardless of the monitor size, cybertext can run forever. The reader can scroll, and scroll, and scroll. Whereas paper sizes are controlled by convention and printing companies, *you* must control the size of a Web page.

On screen, less is best. A successful Web page should limit pertinent information to one screen, approximately 6″ to 7″ long, without requiring the reader to scroll. This equates to approximately 20–22 lines of text, versus the 50–55 lines that can fit on an 8½″ × 11″ piece of paper. Furthermore, a successful Web page should limit lines of text to perhaps two-thirds of the screen, with a graphic, white space, or hypertext links placed in the remaining third. This breaks up the monotony of long lines into manageable chunks. Thus, the reader can maintain focus more easily (Eddings et al.).

Structure

Paper text is sequential. We read from page 1 through the end of the text. Think of a book: you would never consider reading page 102 before you've read page 18 or chapter 7 before you've read chapter 4. In fact, you're expected to read sequentially—page after page after page. That's how it's done; that's how it's always been done. After all, the book (or newsletter or manual) is bound; each page is physically stitched, glued, or stapled to the next. In contrast, Web sites are composed of randomly "stacked" screens. You scan the home page, and then *you* decide what you want to read and when you want to read it. Once you make this decision, you click on hypertext buttons, access any screen, and read content in any order. Thus, you receive "just in time" or "as needed" information. That is, if you want to access the linked text, you can. If you don't want to read this screen, you don't have to. Again, on screen, less is best for the random reader who plans to skim and scan (Goldenbaum and Calvert).

Noise

One reason a Web page should be limited to one screen, with each screen providing a nugget of information, is the problem with computer "noise"—sound and visual distractions. Extended viewing of a computer screen is more demanding than continued reading of paper text. Paper is dull, and it absorbs light; the screen, composed of glass, reflects light, creating visual glare. Most hard-copy text is colorless and motionless; a book is composed of black-and-white sentences quietly lined up like sleeping sentinels. Web sites, in contrast, often contain lots of color, blinking text, animated graphics, frames of layered text, and sound and video. Noise—multiple distractions—inundates the screen. In such a busy communication vehicle as a Web site, the less we give the reader on one screen, the better.

CRITERIA FOR A SUCCESSFUL WEB SITE

Because Web sites are a different type of correspondence than traditional, paper-bound text, you need to employ different criteria when creating your site.

Home Page

The home page on a Web site is your welcome mat. It is the first thing your reader sees; thus, the home page sets the tone for your site. A successful home page should consist of the following components:

1. Identification information. Who are you? What is the name of your company, product, or service? How can viewers get in touch with you if they want to purchase your product? A good home page should clearly name the company, service, or product. Further, a successful home page should provide the reader access to a corporate phone number, an e-mail address, a fax number, an address, and/or customer service contacts. Amazingly, many Web sites omit this information. They provide the reader information about a product or service, but they prevent the reader from making a follow-up purchase or inquiry. That's not good business (Berst).

2. Graphic. Don't just tell the reader who or what you are. Show the reader. An informative, attractive, and appealing graphic depicting your product or service will convey more about your company than mere words will. Check out various Web sites to prove this point. Notice how inviting and family-friendly the McDonalds' Web site is, complete with a sparkling golden arch. Pepsi's Web site is also inviting but avant-garde, symbolizing "Generation Next." Your home page's graphic, like a corporate logo, should represent your company's ideals.

Be careful, however, when you choose the size of your home page graphic. If the graphic is too large, it will take a long time to load. This delay might cause Web surfers to lose patience, stop the HTTP transfer, and visit another site. Then you've lost their business or interest.

3. Lead-in introduction. In addition to a graphic, you might want to provide a one- or two-sentence lead-in. Welcome readers to the site and tell them what they can find in the following screens. At this textbook's writing, the Gateway 2000 home page provided its audience a friendly graphic (the company's spotted cow) and this introduction: "Choose a category and discover products, services, and content that are tailored to you." If you access this site today, you may find a different message.

Your introduction could also provide an overview of your company's goals. For example, in addition to a striking graphic, the home page of Black & Veatch (an international engineering company) provided this introduction:

> Today, Engineering firms need to do much more than just provide engineering design. Increasingly, the industry trend toward design-build and turnkey projects requires that engineering firms be able to deliver a full spectrum of services including engineering, procurement, and construction. Black & Veatch is a company that can do exactly that—provide all the services needed to take a project from concept to completion.

4. Hypertext links/buttons. This is what truly distinguishes a Web site from all other types of technical writing: the ability to click on a hot button and link to another screen. By providing the reader with either hypertext links or graphical user interfaces (GUIs), a home page acts as an interactive table of contents or index. The reader selects the topic he or she wishes to pursue, clicks on that link, and jumps to a new screen. Instead of being forced by the constraints of paper-bound text to read sequentially, line by line, page by page, the reader can follow a more intuitive approach. With hypertext links, you read what you want to read, when you want to read it.

There is no one way to present these hypertext link buttons or GUIs on the home page. You could simply provide a vertical or horizontal list of hot buttons. Graphic links can be presented on the home page in any manner you choose. The only constraints are size and design. Disproportionately large graphics will take up too much room on the screen and delay the Web site's loading. As for design (the placement of the graphics and their "look"), that's up to you, depending on the image you want to create.

Linked Pages

Once your reader clicks on the hypertext links from the home page, he or she will jump to the designated linked pages. These linked pages should contain the following:

1. Headings/subheadings. The reader clicks on the home page link and arrives at a new screen. Will the reader know where he or she is now? Or will the reader be lost in cyberspace? On paper-bound text, when readers turns the page, they know exactly where they are: on the next page. However, there is no "next" page in cyberspace. "Next" connotes sequence, which is clearly evident when pages are stitched together in a book or bound in a manual. In cyberspace, however, "next" could be any screen in any order. That's why the tool bar on your Internet browser reads "Back" and "Forward." Thus, to ensure that readers know where they are in the context of the Web site, you need to use headings. These give the readers visual reminders of their location.

2. Development. As in all correspondence, you need to develop your ideas thoroughly. In Chapter 3, we suggest responding to reporter's questions and specifying. The same applies online. Each linked page will develop a new idea. Prove your points precisely.

Navigation

As mentioned, readers can easily get lost in cyberspace. If a reader is viewing one screen but wants to return to a screen he or she looked at before, how does the reader get there? There are no pages to turn. You need to help the reader navigate through cyberspace. You can do this in two ways:

* *Home buttons.* The reader needs to be able to return to the home page easily from any page of a Web site. Remember, the home page acts as a table of contents or index for all the pages within the site. By returning to the home page, the reader can access any of the other pages. To ensure this easy navigation, you need to provide a hypertext-linked "Home" button on each page.

- *Links between Web pages.* Why make the reader return to the home page each time he or she wants to access other pages within a site? If each page has hypertext links to all pages within the site, then the reader can access any page he or she chooses, in any order of discovery. These hypertext links between pages can be provided in many ways. Many Web sites provide a row of hypertext link GUIs or hot words at the bottom of each page. You select the approach that works best for your Web site. But don't omit this important feature. If readers can't return to the home page or move easily within a Web site, it's almost like getting stuck in a cyberspace undertow. They end up swimming in circles, getting nowhere. Here's a good rule of thumb: a reader should never have to click more than three buttons to get to any page within a Web site.

Document Design

You can do amazing things on a Web site. You can add distinctive backgrounds, colored fonts, different font faces and sizes, animated graphics, frames, and highlighting techniques (such as lines, icons, and bullets). However, just because you can doesn't mean you should! Document design should enhance your text and promote your product or service, not distract from your message. As mentioned before, on a Web site, less is best.

1. Background. On Web sites, you can add backgrounds running the gamut from plain white, to various colors, to an array of patterns: fabrics, marbled textures, simulated paper, wood grains, grass and stone, psychedelic patterns, water and cloud images, and waves. They all look exciting. However, they aren't all effective. When choosing a background, you want to consider your corporate image. If you are an engineering company, do you want a pink background? Perhaps a subtle stone or "brick" background would be more effective in portraying your corporate identity. If you are a child-care center, do you want a dark background? Maybe this will be too dour. A white background with toys as a watermark might more successfully convey your center's mission statement.

In addition, remember that someone will attempt to read your Web site's text. Very few font colors or styles are legible against a psychedelic background—or against most patterned backgrounds, for that matter. Instead, to achieve readability, you want the best contrast between text and background. Despite the vast selection of backgrounds at your disposal, the best contrast is still black text on a white background.

2. Color. When you create a Web site, the default font color will be black. You can change this to any color in the spectrum, but do you want to? What corporate image do you have in mind? The font color should be suitable for your Web site, as well as readable. A yellow font on a blue background will cause headaches for your reader. A red font on a black background is nightmarish. The issue, however, isn't just esthetics. As noted above, a primary concern is contrast. Red, blue, and black font colors on a white background are very legible because their contrast is optimum. Other combinations of color don't offer this contrast; thus, the reader will have trouble deciphering your words (Goldenbaum and Calvert).

Let's also factor in a computer's resolution. Your monitor's resolution is determined by the number of pixels it displays. Here's where we encounter a problem. All monitors aren't the same. Some monitors display 640 pixels across by 480 pixels high. This equals approximately 300,000 pixels. Other monitors, however, can display more than 1 million pixels (Mazur 5). The more pixels a monitor displays, the clearer its resolution. Resolution affects the display of color on a monitor. On monitors with fewer pixels, colors will fade. In contrast, on monitors with more pixels, color will be sharper.

Thus, when you use color on your Web site, the color you see on your monitor will *not* necessarily be the color your reader sees. The WYSIWYG factor does not apply online. As a webmaster, you cannot solve this problem, since you cannot control other people's monitors. All you can do is try to reduce the problem. First, use restraint with color. Stick with black, red, and blue font color for text. Variations such as salmon, lime green, cornflower blue, medium aquamarine, and goldenrod won't be true. Next, test your use of color on several different monitors. You may well be surprised at what you see, but now you can change your font color if needed for clarity (Eddings et al.).

3. Graphics. If different monitors affect font color, imagine what they do to graphics? The same problem of resolution is magnified. The graphic that you see on your monitor will not look the same on all other monitors. You can control this problem somewhat by selecting the correct graphic format. If your graphic is a large image in flat colors without gradations, such as a corporate logo, save it as a gif (graphics interchange format). This is an 8-bit color format. If your image has multiple gradations of color, as photographs would, save the image as a jpeg (joint photographic experts group). A jpeg (or jpg) image supports 24-bit color. Remember, however, that whichever option you choose, the colors won't be true from monitor to monitor (Mazur 6).

Another problem is image size. Your Web site will be affected by the size of your graphics and the number of graphics you have included. For example, a "small" graphic (approximately 90 × 80 pixels, 1½ ″ × 1½″) consumes about 2.5 kilobytes. A "medium" graphic (approximately 300 × 200 pixels, 5″ × 4″) consumes about 25 kilobytes. A "large" graphic (approximately 640 × 480 pixels, 10″ × 8″) consumes about 900 kilobytes. When you use graphics, with varying amounts of color depth, your file size increases. The larger your file, the longer it will take for the images to load. The longer it takes for your file to load, the less interested your reader will be in your Web site. As mentioned already, less is best. Limit the size and number of your graphics.

Furthermore, use suitable graphics. If you work for an accounting firm, do you want graphics depicting a smiling calculator operated by an oversized flamingo? Create and insert your graphics carefully. (We discuss ways in which you can create and/or download Internet graphics in Chapter 9.)

4. Highlighting techniques. Font sizes and styles, lines, icons, bullets, frames, java applets, animation, video, audio! You can do it all on your Web page, but should you? Nothing is more distracting than a Web site full of highlighting techniques. Each design element is like spice. Your Web site is the stew. If you throw in too many different spices, the stew becomes inedible. A little spice goes a long way.

An occasional line (horizontal rule) to separate headings and subheadings from the text might be effective. You should use bullets and icons to enliven your text and to break up the monotony of wall-to-wall words. To create first- and second-level headings, you can change your font size and style. Boldface is effective for headings. Select a readable font (Arial is excellent online).

But other highlighting techniques have their downside. Limit italics, which are not easy to read online, and avoid underlining. Remember, hypertext links will usually appear as underlined words or phrases. Thus, underlining for emphasis could mislead your reader. Java applets take a long time to load. Video requires an add-on users must download before they can enjoy your creation. Audio, of course, requires speakers. Animation can be very sophomoric.

The worst highlighting technique might be frames. Jesse Berst, editorial director of ZDNet considers frames to be one of the "Seven Deadly Web Site Sins." He says, "Too many frames . . . produce a miserable patchwork effect." If you want to achieve the "frame" look, but without the hassle, use tables instead. They are easier to create and revise, they load faster, and they are less distracting to your readers.

In many cases, white space is the best option. Rather than cluttering your site with lines, icons, bullets, animation, audio, video, and frames, make your site more open and inviting. Indent text where appropriate, and leave plenty of vertical and horizontal white space on screen (Yeo 12–14).

Style

1. Conciseness. Conciseness is important in all technical correspondence. (We discuss this in great detail in Chapter 3.) Conciseness is even more important in a Web site. As noted earlier in this chapter, a successful Web page should be limited to one viewable screen, and a line of text should rarely exceed two-thirds of the screen.

2. Personalized tone. The Internet is a very friendly medium of communication, in contrast to many books and journals. If reports are remnants of the stodgy 1950s, Web sites are pure "Generation X." On a Web site, you are allowed, and in fact encouraged, to use pronouns, contractions, and positive words. Create a personalized tone that engages your readers.

Grammar

Perhaps because of the friendly, casual nature of the Internet, many Web sites are poorly proofread. That's a problem. Casualness is one thing; lack of professionalism is another. If your Web site represents your company, you don't want prospective clients to perceive your company as lacking in quality control. In fact, that's what poor grammar online denotes—poor attention to detail.

It's easy to see why many sites are grammatically flawed. First, anyone can go online almost instantaneously. All you do is code in your HTML and transfer the file to your Web server, and you're online. That's great for business, but it's bad for proofreading. Good proofreading takes time, a concept almost antithetical to the Internet. Next, though most word processing packages now have spell checks and grammar checks that tell you immediately when you've made an error, many Web conversion programs do not contain these amenities. Thus, when constructing your Web site, you must be vigilant. The integrity of your Web site demands correct grammar.

SUCCESSFUL WEB SITE CHECKLIST

1. Home Page

✔ Does the home page provide *identification information* (name of service or product, company name, e-mail, address, etc.)?

✔ Does the home page provide an informative and appealing graphic?

✔ Does the home page provide a welcoming and informative introduction?

✔ Does the home page provide hypertext links or buttons connecting the reader to subsequent screens?

2. Linked Pages

✔ Do the linked pages provide headings clearly indicating to the reader which screen he or she is viewing?

✔ Do the linked pages **develop ideas** thoroughly (appropriate amount of detail, specificity, valuable information)?

3. Navigation

✔ Does the Web site allow for easy return from linked pages to the home page?

✔ Does the Web site allow for easy movement between linked pages?

4. Document Design

✔ Does the Web site provide an effective background, suitable to the content and creating effective contrast for reading the text?

✔ Does the Web site use color effectively, in a way suitable to the content and creating effective contrast for reading the text?

✔ Does the Web site use graphics effectively, so that they are suitable to the content, not distracting, and not causing delays for the site to load?

✔ Does the Web site use highlighting techniques effectively (lines, bullets, icons, audio, video, frames, font size and style, etc.) in a way that is suitable to the content and not distracting?

5. Style

✔ Is the Web site concise?
  • Short words (one to two syllables)
  • Short sentences (10–12 words per sentence)
  • Short paragraphs (four typed lines maximum)
  • Line length limited to approximately two-thirds of the screen
  • Text per Web page limited to one viewable screen—minimizing the need to scroll

✔ Is the Web site **personalized**, using positive words and pronouns?

6. Grammar

✔ Does the Web site **avoid grammatical errors**?

PROCESS

Throughout this textbook, we suggest that you write according to a process: a step-by-step approach to writing that is geared toward your success. The same process applies to Web site construction.

Prewriting

Before you begin to construct your Web site, gather data. Answer these reporter's questions.

1. Why are you creating a Web site? Is your corporation creating the Web site to

- Inform the public about products and services?
- Sell products via online order forms?
- Encourage new franchisees to begin their own businesses?
- Advertise new job openings within the company?
- Provide a company profile, including annual reports and employee résumés?
- List satisfied clients?
- Provide contact information?
- All of the above?

2. The answers to several questions tell you *who* your audience is (Wilkinson 15): Are you writing to low-tech clients, prospective employees, the general lay public, stockholders, or high-tech associates in similar businesses? Will your audience be

- Customers looking for product information, such as specifications, pricing information, guarantees, warranties, instructions, and facilities locations?
- Current users of a product or service looking for upgrades, enhancements, changes in pricing, guarantees, warranties, schedules, personnel, facilities locations, and so on?
- New customers wanting to purchase the product or service by way of an online order form?
- Customers looking for online help with instructions or procedures?
- Customers hoping to contact customer service?
- Prospective new hires seeking employment?

3. Answering the above questions will help you decide *what* to include in your site. Will you add forms, résumés, product descriptions, lists of job openings, stock reports, customer contacts?

4. How will you construct the site? Will you use HTML coding or HTML converters and editors? Where will you get the graphics you'll need? Will you download existing graphics, modified to meet legal and ethical standards, or will you create new graphics with the help of your in-house technical artists?

5. Where will your Web site reside? Will it be an Internet site open to the World Wide Web, or will you focus on an intranet or extranet site?

Once you've gathered this information, use storyboarding for a preliminary sketch of your Web site's layout.

1. Sketch the home page. How do you want the home page to look? What color scheme do you have in mind? How many graphics do you envision? Where will you place your graphics and text—margin left, centered, margin right? Using a storyboard to sketch the graphic or graphics, lead-in, and key hypertext buttons will save you time. If you decide up front what your home page will look like, you will be able to research your graphics, determine how many linked pages you'll need, and begin to draft the text.

2. Sketch the linked page(s). How many linked pages do you envision? Will they have graphics, online forms, e-mail links, audio or video, headings and subheadings, or links to other Internet, intranet, or extranet sites? Will you use frames or tables? What will your color scheme for these pages be? Sketch your design for these screens by storyboarding. Doing so will allow you to preview the entire site. This will help you estimate the time required to complete the project, as well as give you a better understanding of what resources (personnel, hardware, software, etc.) you'll need to construct your Web site. Then you can meet deadlines or bill accordingly.

Writing

The second stage of the process encourages you to create a rough draft of your Web site. To do so, follow this procedure:

1. Study the Successful Web Site Checklist. This will remind you what to include in your rough draft and how best to structure the Web site.

2. Review your prewriting. Have your answered all the reporter's questions? How does your Web site storyboarding sketch look? Do you see any glaring omissions or unnecessary items? If so, add what's missing and omit what's unneeded.

3. Draft the Web site. How do you create a rough draft online? That's the ultimate secret of creating a Web site. When you view a Web site, you are not viewing actual text. Go ahead, test this yourself. Click and drag on text online and then hit the delete key. What happened? Nothing! That's because you are viewing an image. The Web pages you see are, in fact, mirror reflections or mirages. What you see *really* isn't there. The real source of the mirage you see online can be viewed only when you peel away the veneer.

Select a Web site from your computer. Then, from the menu bar, click on "View" and scroll down to "Page source." This will reveal to you the actual source of the text and graphics you've been viewing. What you'll see is the HTML coding—a computer language that's used to transfer your text from a word processing package, such as *Notepad*, to the Web.

Figure 10.4 shows the HTML code for a Web page.

```
<HTML>

<TITLE>Projects: Intranets & Web Sites</TITLE>

<BODY background="../images/backprojects.gif" link="#3366cc"
alink="#3366CC" vlink="#009999">

<TABLE cellpadding="4" cellspacing="0" border="1" width="100%"
bgcolor="#339966"> <TR><TD align=right><B> <FONT color=
"#ffffff" size="+1"> Projects  </FONT> </B> </TD>

</TR>

</TABLE>

<P>

<TABLE width="100%">

<TR valign=top>

<TD>

<A href="acg.htm"><IMG src="../images/acgbtn2.gif"
width="131" height="30" vspace="4" border="0" alt="ACG">
</A><BR>

<A href="services.htm"><IMG src="../images/servicesbtn2.gif"
width="131" height="30" vspace="4" border="0" alt="Services">
</A><BR>

<A href="staff.htm"><IMG src="../images/staffbtn2.gif"
width="131" height="30" vspace="4" border="0" alt="Staff">
</A><BR>

<A href="projects.htm"><IMG src="../images/projectsbtn2.gif"
width="131" height="30" vspace="4" border="0" alt="Projects">
</A><BR>

<A href="library.htm"><IMG src="../images/librarybtn2.gif"
width="131" height="30" vspace="4" border="0" alt="Library">
</A><BR>

<A href="contacts.htm"><IMG src="../images/contactsbtn2.gif">
```

FIGURE 10.4 HTML Code for a Web Page

HTML Coding

We know that HTML coding looks daunting. However, it's not as hard as it looks. Actually, HTML is a fairly easy-to-learn language. Table 10.3 provides you the simple rules you'll need to draft your own text. These HTML codes are not case-sensitive.

TABLE 10.3 HTML Codes

	ON	OFF
Document setup	\<html\> Creates a Web page.	\</html\> The forward slash mark (/) turns all the codes off.
Title	\<title\> Title does not appear on Web page.	\</title\>
Body	\<body\> Begins the text.	\</body\>
Emphasis techniques		
Boldface	\<b\>	\</b\>
Italics	\<i\>	\</i\>
Blinking	\<blink\> Text blinks on and off.	\</blink\>
Font sizes		
Largest	\<H1\>	\</H1\>
	\<H2\>	\</H2\>
	\<H3\>	\</H3\>
	\<H4\>	\</H4\>
	\<H5\>	\</H5\>
Smallest	\<H6\>	\</H6\>
Dividers		
Double return	\<P\> Creates a double space.	\</P\>
Line break	\<BR\> Acts like a hard return.	\</BR\>
Horizontal rule	\<HR\> Places a thin line across the page	\</HR\>
Center	\<center\> Centers text or graphics.	\</center\>
Itemized lists		
Ordered list	\<OL\> \<LI\>apples \<LI\>oranges \<LI\>peaches \</OL\>	Looks like this: 1. apples 2. oranges 3. peaches
Unnumbered list	\<UL\> \<LI\>apples \<LI\>oranges \<LI\>peaches \</UL\>	Looks like this: • apples • oranges • peaches

TABLE 10.3 HTML Codes (Cont.)		
	ON	OFF
Backgrounds		
Colors	<body bgcolor="red"> or any color you choose. The "bg" stands for "background."	
	<font color="red"> or any color you choose. This changes the font color from the default black. These colors can also be designated by their numerical hex codes. Black, for example, is numerically written "#000000." Red is "#FF0000," Pink is "#BC8F8F," and Sky Blue is "#3299CC." There are literally thousands of such numerical delineations for different colors.	</font color>
Patterns	<body background="waves.gif"> We've used "waves.gif" as an example. You can find thousands of patterns on the Internet.	
Images/graphics		
	The Internet is perhaps the largest garage sale on Earth. If you want it, you can find it on the Internet. The same holds true for graphics (pictures, icons, lines, bullets, animations, backgrounds, buttons, etc.). Most of it is free. You merely code it in your HTML text as follows: <img src="filename.gif"> or <img src="filename.jpg"> We've used "filename" as an example. Each graphic will have its own name.	
Hypertext links		
	Links are what allow you to jump from linked page to linked page. The code looks like this: <a href="filename.html"> We've used "filename" as an example. You would put within the less than/ greater than symbols the name of the file you plan to link to.	</a>

Do you need to code in HTML in order to create a Web site? No, you don't. You could use one of the many HTML editors and converters on the market. Windows users could try these: *Ant HTML* and *Ant PLUS*, *Front Page*, *HotDog Web Editor*, *Netscape Navigator 3.0* and above, *Microsoft Internet Assistant for Word 6.0 for Windows* and above, *Corel WordPerfect 8.0* and above, and others. Macintosh users could try these: *Ant HTML*, *Alpha-Text*, *WebDoor*, and others (Byrne 9).

However, learning how to code in HTML is valuable for many reasons. First, when you code your own HTML, you control the page. Thus, you decide what you want your site to look like. In contrast, if you depend on an HTML editor or converter, you limit your options to those supported by the program (Scott 16–17). Second, if you use a conversion program and it doesn't do what you had hoped it would, you won't be able to change it—unless you know HTML coding. Finally, if you use an HTML converter, you'll always be dependent on that program. Once you learn to code HTML yourself, however, you can create Web sites on any word processor. All you need is "Notepad in Windows, Teach Text on a Macintosh, or Edit in DOS" (Byrne 9).

Rewriting

The final stage of the writing process is rewriting. Once you've coded in your text using HTML, save the file and open it online in the Internet. What do you see? Is it what you had in mind? If not, revise.

As with all other types of correspondence, you should consider these steps when revising:

1. Add new detail for clarity. Have you said all you need to say? Have you provided contact information, such as phone numbers, e-mail addresses, fax numbers, and a mailing address? Do you need to include a form to help your customers order goods online? Do you want to advertise job openings? Would your Web site profit by including an online catalog in which you describe your products?

Text looks different online than when you code it in. Once you see the first draft of your Web site, you might change your mind about its components. If you do so and want to add new information, the rewriting stage is the time to make those changes.

2. Delete anything unnecessary for conciseness. Do your text and graphics all fit on one viewable screen? Or do you make your readers scroll endlessly to gather all their data? Have you loaded your Web site with too many bells and whistles? These could include java scripting, animation, or audio and video plug-ins. If your readers have to wait too long for your site to load, you could lose their interest—and their business. Review and revise your Web site for conciseness. Delete the unnecessary add-ons. Delete dead words and phrases. Reduce the size of your graphics. More is not always better.

3. Simplify words and phrases. Web sites are very friendly types of correspondence. Multisyllabic legalese is rarely appropriate for this modern medium. Think simple. (See Chapter 3 for help on this point.)

4. Move information for emphasis. Look at several Web sites. Where do other companies place information? You'll notice that most companies list their webmasters (the people or companies who create their Web sites) at the bottom of a screen. Why? Because it would place too much emphasis on the writer and too little emphasis on the product or service to list webmasters at the top of a screen. Similarly, notice where companies mention their product or service: usually at the top of a screen, so they will receive immediate recognition. As in real estate, location, location, location sells. Decide what is most important on your Web site, and then place this information where it can be clearly seen.

5. Reformat for access. No communication medium depends more on layout than a Web site. In fact, rarely do you hear people say, "That Web site has great content." What you hear is, "That's a great-looking Web site" or "That's a terrible-looking Web site." Your Web site's design will make or break it.

Once you load your site, determine how it looks. Is there ample white space? Have you indented text and used bullets or icons to break up the monotony of wall-to-wall words? Where have you placed your graphics? Do the graphics integrate well with the text, and are they pleasing to the eye? What about your headings and subheadings? Are your first-level headings larger than your second-level headings? Should you center them? Would your site profit from a horizontal rule or line separating text from headings? Have you used color effectively? Is there enough contrast between your background and your font color? Where do you want to place your navigational links? Early in the history of the Internet, most companies placed the links at the bottom of the page. Now, however, we see the navigational links placed either along the left or the right margin. These links are viewable instantly, rather than forcing the readers to scroll down to find them.

(For further suggestions regarding document design, see Chapter 8.)

6. Enhance the tone and style of your Web site. As noted, Web sites tend to be rather informal. Don't write a Web site using the same stuffy tone that you would use when writing a report. Enhance the tone of your site by using contractions, positive words, and lots of pronouns. (See Chapter 4 for help.)

7. Correct errors. How many people might read your Web site? Remember, this is the World Wide Web we're talking about. In contrast to a memo or letter, which perhaps one or two people might read, a successful Web site could be seen by millions—worldwide. Now imagine how bad grammatical errors will look when viewed through a magnifying glass the size of the entire earth. People in dozens of countries and on numerous continents will learn that you can't use commas correctly. Don't let this happen to you or to your company. Correct your errors. Proofread your Web site diligently.

8. Make sure it works. Test your Web site. How long does it take to load? How do your graphics and colors look on "multiple browsers, platforms, . . . monitors, and with different transmission speeds" (Yeo 14)? A good idea is to test your site on at least three different computer systems (Eddings et al.). Do your hypertext links work? First, determine whether your readers can easily navigate from one screen to the next within your Web site. Next, if you've created hypertext links to external sites, test these frequently. External sites change constantly. You might need to change your hypertext codes to match these external changes (Yeo 14). Does your e-mail link work? How about the online order form? Can someone actually submit a request for goods on this form? Do your audio and video plug-ins work? No matter how beautiful your Web site looks on your monitor, the site will not be successful unless all the components work.

CHAPTER HIGHLIGHTS

1. To be an effective communicator in business and industry today, you have to communicate electronically.

2. The Internet, the information superhighway, connects computers internationally.

3. The information accessible on the Internet is virtually unrestricted.

4. Web browsers, such as *Netscape Navigator* and Microsoft *Internet Explorer*, allow you to access information on the Web.

5. Companies use intranets to communicate internally.

6. An extranet allows several companies to share information confidentially.

7. E-mail allows you to communicate business and personal information over the Internet.

8. E-mail has limitations, such as a lack of privacy and hard-copy documentation.

9. Not all computers use the same ASCII codes, so e-mail can look different from computer to computer.

10. The instantaneous nature of e-mail demands that you carefully proofread.

11. Interactive online help systems provide "just in time" and "as needed" information to help your audience complete a task.

12. Web sites are created by companies, the government, schools, organizations, and individuals.

13. Use our criteria when creating your Web sites.

14. The process approach to writing applies to Web site construction.

ACTIVITIES

E-Mail

1. Gather examples of e-mail from corporate environments and from your personal home use. Compare and contrast these two types of e-mail. How similar or different is the tone as evident in both types of correspondence? If the tone in both corporate and home-use e-mail is similar, is that appropriate? Could this lead to problems? Explain your answers.

2. Gather examples of e-mail from corporate environments. Discuss the various topics covered in these e-mail samples. Determine if the tone is appropriate, if the page layout is conducive to ease of access, if the correspondence should be saved for future reference, if the communication should have been submitted in a different form (written report, memo, etc.).

3. If your classroom has computer capabilities but no e-mail system, write memos limited to 12–14 lines, and 60 characters per line, to simulate an e-mail screen. Discuss the problems that such a restricted space creates and how these problems could be solved.

4. If your computer classroom has e-mail capabilities, use the system to communicate with other students in your class. Topics of discussion could include peer evaluations of other types of communication or planning strategies for group projects. By performing such an activity, you can practice composing e-mail and writing appropriately to an audience.

5. If your computer classroom has e-mail capabilities, write e-mail messages to students in another class. The topics could be school-related (degree-program objectives, criteria for graduation, challenges to graduation, helpful hints for success in various classes) or more personal chat lines about movies, cars, job opportunities, restaurants, musical tastes, etc.). The value of such an activity is to practice writing, audience recognition, and e-mail techniques.

6. Using the flaming (angry) e-mail example provided in this chapter (Figure 10.2), rewrite the message to soften its tone and to convey the message more positively.

Online Help

1. Study the help menu in your word processing package. Distinguish between the pop-up windows and the online jumps. What are the differences? What are the similarities?

2. In a small group, have each individual ask a software-related question regarding a word processing application. These could include questions such as "How do I print?" and "How do I set margins?" and so on. Then, using the Help menu, find the answers to these questions. Are the answers provided by way of pop-up windows or online jumps? Rewrite any of these examples to improve their conciseness or visual layout.

3. Take an existing document (one you have already written in your technical writing class or writing from your work environment) and rewrite it as online help. To do so, find "hot" topics in the text and expand on them as either pop-up windows or jump screens. Then highlight new hot buttons within the expanded pop-ups or jump screens for additional hypertext documents. Create four or five such layers.

   If you have the computer capability to do so, use your computer system. If you don't have the computer capabilities, use 3 × 5 index cards for the layered pop-ups or jumps.

4. Using our suggestions in this chapter, rewrite Figure 10.5 as an effective online help screen. To do so, reformat the information for access on a smaller screen and create hot-button words or phrases. Determine which of the hot buttons can be expanded with a short pop-up window and which require a longer jump screen. Research any data with which you're unfamiliar to provide additional information for your hypertext surfer.

Because your company is located in the East Side Commercial Park, the outside power is subject to surges and brownouts. These are caused when industrial motors and environmental equipment frequently start and stop. The large copiers in your office also cause power surges that can damage electronic equipment.

As a result of these power surges and outages, your three network servers have become damaged. Power supplies, hard drives, memory and monitors have all been damaged. The repair costs for 1999 damages totaled $55,000.

Power outages at night also have caused lost data and failed backups and file transfers. Not being able to shut down the servers during power failures is leading to system crashes. Data loss requires time-consuming tape restorations. If the failures happen before a scheduled backup, the data is lost, which requires new data entry time. Ten percent of your data loss can not be restored. This will lead to the loss of customer confidence in the reliability of your services.

Finally, staff time losses of $40,000 are directly related to network downtime. Data-entry staff are not able to work while the servers are being repaired. These staff also must spend extra time reentering lost data. Fifteen percent of the data management staff time is now spent recovering from power-related problems.

To solve your problems, you need to install a BACK-UPS 9000 system. It has the configuration and cost-effectiveness appropriate for your current needs.

FIGURE 10.5 Flawed Text That Needs Revision

Web Sites

1. Create a corporate Web site. To do so, make up your own company and its product or service. Your company's service could focus on dog training, computer repair, basement refinishing, vent cleaning, Web site construction, child care, auto repair, personalized aerobic training, or online haute cuisine. Your company's product could be paint removers, diet pills, interactive computer games, graphics software packages, custom-built engines, flooring tiles, or duck decoys. The choice is yours. To create this Web site, follow our writing process. Prewrite by listing answers

to reporter's questions. Then sketch your site through storyboarding. Next, draft your Web site, using HTML coding. Finally, rewrite by adding, deleting, simplifying, moving, reformatting, enhancing, correcting, and making sure your links work.

2. Create a Web site for your technical writing, business writing, or professional writing class. To create this Web site, follow our writing process. Prewrite by listing answers to reporter's questions. Then sketch your site through storyboarding. Next, draft your Web site, using HTML coding. Finally, rewrite by adding, deleting, simplifying, moving, reformatting, enhancing, correcting, and making sure your links work.

3. Create a Web site for your high school, your college, your fraternity or sorority, your church or synagogue, or your professional organization. To create this Web site, follow our writing process. Prewrite by listing answers to reporter's questions. Then sketch your site through storyboarding. Next, draft your Web site using HTML coding. Finally, rewrite by adding, deleting, simplifying, moving, reformatting, enhancing, correcting, and making sure your links work.

4. Create a personal Web site for yourself or for your family. To create this Web site, follow our writing process. Prewrite by listing answers to reporter's questions. Then sketch your site through storyboarding. Next, draft your Web site using HTML coding. Finally, rewrite by adding, deleting, simplifying, moving, reformatting, enhancing, correcting, and making sure your links work.

5. Research several Web sites, either corporate or personal. Use our Successful Web Site Checklist to determine which sites excel and which sites need improvement. Then write a report (see Chapter 15 for report criteria) justifying your assessment. In this report, clarify exactly what makes the sites successful. To do so, you could use a table, listing effective traits and giving examples from the Web sites to prove your point. Next, explain why the unsuccessful sites fail. To do so, again use a table to list effective traits, and give examples from the inferior sites to show why they are unsuccessful. Finally, suggest ways in which the unsuccessful sites could be improved.

6. Reconstruct an existing Web site you consider to be flawed. This might entail adding new information, deleting unnecessary details, simplifying text and graphics, moving information on the site for emphasis, reformatting the text for access, enhancing the tone, correcting errors, and ensuring that all links work.

Case Study

The following short proposal recommends that upper management approve the construction of a corporate Web site. The company's management has agreed with the proposal. You and your team have received a memo directing you to build the Web site. Using the proposal, build your company's Web site.

Date: November 11, 2000
To: Distribution
From: Shannon Conner
Subject: RECOMMENDATION REPORT FOR NEW CORPORATE
           WEB SITE

Introduction

In response to your request, I have researched the impact of the Internet on corporate earnings. Companies are still striving to position themselves on the World Wide Web and maximize their earnings potential. I concluded that from 1994 to 1998, only 16 percent of companies with Internet sites were making a profit from their online services. This fact, however, has changed. From 1998 to 2000, many more corporate Web sites earned a profit for their companies. The time is right for our company, Java Lava, to go online.

Discussion

Why should we go online? The answer is simple: we can maximize our profits by making our local product a global product. How can we accomplish this goal?

1. *International bean sales.* Currently, coffee bean sales account for only 27 percent of our company's overall profit. These coffee bean sales depend solely on walk-in trade. The remaining 73 percent stems from over-the-counter beverage sales. If we go global via the Internet, we can expand our coffee bean sales dramatically. Potential clients from every continent will be able to order our coffee beans online.
2. *International promotional product sales.* Mugs, T-shirts, boxer shorts, jeans jackets, leather jackets, key chains, paperweights, and calendars could be marketed. Currently we give away items imprinted with our company's logo. By selling these items online, we could make money while marketing our company.
3. *International franchises.* We now have three coffeehouses, located at 1200 San Jacinto, 3897 Pecan Street, and 1801 West Paloma Avenue. Let's franchise. On the Internet, we could offer franchise options internationally.
4. *Online employment opportunities.* Once we begin to franchise, we'll want to control hiring practices to ensure that Java Lava standards are met. Through a Web site, we could post job openings internationally and list the job requirements. Then potential employees could submit their résumés online.

In addition to the above information, used to increase our income, we could provide the following:

- A map showing our three current sites.
- Our company's history—founded in 1898 by Hiram and Miriam Coenenburg, with money earned from their import/export bean business. "Hi" and "Mam," as they were affectionately called, handed their first coffeehouse over to their sons (Robert, John, and William) who expanded to our three stores. Now a third generation, consisting of Rob, John, and Bill's six children (Henry, Susan, Andrew, Marty, Blake, and Stefani) could take us into the next millenium.
- Sources of our coffee beans—Guatemala, Costa Rica, Columbia, Brazil, Sumatra, France, and the Ivory Coast.
- Freshness guarantees—posted ship dates and ground-on dates; 100 percent money-back guarantee.
- Corporate contacts (addresses, phone numbers, e-mail, fax numbers, etc.).

Conclusion
Coffee is a "hot" commodity now. The international market is ours for the taking. We can maximize our profits and open new venues for expansion. The Web is our tarmac for takeoff. I'm awaiting your approval.

CHAPTER 11

Technical Description

OBJECTIVES

A technical description is a part-by-part depiction of the components of a mechanism, tool, or piece of equipment. Technical descriptions are important features in several types of correspondence.

OPERATIONS MANUALS

Manufacturers often include an *operations manual* in the packaging of a mechanism, tool, or piece of equipment. This manual helps the end user construct, install, operate, and service the equipment. Operations manuals often include technical descriptions.

Technical descriptions provide the end user with information about the mechanism's features or capabilities. For example, this information may tell the user which components are enclosed in the shipping package, clarify the quality of these components, specify what function these components serve in the mechanism, or allow the user to reorder any missing or flawed components. Here's a brief technical description found in an operations manual:

> The Modern Electronics Tone Test Tracer, Model 77A, is housed in a yellow, high-impact plastic case that measures 1-1/4″ × 2″ × 2-1/4″, weighs 4 ounces, and is powered by a 1604 battery. Red and black test leads

are provided. The 77A has a standard four-conductor cord, a three-position toggle switch, and an LED for line polarity testing. A tone selector switch located inside the test set provides either solid tone or dual alternating tone. The Tracer is compatible with the EXX, SetUp, and Crossbow models.

PRODUCT DEMAND SPECIFICATIONS

Sometimes a company needs a piece of equipment that does not currently exist. To acquire this equipment, the company writes a *product demand* specifying its exact needs:

Subject: Requested Pricing for EDM Microdrills

Please provide us with pricing information for the construction of 50 EDM Microdrills capable of meeting the following specifications:

- Designed for high-speed, deep-hole drilling
- Capable of drilling to depths of 100 times the diameter using .012-inch to .030-inch diameter electrodes
- Able to produce a hole through a 1.000-inch-thick piece of AISI D2 or A6 tool steel in 1.5 minutes, using a .020-inch diameter electrode.

We need your response by January 13, 2000.

STUDY REPORTS PROVIDED BY CONSULTING FIRMS

Companies hire a consulting engineering firm to study a problem and provide a descriptive analysis. The resulting *study report* is used as the basis for a product demand specification requesting a solution to the problem.

One firm, when asked to study crumbling cement walkways, provided the following technical description in its study report:

The slab construction consists of a wearing slab over a 1/2"-thick water-proofing membrane. The wearing slab ranges in thickness from 3-1/2" to 8-1/2," and several sections have been patched and replaced repeatedly in the past. The structural slab varies in thickness from 5-1/2" to 9" with as little as 2" over the top of the steel beams. The removable slab section, which has been replaced since original construction, is badly deteriorated and should be replaced. Refer to Appendix A, Photo 9, and Appendix C for shoring installed to support the framing prior to replacement.

SALES LITERATURE

Companies want to make money. One way to market equipment or services is to describe the product. Such descriptions are common in sales letters, brochures, and proposals and on the Internet. The following technical description is from Hewlett-Packard's Web site:

Key Specifications	DeskJet 720C/722C
Speed, Black	Fast mode: 8 pages per minute Normal mode: 5 pages per minute
Speed, Color	Fast mode: 7 pages per minute Normal mode: 4 pages per minute
Resolution	Fast mode: 300 x 300 dpi Normal mode: 600 x 600 dpi
Print Colors	Black and up to 16.7 million colors
Paper Trays	1
Media Capacity	100 sheets
Media Types	Plain and premium papers, photo papers, transparencies, cards, envelopes, labels, banners, and iron-on transfers
Memory	256 KB
Footprint	W x H x D: 17.5 x 7.3 x 14 in (446 x 185 x 355 mm) Weight: 12 lb (5.5kg)

Not all technical writing is photographic and quantifiable information. Some descriptions are impressionistic. Note that the following sales piece uses impressionistic adjectives and adverbs (italicized), allowing your imagination to depict the product's components:

> Inside the *elegant* Zephyr you'll find *comfortably designed* seats done in *rich, tailored* cloths. The *special* suspension system surrounds you with a driving experience quieter than Zephyr's past. Outside you'll find a *new bright* grille, sport mirrors, and a *smart* dual pinstripe. Finally, underneath all this comfort and luxury beats the *road-hugging* heart of an American classic.

Whatever your purpose in writing a technical description, the following will help you write more effectively:

- Criteria
- Process (with sample process log)
- Technical description examples

CRITERIA FOR WRITING TECHNICAL DESCRIPTIONS

As with any type of technical writing, there are certain criteria for writing technical descriptions.

TITLE

Preface your text with a title precisely stating the topic of your description. This could be the name of the mechanism, tool, or piece of equipment you're writing about.

OVERALL ORGANIZATION

In the *introduction* you specify what *topic* you are describing, explain the mechanism's *functions or capabilities*, and list its *major components*.

EXAMPLE 1
The Apex Latch (#12004), used to secure core sample containers, is composed of three parts: the hinge, the swing arm, and the fastener.

EXAMPLE 2
The DX 56 DME (Distance Measuring Equipment) is a vital piece of aeronautical equipment. Designed for use at altitudes up to 30,000 feet, the DX 56 electronically converts elapsed time to distance by measuring the length of time between your transmission and the reply signal. The DX 56 DME contains the transmitter, receiver, power supply, and range and speed circuitry.

In the *discussion*, you use highlighting techniques (itemization, headings, underlining, white space) to describe each of the mechanism's components.

Your *conclusion* depends on your purpose in describing the topic. Some options are as follows:

- *Sales.* "Implementation of this product will provide you and your company . . ."
- *Uses.* "After implementation, you will be able to use this XYZ to . . ."
- *Guarantees.* "The XYZ carries a 15-year warranty against parts and labor."
- *Testimony.* "Parishioners swear by the XYZ. Our satisfied customers include . . ."
- *Comparison/contrast.* "Compared to our largest competitor, the XYZ has sold three times more . . ."
- *Reiteration of introductory comments.* "Thus, the XYZ is composed of the aforementioned interchangeable parts."

Use *graphics* in your technical descriptions. Today, many companies use the *super comic book look*—large, easy-to-follow graphics that complement the text. You can use line drawings, photographs, clip art, exploded views, or sectional cutaway views of your topic, each accompanied by *call-outs* (labels identifying key components of the mechanism). These types of graphics are discussed in Chapter 9. When using graphics, try one of the techniques shown in Figures 11.1, 11.2, and 11.3.

Title

Introduction
_____
_____

Discussion

    Handle _____
    _____
    _____

    Shaft _____
    _____
    _____

    Blade _____
    _____
    _____

Conclusion
_____
_____
_____

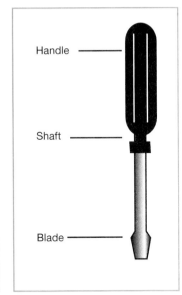

Handle ——

Shaft ——

Blade ——

FIGURE 11.1 Text and Graphics Separate

Title

Introduction
_____
_____
_____

Discussion

    1. Handle _____
    _____
    _____

    2. Shaft _____
    _____
    _____

    3. Blade _____
    _____
    _____

Conclusion
_____
_____
_____

    —— ①

    —— ②

    —— ③

FIGURE 11.2 Text and Graphics Merged

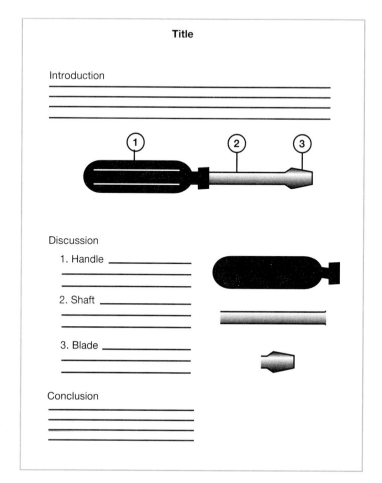

FIGURE 11.3 Text and Graphics Merged with Exploded View

INTERNAL ORGANIZATION

When describing your topic in the discussion portion of the technical description, itemize the topic's components in some logical sequence. Components of a piece of equipment, tool, or product can be organized by importance (as discussed in Chapter 3).

However, *spatial* organization is better for technical descriptions. When a topic is spatially organized, you literally lay out the components as they are seen, as they exist in space. You describe the components as they are seen either from left to right, from right to left, from top to bottom, from bottom to top, from inside to outside, or from outside to inside.

DEVELOPMENT

To describe your topic clearly and accurately, detail the following:

Weight	Materials (composition)
Size (dimensions)	Identifying numbers

Color	Make/model
Shape	Texture
Density	Capacity

WORD USAGE

Your word usage, either photographic or impressionistic, depends on your purpose. For factual, objective technical descriptions, use photographic words. For subjective, sales-oriented descriptions, use impressionistic words. Photographic words are denotative, quantifiable, and specific. Impressionistic words are vague and connotative. Table 11.1 shows the difference.

Another way to note the difference between impressionistic words and photographic detail is through the ladder of abstraction, shown in Figure 11.4.

PROCESS

Now that you know what should be included in a technical description, your next question is, "How do I go about writing this type of document?" As always, effective writing follows a systematic, step-by-step process. To write your technical description, (a) *prewrite* to gather data and determine objectives; (b) *write* a draft of the description; and (c) *rewrite* by revising the draft, thereby making it as perfect as possible.

PREWRITING

We've discussed reporter's questions and clustering/mind mapping as techniques for gathering data and determining objectives. Either of these two prewriting tools can be used for your technical description. However, another prewriting technique can be helpful: *brainstorming/listing*.

Brainstorming/listing, either individually or as a group activity, is a good, quick way to sketch out ideas for your description. To brainstorm/list, do the following:

1. *Title your activity*, at the top of your page, to help you maintain your focus (for example, *Description: XYZ model 2267 Light Pen*).

TABLE 11.1 Photographic versus Impressionistic Word Usage

PHOTOGRAPHIC	IMPRESSIONISTIC
6′9″	tall
350 lb	fat
gold	precious metal
6,000 shares of United Can	major holdings
700 lumens	bright
.030 mm	thin
1966 XKE Jaguar	impressive car

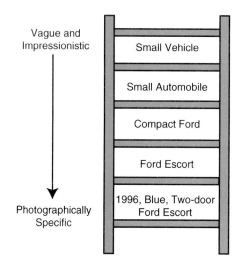

Vague and
Impressionistic

Small Vehicle

Small Automobile

Compact Ford

Ford Escort

1996, Blue, Two-door
Ford Escort

Photographically
Specific

FIGURE 11.4 Ladder of Abstraction

2. *List any and all ideas* you might have on the topic. Don't editorialize. Avoid criticizing ideas at this point. Just randomly jot down as many aspects of the topic as possible, without making value judgments.

3. *Edit your list.* Reread the list and evaluate it. To do so, (a) select the most promising features, (b) cross out any that don't fit, and (c) add any obvious omissions.

WRITING

Once you've gathered your data and determined your objectives, the next step is writing your description.

1. *Review your prewriting.* After a brief gestation period (an hour, a half-day, or a day) in which you wait to acquire a more objective view of your prewriting, go back to your brainstorming/listing and reread it. Have you omitted all unneeded items and added any important omissions? If not, do so.

2. *Organize your list.* Make sure the items in the list are organized effectively. To communicate information to your readers, organize the data so readers can follow your train of thought. This is especially important in technical descriptions. You not only want your readers to understand the information, but also to visualize the mechanism or component. This visualization can be achieved by organizing your data spatially.

3. *Title your draft.* At the top of the page, write the name of your piece of equipment, tool, or mechanism.

4. *Write a focus statement.* In one sentence, write (a) the name of your topic (plus any identifying numbers), (b) the possible functions of your topic or your reason for writing the description, and (c) the number of parts composing your topic. This will eventually act as the introduction.

For now, however, it can help you organize your draft and maintain your focus.

5. *Draft the text.* Use the sufficing technique discussed in earlier chapters. Write quickly without too much concern for grammatical or textual accuracy, and let what you write suffice for now. Just get the information on the page, focusing on overall organization and a few highlighting techniques (headings, perhaps). The time to edit is later.

6. *Sketch a rough graphic of your equipment, component, or mechanism.*

REWRITING

Rewriting is the most important part of the process. This is the stage during which you fine-tune and polish your technical description, making it as perfect as it can be. A perfected description ensures your credibility. An imperfect description makes you and your company look bad.

Revise your draft as follows:

1. *Add any detail required to communicate more effectively.* This can involve brand names, model numbers, and/or specificity of detail (size, material, density, dimensions, color, weight, etc.).

2. *Delete dead words and phrases for conciseness.* (Refer to our discussion of effective technical writing style in Chapter 3.).

3. *Simplify long-winded, old-fashioned words and phrases for easier understanding.* Chapter 3 will help you with this important aspect of your writing.

4. *Move (cut and paste) information to ensure spatial organization.* You don't want to give your reader a distorted view of your mechanism. To avoid doing so, the technical description must maintain a spatial order—left to right, right to left, and so on. For example, let's say you are describing a pencil, graphically depicted horizontally so the eraser end is to the left and the graphite tip is to the right. This pencil consists of the eraser, metal ferrule, wooden body, and sharpened point (that's the correct spatial order). If you describe the eraser first, then the tip, then the body, and then the ferrule, you've distorted the spatial order.

5. *Reformat your text for ease of access through highlighting techniques— headings, boldface type, font size and style, itemization, and so forth.* Rewriting is also the time to perfect your graphics. When revising your description, make sure that your graphics are effective. To do so,

   • Add a figure number and title. Place these beneath the graphic (for example, "Figure 11.5. Exhalation Valve with Labeled Call-Outs").

   • Draw or reproduce graphics neatly.

   • Place the graphics in an appropriate location, near where you first mention them in the text.

   • Make sure that the graphics are an effective size. You don't want the graphics to be so small that your reader must squint to read them, nor do you want them to be so large that they overwhelm the text.

• Label the components. Use call-outs to name each part, as in Figure 11.5.

6. *Enhance the style of your text.* Make sure that you've avoided impressionistic words. Instead, use photographic words, which are precise and specific. In addition, strive for a personalized tone. It is difficult to incorporate pronoun usage in technical descriptions. However, even when writing about a piece of equipment, remember that "people write to people." Although you're describing an inanimate object, you're writing to another human being. Therefore, use pronouns to personalize the text. (Examples later in this chapter clarify how you can accomplish this.)

7. *Correct your draft.* Proofread to correct grammar errors as well as errors in content: measurements, shapes, textures, materials, colors, and so on.

8. *Avoid sexist language.*

The following checklist will help you in writing technical descriptions.

TECHNICAL DESCRIPTION CHECKLIST

✔ Does the technical description have a title noting your topic's name (and any identifying numbers)?

✔ Does the technical description's introduction (a) state the topic, (b) mention its functions or the purpose of the mechanism, and (c) list the components?

✔ Does the technical description's discussion use headings to itemize the components for reader-friendly ease of access?

✔ Is the detail within the technical description's discussion photographically precise? That is, does the discussion portion of the description specify the following?

Colors	Capacities
Sizes	Textures
Materials	Identifying numbers
Shapes	Weight
Density	Make/model

✔ Are all of the calculations and measurements correct?

✔ Do you sum up your discussion using any of the optional conclusions discussed in this chapter?

✔ Does your technical description provide graphics which are correctly labeled, appropriately placed, neatly drawn or reproduced, and appropriately sized?

✔ Do you write using an effective technical style (low fog index) and a personalized tone?

✔ Have you avoided sexist language and multicultural biases?

✔ Have you avoided grammatical and mechanical errors?

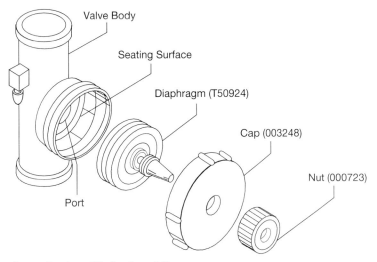

Valve Body

Seating Surface

Diaphragm (T50924)

Cap (003248)

Nut (000723)

Port

Source: Courtesy of Puritan Bennett Corp.

FIGURE 11.5 Exhalation Valve with Labeled Call-Outs

PROCESS LOG

The following student-written process log clarifies the way process is used in writing technical descriptions.

PREWRITING

We asked students to describe a piece of equipment, tool, or mechanism of their choice. One student chose to describe a cash register pole display. To do so, he first had to gather data using listing/brainstorming. He provided the following list:

1. Three parts
2. Pole printed circuit board (PCB)
3. Case assembly
4. Filter
5. Plastic materials

Next, we asked students to go back to their initial lists and edit them. Since this student's list was so sketchy, he didn't need to omit any information. Instead, he had to add omissions and missing detail, as in Figure 11.6.

WRITING

After the prewriting activity, we asked students to draft a technical description. Focusing on overall organization, highlighting, detail, and a sketchy graphic, the aforementioned student wrote a rough draft (Figure 11.7).

Pole PCB

- length—15 mm
- width—5.1mm
- tube length—10.8 mm
- face plate width—2.3 mm
- thick—1.7 mm
- PCB thickness—0.2 mm
- 10" stranded wire with female connectors
- fiberglass and copper construction

Pole Case Assembly

- long—15.5 mm
- bottom width—2.5 mm
- top width—0.9 mm
- mounting pole—5 mm high × 3.2 mm diameter
- tounge for mounting—3.1 mm
- lower mounting tounge—1.5 mm
- side mounting tounge—0.8 mm high
- high—6.1 mm
- almond-colored plastic

Filter

- long—15.6 mm
- high—6.2 mm
- thick—0.7 mm
- plastic, blue

FIGURE 11.6 Student's Listing/Brainstorming

REWRITING

Once students completed their rough drafts, we used peer review groups to help each student revise his or her paper. In the following revision, students (a) added new detail for clarity, (b) corrected any errors, and (c) perfected graphics.

The preceding rough draft, complete with student comments, is shown in Figure 11.8.

The student incorporated these suggestions and prepared the finished copy (Figure 11.9).

The QL 169 Customer Pole Display provides the viewing of all transaction data for the customer. The display consist of a printed circuit board, a case assembly, and a filter display.

Display Circuit Board

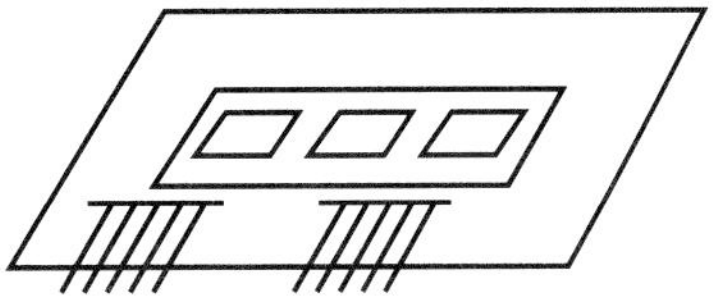

- length—15 mm
- width—5.1 mm
- tube length—10.8 mm
- face plate width—2.3 mm
- thick—1.7 mm
- PCB thickness—0.2 mm
- 10" stranded wire with female connectors
- fiberglass and copper construction

Display Case Assembly

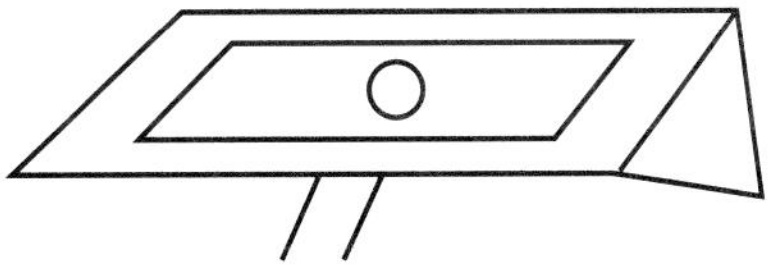

- long—15.5 mm
- bottom width—2.5 mm
- top width—0.9 mm
- mounting pole—
  5 mm high x 3.2 mm diameter
- tounge for mounting—3.1 mm
- lower mounting tounge—1.5 mm wide
- side mounting tounge—0.8 mm high
- high—6.1 mm
- almond-colored plastic

Display Filter

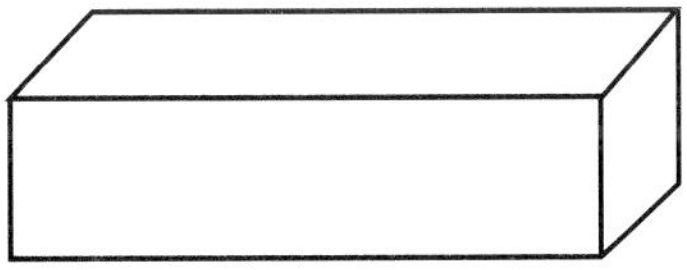

- long—15.6 mm
- high—6.2 mm
- thick—0.7 mm
- plastic, blue

FIGURE 11.7 Student's Rough Draft

No Title   —— *awkward* ——

The QL 169 Customer Pole Display provides the viewing of all transaction data for the customer. The display consist of a printed circuit board, a case assembly, and a filter display.   *SP*

Display Circuit Board   *Make All Caps*
- length—15 mm
- width—5.1 mm   *Number your*
- tube length—10.8 mm   *components*
- face plate width—2.3 mm
- thick—1.7 mm   *thickness*
- PCB thickness—0.2 mm
- 10" stranded wire with female connectors
- fiberglass and copper construction

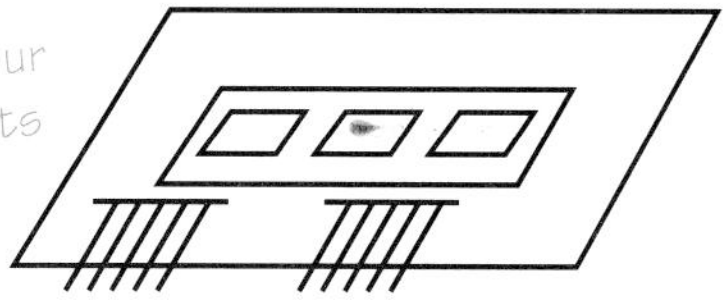

Too vague (specify)

Display Case Assembly   *Make All Caps*   *Number components*
- long—15.5 mm   *length*
- bottom width—2.5 mm
- top width—0.9 mm
- mounting pole—
  5 mm high x 3.2 mm diameter
- tounge for mounting—3.1 mm
- lower mounting tounge—1.5 mm wide   *SP*
- side mounting tounge—0.8 mm high
- high—6.1 mm   *height*
- almond-colored plastic

vague

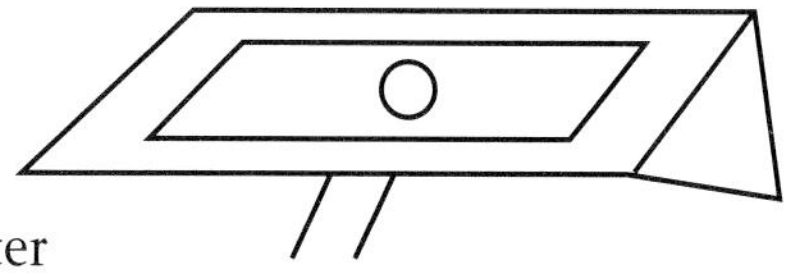

Display Filter
- long—15.6 mm   *length*
- high—6.2 mm   *height*
- thick—0.7 mm   *thickness*
- plastic, blue

what

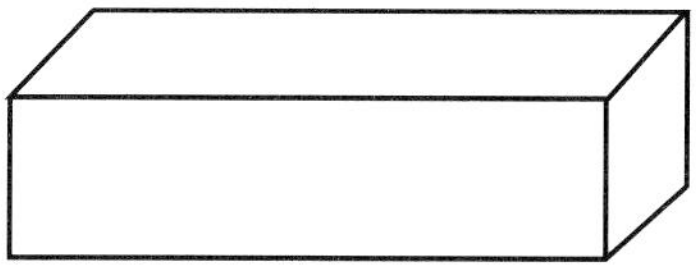

add a conclusion

FIGURE 11.8 Student's Suggested Revisions

QL169 Pole Display

The QL169 pole display provides an alphanumeric display for customer viewing of cash register sales. The display consists of a printed circuit board (PCB) assembly, a case assembly, and a display filter.

Item Description

1. PCB ASSEMBLY
2. CASE ASSEMBLY
3. DISPLAY FILTER

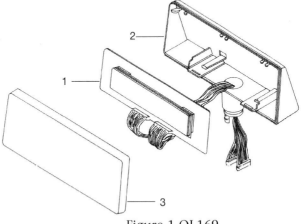

Figure 1 QL169

1.0 PCB Assembly

The printed circuit board, containing the display's electrical circuitry, is constructed of fiberglass with copper etchings.

1.1 Length—15 mm
1.2 Width—5.1 mm
1.3 Tube length—10.8 mm
1.4 Tube faceplate
     width—2.3 mm
1.5 Tube total width—2.8 mm
1.6 Tube thickness—1.7 mm
1.7 PCB thickness—0.2 mm
1.8 20 conductor 10" 22 gauge
     stranded wire with 2 AECC female connectors
     (AECC part #7214-001)
1.9 American Display Company blue phosphor display tube
     (ADC part #1172177)

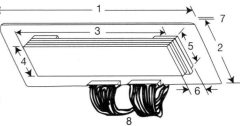

Figure 2 PCB

2.0 Case Assembly

Almond-colored ABS plastic, used to construct the case assembly, protects the PCB.

2.1 Length—15.5 mm
2.2 Bottom width—2.5 mm
2.3 Top width—0.9 mm
2.4 Mounting pole—5 mm high and 3.2 mm in diameter
2.5 Mounting tongue inside width—3.1 mm from side of assembly
2.6 Lower mounting tongue— 1.5 mm wide
2.7 Side mounting tongue— 0.8 mm high
2.8 Tongue thickness—0.2 mm
2.9 Height—6.1 mm

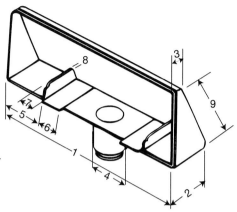

Figure 3 Case Assembly

3.0 Display Filter

3.1 Length—15.6 mm
3.2 Height—6.2 mm
3.3 Thickness—0.7 mm

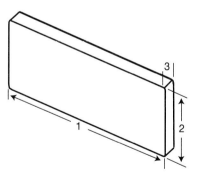

Figure 4 Display Filter

Transparent blue plastic, used to construct the display filter, allows the customer to view the readings. The QL169 display provides easy viewing of clerk transactions and ensures cashier accuracy.

FIGURE 11.9 Student's Finished Copy (Cont.)

SAMPLE TECHNICAL DESCRIPTION

On page 274 is a professionally written technical description merging graphic and text (Figure 11.10). This is a highly effective technical description with these characteristics:

- A title
- An excellent introduction that lists topic, function, and components
- An effectively drawn and placed graphic
- A reader-friendly discussion with headings exactly corresponding to the call-outs in the drawing
- A precisely detailed discussion focusing on materials, color, dimensions, and so on
- An effective conclusion

CHAPTER HIGHLIGHTS

1. Technical descriptions are often used in user manuals, product specifications, reports, and sales literature.
2. The introduction of a technical description explains the mechanism's function and major components.
3. Graphics make a technical description useful to the reader.
4. Follow a logical pattern of organization in a technical description.
5. Carefully consider where you place text and visual aids.
6. Word usage can be either photographic or impressionistic.

ACTIVITIES

1. Write a technical description, either individually or as a team. To do so, first select a topic. You can describe any tool, mechanism, or piece of equipment. However, don't choose a topic too large to describe accurately. To provide a thorough and precise description, you'll need to be exact and minutely detailed. A large topic, such as a computer, an oscilloscope, a respirator, or a Boeing airliner, would be too demanding for a two to four-page description. On the other hand, don't choose a topic that is too small, such as a paper clip, a nail, or a shoestring. Choose a topic that provides you with a challenge but that is manageable. You might write about any of the following topics:

Hammer	Computer disk	VCR remote control
Wrench	Computer mouse	X-acto knife
Screwdriver	Light bulb	Ballpoint pen
Pliers	Calculator	Pencil
Wall outlet	Automobile tire	Credit card

Electronic A1 Feed Switch

The Electronic A1 Feed Switch provides an automatic solid-state closure
(normally open type) after sensing the lack of presence of food. Figure 1
shows the three principal parts of the feed switch.

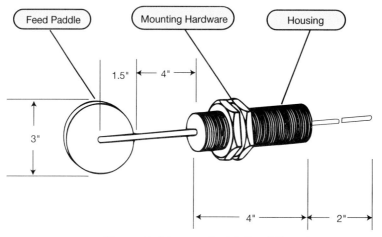

Figure 1. Electronic A1 Feed Switch

- HOUSING: The housing consists of a blue molded polycarbonate cylin-
  der, 4" in length and 1" in diameter. Encircling the complete length of
  the housing are 1"–18 threads. At the middle of one end of the housing
  is a 2" Belden 6906 cable. Attached to the middle of the other end of
  the housing is a feed paddle.
- MOUNTING HARDWARE: The mounting hardware consists of two
  white, flat, hexagonal, plastic Delryn locking nuts, Grade 2, 1" ID, 1-
  1/4" OD, 1/8" thick, and 1"–18 EF threads per inch.
- FEED PADDLE: The feed paddle consists of a 3" diameter, 16 Gauge
  (.060") No. 304 galvanized steel, Finish 2D or better, welded to a No.
  101 Finish 3D or better galvanized steel shaft 5 1/2" in length and
  1/4" in diameter. Exposed threads 1/4"–20 and 1/2" in length pro-
  trude from the free end of the shaft. Welded on three sides to the
  middle of the 3" diameter steel is the 1-1/2" of the unthreaded shaft
  end. The feed paddle is white Teflon coated to a thickness of 1/64".

The Electronic A1 Feed Switch, with its five-year warranty, is reliable in
dusty and dirty environments. The easy-to-install, short-circuit-protected
switch is internally protected for transient voltage peaks to a maximum
of 5 KV, for up to 10 msec duration. Our feed switch can detect dry con-
centrate, meal, rolled barley, and fine and coarse feeds, as well as special
feeds such as soybean meal and high-moisture corn.

FIGURE 11.10 Professionally Written Technical Description

Once you or your team has chosen a topic, prewrite (listing the topic's components), write a draft (abiding by the criteria provided in this chapter), and rewrite (revising your draft).

2. Find examples of professionally written technical descriptions in technical books and textbooks, professional magazines and journals, or on the Internet. Bring these examples to class and discuss whether they are successful according to the criteria presented in this chapter. If the descriptions are good, specify how and why. If they are flawed, state where and suggest ways to improve them. You might even want to rewrite the flawed descriptions.

3. Select a simple topic for description, such as a pencil, coffee cup, toothbrush, or textbook (you can use brainstorming/listing to come up with additional topics). Describe this item without mentioning what it is or providing any graphics. Then, read your description to a group of students/peers and ask them to draw what you've described. If your verbal description is good, their drawings will resemble your topic. If their drawings are off base, you haven't succeeded in providing an effective description. This is a good test of your writing abilities.

12 Instructions and User's Manuals

OBJECTIVES

Instructions and user's manuals are important in technical writing. Every manufactured product comes complete with either a short (one- to four-page) instruction booklet or with a more detailed user's manual (user's manuals can be anywhere from four pages to a book-length text).

You can receive short instructions for baking brownies; making pancakes; putting together children's toys; or changing a tire, the oil in your car, or the coolant in your engine. Longer user's manuals help people set up stereo systems, construct electronic equipment, maintain computers, and operate fighter planes.

There are even instructions for making chicken noodle soup. One major soup manufacturer recently stopped printing instructions for soup preparation on its cans of soup. Executives assumed that such information was unnecessary. They were mistaken. The company received thousands of calls from consumers asking how to prepare the soup. Responding to befuddled customers, the company reverted to printing these instructions. "Pour soup in pan. Slowly stir in one can of water. Heat to simmer, stirring occasionally."

Even pop-top cans of soda include instructions: "1. Lift up. 2. Pull back. 3. Push down." When such simple items as soup and soda need instructions, you can imagine how necessary instructions are for complex products and procedures. Include instructions or user's manuals whenever your audience needs to know how to

Operate a mechanism	Restore a product
Install equipment	Correct a problem
Manufacture a product	Service equipment
Package a product	Troubleshoot a system
Unpack equipment	Care for a plant
Test components	Use software
Maintain equipment	Set up a product
Clean a product	Implement a procedure
Monitor a system	Construct anything
Repair a system	Assemble a product

Whatever your objectives are, this chapter will help you write your short instructions or help you construct a longer user's manual.

CRITERIA FOR WRITING INSTRUCTIONS

Follow these criteria for writing an effective instruction.

TITLE

Preface your text with a title that explains two things: *what* you're writing about (the name of your product or service) and *why* you're writing (the purpose of the instruction). For example, to merely title your instruction "Overhead Projector" would be uninformative. This title names the product, but it doesn't explain why the instruction is being written. Will the text discuss maintenance, setup, packing, or operating instructions? A better title would be "Operating Instructions for the XYZ Overhead Projector."

ORGANIZATION

As the writer of the instruction, you know what your objectives are. You know why you're writing and what you want to accomplish. However, your readers, upon picking up the instruction, are unaware of your intentions. Whereas you know where you're going, they must patiently await further information regarding the destination. To help your readers follow your train of thought, provide them with a road map. One way to do this is through organization. A neatly organized instruction can orient your readers, thereby helping them follow your directions. To organize your instruction effectively, include an *introduction*, a *discussion*, and a *conclusion*.

Introduction

In the first paragraph of your instruction, tell your readers three things: what *topic* you'll be discussing, your *reasons* for writing, and the *number of steps* involved in the instruction. The topic names the product or service. Your reason for writing either explains the purpose of the instruction ("maintaining the machine will increase its longevity") or comments about the product's capabilities or ease of use. Stating the number of steps allows your readers to plan ahead and organize their thoughts and their time. Look at the following examples:

GOOD
The following seven steps will help you operate the Udell PQ 4454 Overhead Projector.

BETTER
The Udell PQ 4454 is an overhead projector for business and school. It's easy to use and requires little maintenance. Follow these eight steps for operating the machine.

BEST
The Udell PQ 4454 Overhead Projector is used to project written or graphic material onto a screen or wall. Because of its capability to enlarge and project, it is ideal for use in schools or businesses. Six simple steps will help you operate the projector.

The first of the preceding examples includes product plus steps; the second adds ease of use. The third example, however, is best because it mentions product, steps, ease of use, plus capabilities.

In addition to this introductory overview focusing on the topic and steps to be performed, you might want to tell your readers what tools or equipment they'll need to perform the procedures. You can provide this information through a simple list, or perhaps you'll want to add graphics (the type of tool needed plus a picture of that tool). This depends on whether your readers are high tech, low tech, or lay.

EXAMPLE 1
List of Required Tools
Tools Required
1—2 mm hex wrench
1—pliers

EXAMPLE 2
List of Required Tools
    plus Graphic Depiction
Tools Required
1—2 mm hex wrench

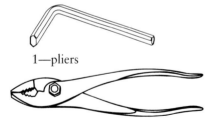

1—pliers

Finally, decide whether you should preface your instructions with dangers, warnings, cautions, or notes. This is an essential consideration to avoid costly lawsuits and to avoid potentially harming an individual or damaging equipment.

When including hazard alert messages, you need to consider the following:

- *Placement.* Where should the caution or warning notice go? Do you place the information before or after the appropriate text? If you place the information after the text, *after the reader has already performed the task,* the damage is done. Your readers already will have hurt themselves or harmed the equipment. The hazard alert is placed before the step. In fact, you might want to state your warnings or cautions in the introduction, before the audience even begins to perform a task.

- *Access.* If your hazard alert is printed before the appropriate step, but you haven't used any highlighting techniques to make it obvious, your readers won't see the danger. You need to make the warning or caution notice obvious. To do so, vary your typeface and type size, use white space to separate the warning or caution from surrounding text, and/or box the warning or caution.

- *Definitions.* What does *caution* mean? How does it differ from *warning, danger,* or *note?* Unfortunately, these terms are not defined consistently.

  Three primary institutions that seek to provide a standardized definition of terms are ANSI (American National Standards Institute), the U.S. military (MILSPEC), and OSHA (Occupational Safety and Health Administration). However, these institutions often define terms differently. For example, both ANSI and MILSPEC define *note* similarly as essential operating information. OSHA defines *notice* as general instructions relative to safety measures. ANSI defines *caution* as a hazardous situation that might cause minor or moderate injury, whereas MILSPEC defines *caution* as something that could result in damage or destruction to equipment. OSHA uses *caution* to designate potential hazards regarding unsafe practices. Personal injury, damage to equipment, and hazardous unsafe practices should be distinguished more clearly. Both ANSI and MILSPEC define both *warning* and *danger* as an activity possessing the potential for serious injury or death. OSHA defines *danger* as immediate danger. The words overlap and don't distinguish between death and serious harm.

  To avoid confusion, we suggest the following hierarchy of definitions, which clarifies the degree of hazard:

  1. *Note.* Important information, necessary to perform a task effectively or to avoid loss of data or inconvenience.

  2. *Caution.* The potential for damage or destruction of equipment.

  3. *Warning.* The potential for serious personal injury.

  4. *Danger.* The potential for death.

  Our goal is to avoid confusion, which can occur if one writes, "Cautionary Warning Notice!"

- *Colors.* Another way to emphasize your hazard message is through a colored window (box) around the word. Usually, *Note* is printed in blue or black, *Caution* in yellow, *Warning* in orange, and *Danger* in red.
- *Text.* To further clarify your terminology, provide the readers text to accompany your hazard alert. Your text should have the following three parts:
  1. *A one- or two-word identification alerting the reader. High Voltage, Hot Equipment, Sharp Objects*, or *Magnetic Parts*, for example, will warn your reader of potential dangers, warnings, or cautions.
  2. *The consequences of the hazards, in three to five words.* "Electrocution can kill," "Can cause burns," "Cuts can occur," or "Can lead to data loss," for example, will tell your readers the results stemming from the dangers, warnings, or cautions.
  3. *Avoidance steps.* In three to five words, tell the readers how to avoid the consequences noted above: "Wear rubber shoes," "Don't touch until cool," "Wear protective gloves," or "Keep disks away."
- *Icons.* Equipment is manufactured and sold globally; people speak different languages. Your hazard alert should contain an *icon*—a picture of the potential consequence—to help everyone understand the caution, warning, or danger. (Icons are discussed in Chapter 9.)

Figure 12.1 shows an effective page layout and the necessary information to communicate hazard alerts.

Discussion

Itemize and thoroughly discuss the steps in your instruction. Organize them *chronologically*—as a step-by-step sequence. Obviously, you cannot tell your readers to do step 6, then go back to step 2, then accomplish step 12, then do step 4. Such a distorted sequence would fail to accomplish your goals. To operate machinery, monitor a system, or construct equipment, your readers must follow a chronological sequence. Be sure that your instruction is chronologically accurate.

Conclusion

As with a technical description, you can conclude your instruction in various ways. You can end your instruction with a (a) comment about warranties, (b) sales pitch highlighting the product's ease of use, (c) reiteration of the product's applications, or (d) summary of the company's credentials. (For examples, see the discussion on conclusions in Chapter 11.)

Another valid way to conclude, however, is to focus on *disclaimers*, as in the following example:

> This operating guide to the Udell PQ 4454 Overhead Projector is included primarily for ease of customer use. Service and application for anything other than normal use or replacement parts must be performed by a trained and qualified technician.

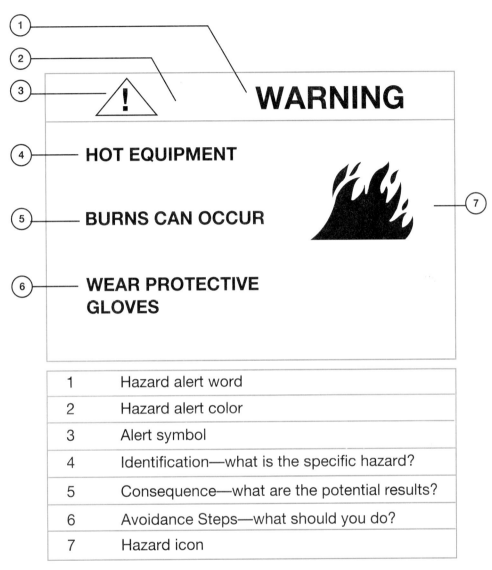

1	Hazard alert word
2	Hazard alert color
3	Alert symbol
4	Identification—what is the specific hazard?
5	Consequence—what are the potential results?
6	Avoidance Steps—what should you do?
7	Hazard icon

FIGURE 12.1 Hazard Alert

AUDIENCE RECOGNITION

How many times have you read an instruction and been left totally confused? You were told to "place the belt on the motor pulley," but you didn't know how. Or you were told to "discard the used liquid in a safe container," but you didn't know what was safe for this specific type of liquid. You were told to "size the cutting according to regular use," but you had never regularly performed this activity.

Here's a typical instruction written without considering the audience:

To overhaul the manual starter, proceed as follows: Remove the engine's top cover. Untie starter rope at anchor and allow starter rope to slowly wind onto pulley. Tie a knot on the end of starter rope to prevent it from being pulled into housing. Remove pivot bolt and lift manual starter assembly from power head.

Although many high-tech readers might be able to follow these instructions, many more readers will be confused. How do you remove the engine's top cover? Where's the anchor? Where's the pivot bolt and how do you remove it? What's the power head?

The problem is caused by writers who assume that their readers have high-tech knowledge. This is a mistake for several reasons. First, even high-tech readers often need detailed information because technology changes daily. You can't assume that every high-tech reader is up to date on these technical changes. Thus, you must clarify. Second, low-tech and lay readers—and that's most of us—carefully read each and every step, desperate for clear and thorough assistance.

As writer, you should provide your readers with the clarity and thoroughness they require. To do this, recognize accurately who your readers are and give them what they want, whether that amounts to technical updates for high-tech readers or precise, even simple information for low-tech or lay readers. The key to success as a writer of instructions is, "Don't assume anything. Spell it all out—clearly and thoroughly!"

GRAPHICS

As with technical descriptions, clarify your points graphically using the super comic book look. Use drawings and/or photographs that are big, simple, clear, keyed to the text, and labeled accurately. Not only do these graphics make your instructions more visually appealing, but also they help your readers and you. What the reader has difficulty understanding, or you have difficulty writing clearly, your graphic can help explain pictorially.

A company's use of graphics often depends on audience. With a high-tech reader, the graphic might not be needed. However, with a low-tech reader, the graphic is used to help clarify. Look at the following examples of instructions for the same procedure. Figure 12.2 uses text with the super comic book look; Table 12.1 uses text without graphics.

Note that Figure 12.2 uses large, bold graphics to supplement the text, clearly depicting what action must be taken. This is appropriate for a low-tech end user. On the other hand, Table 12.1, written for a manufacturer's technician (high-tech), assumes greater knowledge and, therefore, deletes the graphic.

STYLE

1. Number your steps. Don't use bullets or the alphabet. Numbers, which you can never run out of, help your readers refer to the correct step. In contrast, if you used bullets, your readers would have to count to locate steps—seven bullets for step 7, and so on. If you used the alphabet, you'd be in trouble when you reached step 27.

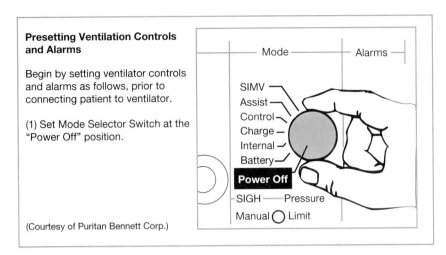

Presetting Ventilation Controls and Alarms

Begin by setting ventilator controls and alarms as follows, prior to connecting patient to ventilator.

(1) Set Mode Selector Switch at the "Power Off" position.

(Courtesy of Puritan Bennett Corp.)

FIGURE 12.2 Text with Super Comic Book Graphics

TABLE 12.1	Text Without Graphics for High-Tech Readers	
LOCATION ITEM	**ACTION**	**REMARKS**
	Note For additional operating instructions, refer to Companion 2800 operator's manual.	
1. MODE switch	Set to POWER OFF.	
2. Rear panel	Ensure that instructions on all labels are observed.	
3. Patient tubing circuit	Assemble and connect to unit.	See Table 2, step 3.
4. Power cord	Connect to 120/220 V ac grounded outlet.	If power source is external battery, connect external battery cable to unit per Table 2, step 4.
5. Cascade I humidifier	Connect to unit.	See Table 2, step 2.

(Courtesy of Puritan Bennett Corp.)

2. *Use highlighting techniques.* Boldface, different font sizes and styles, emphatic warning words, color, italicization, and so on call attention to special concerns. A danger, caution, warning, or specially required technique must be evident to your reader. If this special concern is buried in a block of unappealing text, it will not be read. This could be dangerous to your reader or costly to you and your company. To avoid lawsuits and/or to help your readers see what is important, call it out through formatting.

3. *Limit the information within each step.* Don't overload your reader by writing lengthy steps:

OVERLOADED STEPS

1. Start the engine and run it to idling speed while opening the radiator cap and inserting the measuring gauge until the red ball within the glass tube floats either to the acceptable green range or to the dangerous red line.

Instead, separate the distinct steps:

SEPARATED STEPS

1. Start the engine and run it to idling speed.
2. Open the radiator cap and insert the measuring gauge.
3. Note whether the red ball within the glass tube floats to the acceptable green range or up to the dangerous red line.

4. *Develop your points thoroughly.* Don't say, "After rotating the discs correctly, grease each with an approved lubricant." Instead, clarify what you mean by correct rotating and specify the approved lubricant, as in the following example. Unless you tell your readers what you mean, they won't know.

WELL-DEVELOPED STEPS

1. Rotate the disks clockwise so that the tabs on the outside edges align.
2. Lubricate the discs with 2 oz. of XYZ grease.

5. *Use short words and phrases.* Remember the fog index. (See Chapter 3.)

6. *Begin your steps with verbs—the imperative mood.* Note that each of the numbered steps in the following example begins with a verb:

VERBS BEGIN STEPS

1. *Number* your steps.
2. *Use* highlighting techniques.
3. *Limit* the information within each step.
4. *Develop* your points thoroughly.
5. *Use* short words and phrases.
6. *Begin* your steps with verbs.

7. Personalize your text. Remember, people write to people. Involve your readers in the instruction by using pronouns. (See Chapter 4.)

8. Do not omit articles. Articles—*a, an,* and *the*—are part of our language. Although you might see instructions that omit these articles, please don't do so yourself. Articles do not take up much room in your text, but they do make your sentences read more fluidly.

PROCESS

Now that you know what should be included in an instruction, it's time to write. But how do you start? Where do you begin?

As in other technical writing activities, follow a step-by-step process: *prewriting* to gather data and determine objectives, *writing* to draft your correspondence, and *rewriting* to revise and perfect.

PREWRITING

You've already been introduced to three prewriting techniques: reporter's questions, clustering/mind mapping, and brainstorming/listing. For writing an instruction, reporter's questions would be a good place to start gathering data and determining objectives. Ask yourself:

- *Who* is my reader?
  - —High-tech?
  - —Low-tech?
  - —Lay?
- *What detailed* information is needed for my audience? For example, a low-tech or lay reader will be content with "inflate sufficiently" whereas an instruction geared toward high-tech readers will require "inflate to 25 psi."
- *How* must this instruction be organized?
  - —Chronologically
- *When* should the procedure occur?
  - —Daily, weekly, monthly, quarterly, biannually, or annually?
  - —As needed?
- *Why* should the instruction be carried out?
  - —To repair, maintain, install, operate, and so on. (See the list of reasons for using an instruction at the beginning of this chapter.)

Once you've gathered this information, you can use brainstorming to sketch out a rough list of the steps required in the instruction and the detail needed for clarity. After that, your major responsibility is to sequence your steps chronologically. To accomplish this goal, use a prewriting technique called *flowcharting.*

Flowcharts chronologically trace the stages of an instruction, visually revealing the flow of action, decision, authority, responsibility, input/output, preparation, and termination of process. Flowcharting is not just a graphic way to help you gather data and sequence your instruction, however. It is two-dimensional writing. It provides your reader content as well as a panoramic view of an entire sequence. Rather than just reading an instruction step by step, having little idea what's around the corner, with a flowchart your reader can figuratively stand above the instruction and see where it's going. With a flowchart, readers can anticipate cautions, dangers, and warnings, since they can see at a glance where the steps lead.

To create a flowchart, use the International Standardization Organization (ISO) flowchart symbols shown in Figure 12.3.

Figure 12.4 shows an excellent example of a flowchart, visually depicting an instruction's sequence.

WRITING

Once you've graphically depicted your instruction sequence using a flowchart, the next step is to write a rough draft of the instruction.

1. *Review your prewriting.* In doing so, you will accomplish the following:

   • *Clarify your audience.* Now is a good time to review your reporter's questions and to determine whether your initial assumptions regarding audi-

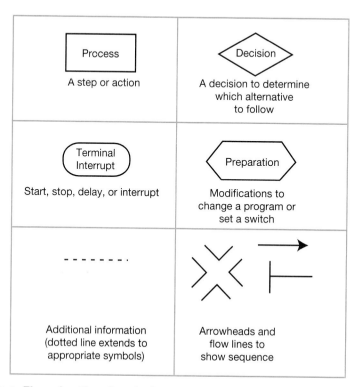

FIGURE 12.3 Flowcharting Symbols

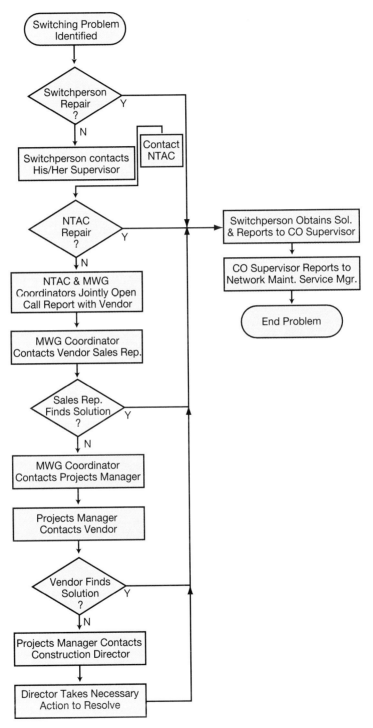

FIGURE 12.4 Flowchart of Handling Service Problems

ence are still correct. If your attitude toward your audience has changed, redefine your readers and their needs.

- *Determine your focus.* In your prewriting, you answered the question *why*. Are you still certain of your objectives for writing the instruction? If so, write a one-sentence thesis or statement of objectives to summarize this purpose. For example, you could write, "This instruction should be followed so that glitches in the system can be discovered and corrected." This thesis or topic sentence will help you maintain your focus throughout the instruction. If you are uncertain about your purpose for writing the instruction, however, rethink the situation. You will not be able to write an effective instruction until your objectives are clearly stated.

- *Organize your steps.* Review your flowchart to determine whether any steps in the instruction have been omitted or misplaced. If the chronological sequence is incorrect, reorganize your detail. If the chronology is correct, you're ready to move on to the next step in drafting.

2. Review the instruction criteria. Before you write the rough draft, remind yourself what the instruction will consist of. You'll need an introduction, a discussion with itemized steps, graphics to help your reader perform the requested procedures, and a conclusion.

3. Write the draft. You've gathered your data and determined your objectives. You've correctly sequenced the steps and reminded yourself of the instruction criteria. At this point, you should feel comfortable with the prospect of writing. Follow the sufficing technique again, roughly drafting your instruction. Don't worry about perfect grammar or graphics. Perfecting the text comes in the next step, rewriting.

REWRITING

Rewriting, the last stage in the process, is the time to perfect your instruction. For manuals, this stage calls for *usability testing*. You've written the instruction, but is it usable? If the instruction doesn't help your audience complete the task, what have you accomplished? To determine the success of your manual, test the instruction.

Peer evaluators are ideal for this purpose. Have a team of readers review your manual. An ideal team consists of readers with different levels of knowledge regarding the topic. A review team consisting of only high-tech peers would skew your findings. In contrast, low-tech and lay peer evaluators would give you more informative feedback. Have your audience assess the manual based on the Usability Testing Features shown in Table 12.2.

Your peer evaluators do not have to suggest ways to fix any problems encountered. Their goal is just to give you feedback. If they say, "I don't understand this word," "Step 3 seems too long," "I could use a graphic here to clarify your meaning," or "More white space would keep me from feeling overwhelmed," you can revise the text accordingly. To do so, reread the instruction, consider the suggestions made during usability testing, and improve the draft by using the eight-step rewriting procedure explained here.

TABLE 12.2 Usability Testing Features		
FEATURE	**REQUIREMENTS**	**COMMENTS**
Audience recognition	• Technical terms are defined. • Examples explain difficult steps at the reader's level of understanding. • Graphics depict correct completion of difficult steps at the reader's level of understanding. • The tone and word usage are appropriate for the intended audience.	
Development	• Steps are precisely developed. • All required information is provided, including hazards and required equipment/tools. • No irrelevant or rarely needed information is included.	

Conciseness (memory load)	• Words, sentences, and paragraphs are kept to a minimum.
	• Readers are not overloaded.
	  1. They do not have to remember important information from step to step.
	  2. Numbered steps are not too long.
	  3. Cross referencing helps the readers refer to information provided elsewhere.	

Consistency	• A consistent hierarchy of headings is used. • Graphics are presented consistently (same location, same use of figure titles, same size, etc.). • Words mean the same thing throughout (technical terms, cautions, warnings, etc.). • The same system of numbering steps is used throughout.	

Ease of use	• Readers can easily find what they want because instructions include
	  1. Table of contents
	  2. Hierarchical headings
	  3. Headers and footers
	  4. Index
	  5. Cross referencing or hypertext links	

Layout	• Graphics depict how to perform steps. • White space makes information accessible. • Color emphasizes hazards or specific aspects of a step. • Numbers and bullets divide steps into manageable chunks. • Boldface and italics emphasize important information.	

1. Add detail for clarity. Don't hope you've said enough to clarify your steps. Tell your readers *exactly* how to do it or why to do it. Part of detail involves warnings and cautions, for example. Make sure that you've added all necessary dangers or concerns. In addition, assess each step. Have you added sufficient detail to clarify your intent? For instance, if you are writing to a lay reader, you don't want to say merely "Insert the guide posts." You must say, "Before inserting the guide post, dig a 2-ft.-deep-by 2-ft.-wide hole. Pour the ready-mixed concrete into the hole. Insert the post into the hole and hold it steady until the concrete starts to set (approximately one minute)."

2. Delete. Omit unnecessary words and phrases to achieve conciseness. In addition, delete extraneous information which a high-tech reader might not find useful. This, however, could be dangerous. What you might see as extraneous, your reader might see as necessary. If you delete any information, do so carefully.

3. Simplify. Don't overload your steps. If you have too much information in any one stage, divide this detail into smaller units, adding new steps. Doing so will help your readers follow your instructions. Simplify your word usage; refer to the fog index.

4. Move information. Make sure that the correct chronological sequence has been followed.

5. Reformat using highlighting techniques. If you include dangers, cautions, warnings, or additional information for clarity, set these off (make them emphatic) through formatting. Use color, windowing, font size and style, emphatic words, graphics, etc.

6. Enhance your text. Use the imperative mood for each step in your instruction. Whenever an action is required, begin with a verb to emphasize the action. In addition, you can enhance your text through personalization. Remember, people write to people. To achieve this person-to-person touch, add pronouns to your instruction.

7. Correct your instruction for accuracy. Proof your text for grammatical correctness as well as for contextual accuracy.

8. Avoid sexist language.

Finally, before you send the instruction to your end user, refer to the Effective Instruction Checklist.

(EFFECTIVE INSTRUCTION CHECKLIST)

✔ *Does the instruction have an effective title?*
- Is the topic mentioned?
- Is the reason for performing the instruction mentioned?

✔ *Does the instruction have an effective introduction?*
- Is the topic mentioned?
- Is the reason for performing the instruction mentioned?

- Is the ease of use or capabilities mentioned?
- Have the number of steps been listed?
- Has a list of required tools or equipment been provided?

✔ *Are hazard alert messages used effectively?*

- Are the hazard alert messages placed correctly?
- Is the correct term used (Danger, Warning, Caution, or Note)?
- Is the correct color used with the appropriate term?
- Does the hazard alert text identify the hazard, provide the consequences, and suggest avoidance steps?
- Is an effective icon used to depict the hazard?

✔ *Is the instruction's discussion effective?*

- Is the instruction organized chronologically?
- Does each step avoid overloading by presenting one clearly defined action?
- Does each step begin with a verb?
- Does the instruction's discussion contain the same number of steps mentioned in the introduction?

✔ *Is the instruction well developed?*

- Does the instruction avoid assuming that readers will understand and, instead, clearly explain each point?

✔ Does the instruction recognize audience effectively?

- Is the instruction written for a high-tech, low-tech, or lay audience?
- Has sexist language been avoided?
- Does the text include language appropriate for a multicultural audience?
- Is the text personalized using pronouns?

✔ *Is the conclusion effective?*

- Have warranties been mentioned?
- Is a sales pitch provided showing ease of use?
- Has the conclusion reiterated a reason for performing these steps?
- Are disclaimers provided?

✔ *Is the instruction's document design effective?*

- Are highlighting techniques effectively used? These include graphics (such as comic book look photographs or line drawings), varied typefaces and type sizes, color, and white space.

✔ *Is correct technical writing style used, abiding by the fog index?*

✔ *Have grammatical and textual errors been avoided?*

PROCESS LOG

Following is an example of an instruction collaboratively written by a group of students. To write this instruction, the students gathered data, determined objectives and sequenced their instruction through prewriting, drafted an instruction in writing, and finally revised their draft in rewriting.

Their assignment was to write an instruction on how to use an overhead projector, an item readily available in most classrooms.

PREWRITING

Figure 12.5 illustrates reporter's questions and students' flowcharting.

WRITING

After completing the prewriting, the students composed a draft, complete with sketchy graphics (Figure 12.6).

REWRITING

The students revised the rough draft, considering all the postwriting techniques discussed in this text. The revision is shown in Figure 12.7. The students then prepared their finished copy (Figure 12.8).

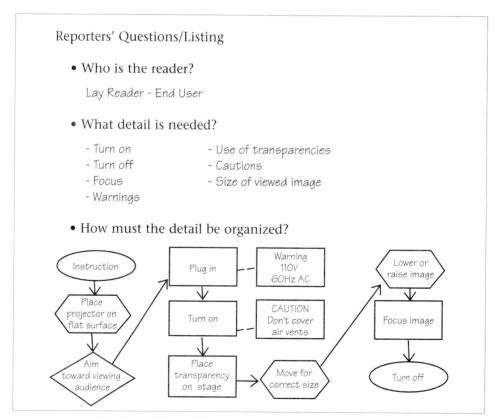

FIGURE 12.5 Students' Prewriting

The Udell Overhead Projector projescts material on a flat surface. Follow these five steps to do so.

1. Place the machine on a flat surface within viewing range of the screen.
2. Plug power plug into a wall outlet and turn the machine on.
3. Put the transparency on the projectors horizontal viewing surface.
4. Move the image up or down for better viewing and focus image by adjusting the focus knob.
5. Turn machine off when completed and umplug.

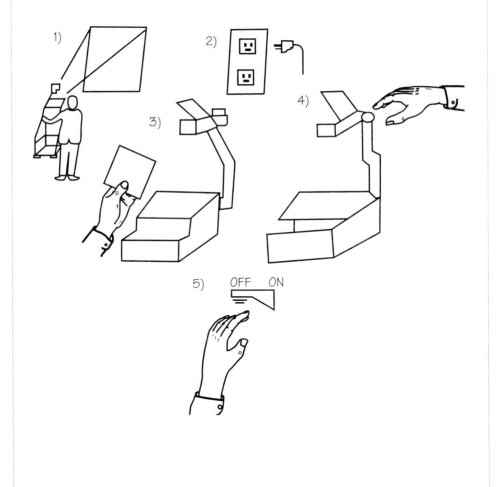

FIGURE 12.6 Students' Rough Draft

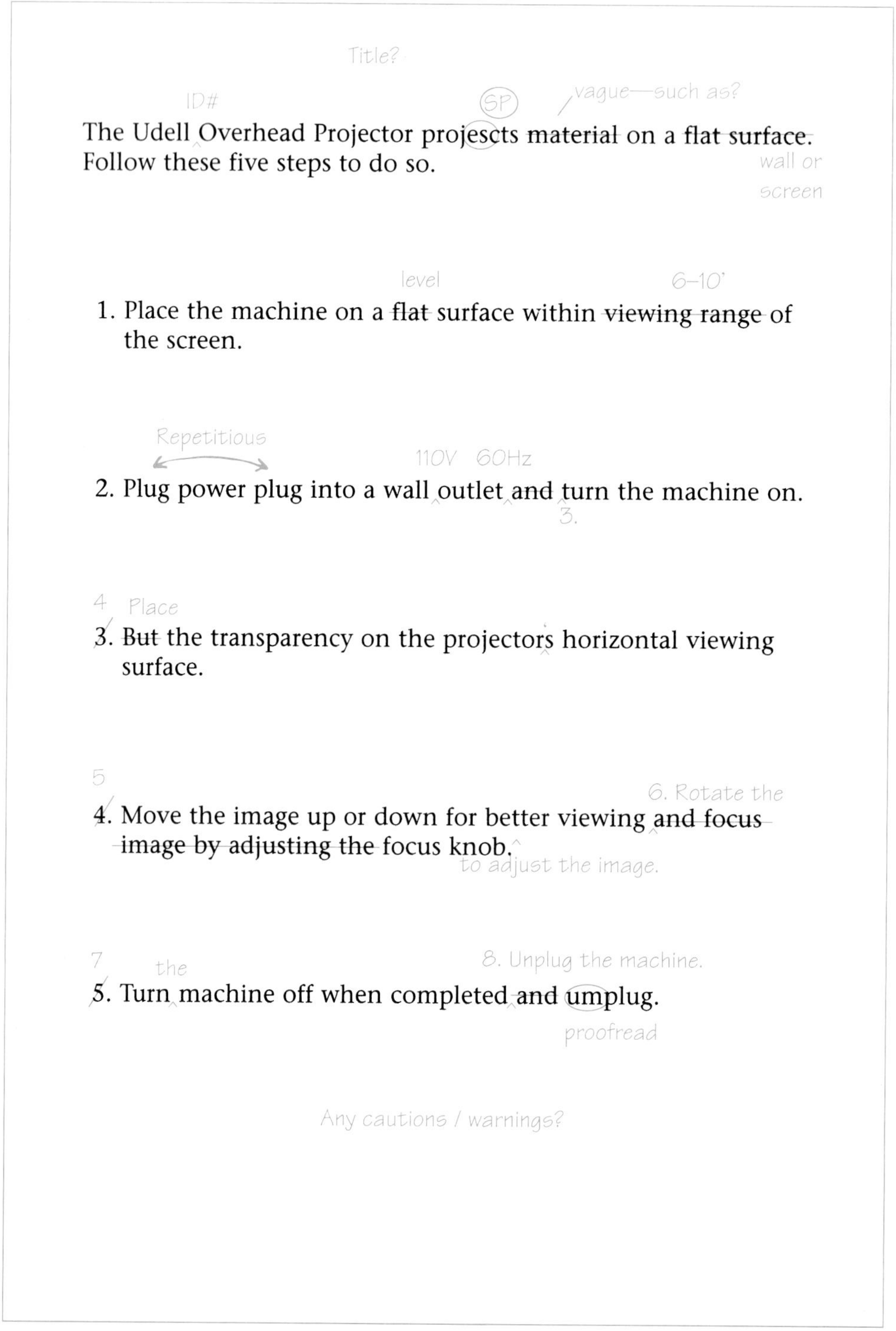

Title?

ID#　　　　　　　　　*(SP)*　*vague—such as?*

The Udell Overhead Projector projescts material on a flat surface.
Follow these five steps to do so.

wall or screen

level　　　　　　　　*6–10'*

1. Place the machine on a flat surface within viewing range of the screen.

Repetitious

110V 60Hz

2. Plug power plug into a wall outlet and turn the machine on.

3.

4. Place

3. But the transparency on the projectors horizontal viewing surface.

5

6. Rotate the

4. Move the image up or down for better viewing and focus image by adjusting the focus knob,

to adjust the image.

7   the　　　　　　　*8. Unplug the machine.*

5. Turn machine off when completed and umplug.

proofread

Any cautions / warnings?

FIGURE 12.7 Students' Suggested Revisions

THE UDELL 4454 OVERHEAD PROJECTOR

The Udell 4454 Overhead Projector is used to project written or graphic material onto a screen or wall. Because of its capability to enlarge and project, the Udell is ideal for use in school and business.

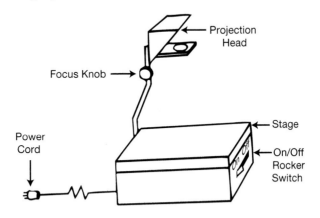

> **WARNING**
> The projector bulb heats up during operation.
> DO NOT TOUCH.

> **CAUTION**
> Keep air vents uncovered to allow for proper cooling.

To operate the Udell, follow these eight simple steps.

1. Place the projector on a level surface, approximately 6–10 feet from a projection screen or blank wall.

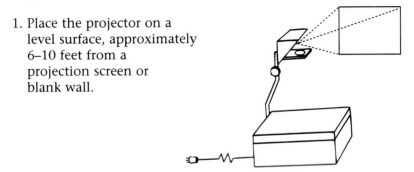

FIGURE 12.8 Students' Finished Copy

2. Plug the power supply cord into a 110 V 60 Hz AC wall outlet.

3. Push the rocker switch to the "on" position.

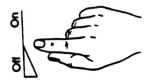

CAUTION
Keep air vents uncovered to allow for proper cooling.

4. Place the material to be viewed squarely on the projector's stage.

5. Adjust the height of the projected image by lowering or raising the projector's lens head.

FIGURE 12.8 Students' Finished Copy (Cont.)

6. Rotate the focus knob for clear viewing of the projected image.

7. Push the rocker switch to the "off" position when you are through viewing your material.

8. Unplug the unit's power cord.

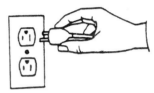

Following these eight simple steps will help you use the Udell overhead projector.

FIGURE 12.8 Students' Finished Copy (Cont.)

SAMPLE INSTRUCTION

Figure 12.9 is an excellent student-written instruction.

INSTRUCTIONS FOR DRY MOUNTING PHOTOGRAPHS

Dry mounting photographs keeps them safe for your family and generations to come. To successfully dry mount your photographs, follow the seven steps we've provided.

Dry mounting requires the following materials:

Materials

1. Press
2. Tacking iron
3. Mat knife
4. Mat board
5. Dry-mount tissue
6. Ruler
7. Print (photographed image)

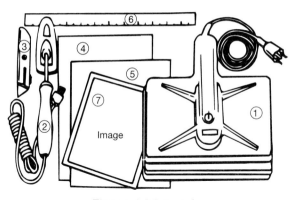

Figure 1 Materials

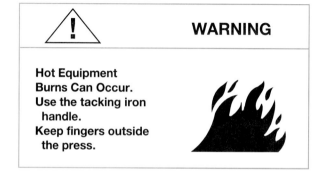

WARNING

**Hot Equipment
Burns Can Occur.
Use the tacking iron
   handle.
Keep fingers outside
   the press.**

FIGURE 12.9 Dry Mounting Instruction

1.0   DRYING THE MATERIAL

1.1   Plug in the press and turn it on. Wait to see the red light.
1.2   Heat the press to 250 degrees by adjusting the dial on the lid.
1.3   Predry the mount board and the print.
1.4   Place them in the press and hold the lid down for 30 seconds.

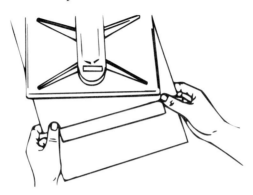

Figure 2

2.0   WIPING THE MATERIALS CLEAN

2.1   After the materials have been baked, let them cool for two to three minutes.
2.2   Wipe away any loose dust or dirt with a clean cloth.

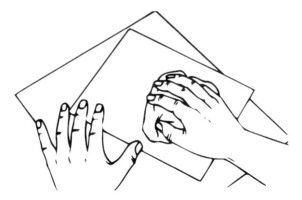

Figure 3

FIGURE 12.9 Dry Mounting Instruction (Cont.)

3.0 TACKING THE MOUNTING TISSUE TO THE PRINT

3.1 Plug in the tacking iron and turn it on.
3.2 With the print face down and the mounting tissue on top, tack the tissue to the center of the print.

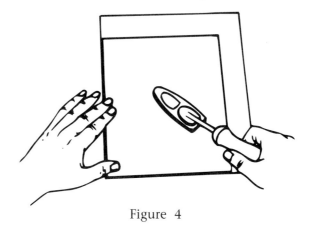

Figure 4

4.0 TRIMMING THE MOUNTING TISSUE

4.1 Place a piece of smooth cardboard under the print so that you won't cut your table.
4.2 Trim off excess mounting tissue using the ruler.

> CAUTION
> Press firmly on the ruler to prevent slipping!
> Don't cut into the image area of the print!

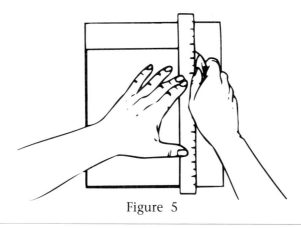

Figure 5

FIGURE 12.9 Dry Mounting Instruction (Cont.)

5.0 ATTACHING THE MOUNTING TISSUE AND PRINT
TO THE BOARD

> CAUTION
> The tissue must lie completely flat on the board
> to prevent wrinkles under the print!

5.1 Position the print and the tissue, face up, on the top of the mount board.
5.2 Slightly raise one corner of the print and touch the tacking iron to the tissue on top of the board.
5.3 Press the corner down lightly, taking care to avoid wrinkles.

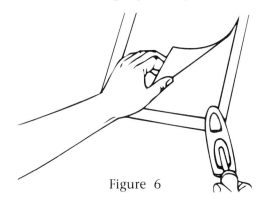

Figure 6

6.0 MOUNTING THE PRINT

6.1 Put the board, tissue, and print (with cover sheet on top) into the press.
6.2 Hold down the press lid firmly for 30 seconds.

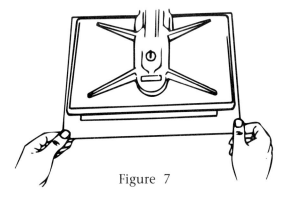

Figure 7

FIGURE 12.9 Dry Mounting Instruction (Cont.)

7.0   TRIMMING THE MOUNTED PRINT

7.1   Pressing down firmly on the ruler to avoid slipping, trim the edges of the mounted print with a sharp mat knife.

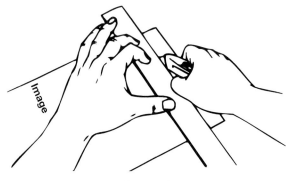

Figure 8

By carefully dry mounting your photographs, they will look beautiful and last longer. Enjoy your artistry.

FIGURE 12.9 Dry Mounting Instruction (Cont.)

CRITERIA FOR WRITING A USER'S MANUAL

Your user's manual might focus on installation, operation, maintenance, or troubleshooting. It might be packed with the product or produced separately to be sold at a later date. The user's manual might be sized to fit in your car's glove compartment, or it might be as large as an encyclopedia. Whatever the size, shape, or purpose of your user's manual, it will include an instruction or several instructions, abiding by the criteria already provided in this chapter.

However, in contrast to a short instruction, user's manuals contain additional information, including *any* or *all* of the following components (the order of these components will vary):

1. Cover page. The cover page accomplishes at least two goals. First, the cover page will name the product or service being discussed *and* explain the purpose of the manual.

EXAMPLE 1
All CALLS Cordless Telephone
   Instruction Manual

EXAMPLE 2
AutoCall Cordless Telephone
   Use and Care Guide

Second, in addition to naming the product or service and stating the purpose of the manual, a cover page might also graphically depict the product or service. This could be accomplished, either with a photograph or a line drawing. Doing so allows the readers to identify the product or service more readily.

2. *Hazard alerts*. Usually, when readers turn the cover page to open the manual, they see hazard alerts or safety guidelines. Companies want to protect their customers from harm, minimize potential damage to equipment, and avoid costly lawsuits. Hazard alerts could be a list of safety tips, such as the following guidelines from a roofing user's manual:

- Please read all instructions.
- Use care with ladders.
  - —Rest ladders on steady ground.
  - —Lean ladders against a wall.
  - —Allow only one person on a ladder.
- Do not touch electric lines.
- Stay off your roof during strong winds.
- Use a scaffolding system for steep roofs.

Better than a list, however, would be a hazard alert like those already discussed in this chapter. A more effective hazard alert would include the appropriate word ("Danger," "Warning," "Caution," or "Note"), color, hazard identification, consequence, avoidance steps, and icon depicting the hazard.

3. *Table of contents*. Your user's manual will have several sections. As already noted, you'll include numerous instructions (setup, installation, maintenance, operation, troubleshooting, for example). In addition, you might include specifications, warranties, guarantees, and customer service contact numbers.

However, readers might want to access only one of these sections. An effective table of contents will allow your readers to find the one or two items they specifically want. (We discuss successful tables of content in Chapter 16, "Proposals.")

4. *Introduction*. Companies need customers. A user's manual might be the only contact a company has with its customer. Therefore, user's manuals often are reader-friendly and seek to achieve audience recognition and audience involvement. The manuals try to reach out and touch the customer.

Look at these introductions from three user's manuals:

EXAMPLE 1
Thank you for purchasing our Drip Coffeemaker. Before your first use,
please take a few minutes to read our Use and Care manual. Then find a safe
place to keep it for future reference.

EXAMPLE 2
Installation and operating procedures are contained in this booklet about your new color TV. The minutes you spend reading this book will contribute to hours of viewing pleasure.

EXAMPLE 3
Welcome to the world of modern electronics. The cordless phone you have purchased is the product of advanced technology. This manual will help you install your new phone easily.

Each of the above introductions uses pronouns ("you," "your," and "our") to personalize the manual. The introductions also use positive words, such as "Welcome," "Thank you," and "pleasure" to achieve a positive customer contact. This is important because companies know that without customers, they're out of business. An effective introduction promotes good customer-company relationships.

5. Definitions of terminology. If your manual uses the abbreviations "BDC," "CCW," or "CPR," will the readers know that you are referring to "bottom dead center," "counterclockwise," or "continuing property records"? Probably not.

If the readers aren't familiar with your terminology, they might miss important information and perform an operation incorrectly.

To avoid this problem, define your abbreviations, acronyms, and/or symbols. Since "Danger," "Warning," "Caution," and "Note" are rarely defined the same way in different manuals, an effective manual defines what it means by these hazard alerts. You can give your definitions either early in the user's manual or in a glossary located at the end of your manual (we discuss glossaries in Chapter 14 and 16). The following example defines terms alphabetically:

⚠	This symbol designates an important operating or maintenance step.
BDC	Bottom dead center
CCW	Counterclockwise
Danger	This hazard alert designates the possibility of death. Be extremely careful when performing an operation.
RMS	Root mean square

6. Technical descriptions. In addition to instructions, many user's manuals contain technical descriptions of the product or system. A description could be a part-by-part explanation or labeling of a product or system's components. Such a description helps readers recognize parts when they are referred to in the instruction. For example, if the user's manual tells the reader to lay shingles with the tabs pointing up, but the reader doesn't know what a "tab" is, the step cannot be performed. In contrast, if a description with appropriate call-outs is provided, then the reader's job has been simplified.

Perhaps the user's manual will contain a list of the product's specifications, such as its size, shape, capacity, capability, and materials of construction. Table 12.3 is an example of a specification from a cordless telephone user's manual. Such a specification allows the user to decide whether the product meets his or her needs. Does it state the desired frequency? Is it the preferred weight and size? The specification answers these questions.

Finally, a user's manual might include a schematic, depicting the product or system's electrical layout. This would help the readers troubleshoot the mechanism. (Chapter 9 provides a sample schematic.)

7. *Warranties.* Warranties protect the customer and the manufacturer. Many warranties tell the customer, "This warranty gives you specific legal rights, and you may also have other rights which vary from state to state." A warranty protects the customer if a product malfunctions sooner than the manufacturer suggests it might: "This warrants your product against defects due to faulty material or workmanship." In such a case, the customer usually has a right to free repairs or a replacement of the product.

However, the warranty also protects the manufacturer. No product lasts forever, under all conditions. If the product malfunctions after a period of time designated by the manufacturer, then the customer is responsible for the cost of repairs or a replacement product. The designated period of time differs from product to product. Furthermore, many warranties tell the customers, "This warranty does not include damage to the product resulting from normal wear, accident, or misuse."

Disclaimers are another common part of warranties that protect the manufacturer. For example, some warranties include the following disclaimers:

TABLE 12.3 Cordless Telephone Specifications	
BASE UNIT	**SPECIFICATIONS**
Transmit Frequency	46.6 to 46.9 MHz
Receive Frequency	49.6 to 49.9 MHz
Power Requirements	117V, 60Hz, 6W
Size	84mm(W) × 60m(H) × 246mm(D)
Weight	755g
HANDSET	**SPECIFICATIONS**
Transmit Frequency	49.6 to 49.9 MHz
Receive Frequency	46.6 to 46.9 MHz
Power Requirements	Rechargeable nickel cadmium batteries
Size	60mm(W) × 210mm(H) × 44mm(D)
Weight	345g

Note: Any changes or modifications to this system not expressly approved in this manual could void your warranty.

Proof of purchase by way of a receipt clearly noting that this unit is under warranty must be presented.

The warranty is only valid if the serial number appears on the product.

The manufacturer of this product will not be liable for damages caused by failure to comply with the installation instructions enclosed within this manual.

8. Accessories. A company always tries to increase its income. One way to do so is by selling the customer additional equipment. A user's manual promotes such equipment in an accessories list. This equipment isn't mandatory, required for the product's operation. Instead, accessories lists offer customers equipment such as extra-long cables or cords, carrying cases, long-life rechargeable batteries, extendable antennas, a modem, CD-ROM capabilities, and video instruction packages. Often the specifications are also provided for these accessories.

9. Corporate contact information. The user's manual helps customers purchase accessories, answer customer questions, and solve customer complaints. Therefore, most user's manuals conclude by giving telephone numbers and/or addresses for local and regional service locations, 24-hour service hotlines, and consumer information bureaus.

Figure 12.10 is a sample user's manual.

CHAPTER HIGHLIGHTS

1. To organize your instruction effectively, include an introduction, a discussion of sequenced steps, and a conclusion.

2. Be sure to number your steps and start each step with a verb.

3. Hazard alerts are important to protect your reader and your company. Place these alerts early in your instruction and/or before the appropriate step.

4. In the instruction, follow a chronological sequence.

5. Your readers do not know how to perform the procedure; that's why they are reading the instruction. Therefore, be precise.

6. Graphics will help your reader see how to perform each step.

7. Highlighting techniques emphasize important points, thus minimizing damage to equipment or injury to its users.

8. Don't compress several steps into one excessively long and demanding step.

9. The writing process, complete with usability testing, will help you construct an effective instruction.

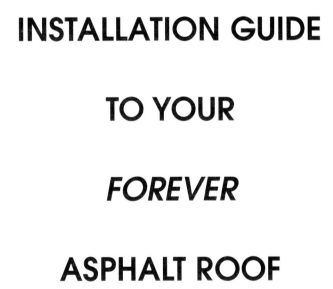

FIGURE 12.10 Installation Guide

IMPORTANT SAFETY TIPS

 Please read the entire manual before starting installation of *FOREVER* Asphalt Shingles.

1. DO NOT touch electric lines above your house.

2. Use care in climbing your ladder.
   • Rest ladders on steady ground.
   • Lean ladders against a wall.
   • Allow only one person on a ladder.

3. On steep roofs, use a scaffolding system.

4. Stay off your roof during strong winds or thunderstorms.

2

FIGURE 12.10 Installation Guide (Cont.)

TABLE OF CONTENTS

3

FIGURE 12.10 Installation Guide (Cont.)

TOOLS REQUIRED FOR INSTALLATION AND SPECIFICATIONS

TOOLS	SPECIFICATIONS
Ladder	Appropriate size
Hammer	Claw-type, smooth face
Utility Knife	6" retractable safety blade
Tape Measure	10' minimum
Chalk Line	Light-colored chalk
Shovel	Flat-end, long-handled tear-off shovel
Underlayment	#15 asphalt-saturated roofing felt One roll covers 200 square ft. of roof.
Three-tab *FOREVER* Asphalt Shingles	One square of shingles covers 100 square ft of roof. Add 10% for capping, borders, and waste.
Nails	2 1/2 lbs., 11 or 12 gauge, 3/8" diameter hot-galvanized nails will cover 1 square of shingles.

4

FIGURE 12.10 Installation Guide (Cont.)

DEFINITION OF TERMS

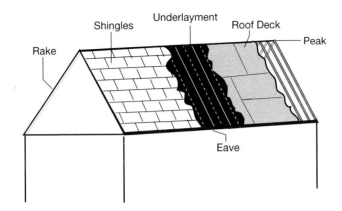

Figure 1 Parts of the Roof

- Rake—The vertical edge or side of the roof
- Shingles—Asphalt coverings for your roof
- Underlayment—Saturated felt to protect your roof deck from rot and mildew
- Peak—The upper horizontal edge of the roof
- Roof deck—The plywood base that covers your rafters
- Eave—The lower horizontal edge of the roof
- Tab—The cutout side of the shingle

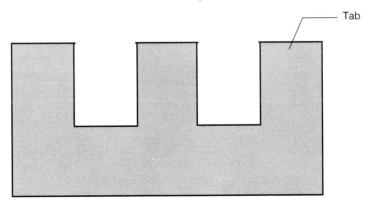

Figure 2 Shingle

5

FIGURE 12.10 Installation Guide (Cont.)

INTRODUCTION:
Benefits of Choosing
FOREVER
Asphalt Shingles

Congratulations! You have made an outstanding decision by choosing to install *FOREVER* Asphalt Shingles. Our shingles are superior to other types of roofing materials for the following reasons:

Table 1 Benefits of Asphalt Shingles

ITEM	Asphalt	Wood	Slate	Ceramic
INEXPENSIVE	X	X		
BEAUTIFUL	X	X	X	X
DURABLE	X		X	
EASY TO INSTALL	X			

6

FIGURE 12.10 Installation Guide (Cont.)

PREPARING YOUR ROOF DECK

Before you can install your new *FOREVER* Asphalt Shingles, you need to prepare your roof deck. To do so, follow these steps.

TOOLS REQUIRED:
Hammer
Shovel

1. Use the hammer to pry up old shingles, the old underlayment, and old nails.
2. Scoop up and remove the roof debris with the shovel.
3. After removing the old roofing material, check the roof deck for sagging, rotting wood.

NOTE: If you find any sagging or rotting wood, you will need to replace those sections with new plywood. Please call 1-800-FOREVER for information about our quality plywood products.

7

FIGURE 12.10 Installation Guide (Cont.)

LAYING AND MARKING THE UNDERLAYMENT

Once you have prepared the roof deck, next you need to install the underlayment. To do so, follow these steps:

TOOLS/EQUIPMENT REQUIRED
Underlayment
Chalk

NOTE: Underlayment roofing felt comes in rolls 4′ wide X 50′ long. Each roll covers 200 sq. ft. You will need half as many rolls as squares of shingles.

1. Install the underlayment by rolling it out perpendicular to the peak and the eave.

NOTE: Be sure to overlap each piece of underlayment by 2″ to provide weatherproofing coverage.

2.0 Use the chalk to put horizontal guidelines on the underlayment.
2.1 Make the first line 12″ from the eave and parallel to it.
2.2 Mark subsequent lines every 5″ to the roof peak (see Figure 3, "Marking the Underlayment").

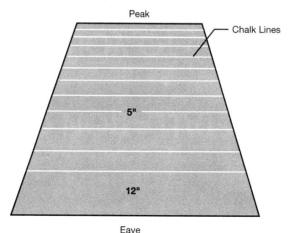

Figure 3 Marking the Underlayment

8

FIGURE 12.10 Installation Guide (Cont.)

INSTALLING YOUR *FOREVER* ASPHALT SHINGLES

Starter Course

The starter course is the first row of shingles installed. This row is closest to the eave. To lay this starter course, follow these steps:

EQUIPMENT/TOOLS REQUIRED
Hammer
Nails
Shingles

1. Starting at the left rake, lay the shingles side by side on the roof with the tabs pointing up to the roof peak.
2. Nail the shingles to the roof using three nails per shingle (see Figure 4, "Starter Course," for nail placement).

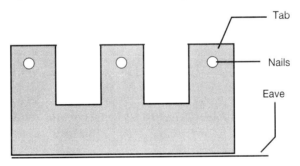

Figure 4 Starter Course

First Course

> NOTE: This course and all subsequent courses will be laid with the <u>tabs pointing down</u> to the eave, not up to the peak.

1. Beginning at the left rake eave, lay a <u>full</u> shingle directly over the starter course shingle.
2. Nail this shingle to the starter course using three nails per shingle.
3. Continue to the right two to three full shingles.

9

FIGURE 12.10 Installation Guide (Cont.)

Second and Other Courses
1. Cut off a tab on the left side of a shingle.
2. Overlapping the starter course halfway, place this shingle over the first shingle in the first course.
3. Nail the shingle to the first course using two nails per shingle.
4. Alternating full shingles with shingles missing a tab, install all subsequent courses as noted above. (See Figure 5, "Shingle Placement for Second and Other Courses.")

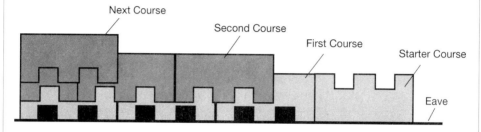

Figure 5 Shingle Placement for Second and Other Courses

10

FIGURE 12.10 Installation Guide (Cont.)

WARRANTY

FOREVER Asphalt Shingles are warranted to give normal service for 20 years or as long as you own your home. We will replace any shingle which cracks, chips, flakes, or peels under normal conditions. As with all roofing material, some nonwarranted damage may occur in severe hail or windstorms or if the roof is not properly installed. To receive warranty service, save your dated sales receipt and proofs of purchase from each square of shingles you purchase.

For tips on installation, call our 24-hour service hotline at
1–800–FOREVER

FOREVER Ashpalt Shingles
8942 Hancock Center
Clearwater, FL 40043

11

FIGURE 12.10 Installation Guide (Cont.)

ACTIVITIES

1. Write an instruction. To do so, first select a topic. You can write an instruction telling how to monitor, repair, test, package, plant, clean, operate, manage, open, shut, set up, maintain, troubleshoot, install, use software, and so on. Choose a topic from within your field of expertise or one that interests you. Follow the writing process techniques to complete your instruction. Prewrite (using a flowchart), write a draft (abiding by the criteria for instructions presented in this chapter), and rewrite to perfect your text.

2. Write a user's manual. To do so, first select a topic. For example, you could write a user's manual which explains how to operate a portable CD player, a cassette tape player, or a combination radio–alarm clock. The user's manual could focus on the installation and operation of any one of the following computer accessories: printer, modem, CD-ROM unit, or software package. In addition, TV sets, VCRs, ceiling fans, wall telephone units, or classroom overhead projectors could serve as the topic of a user's manual that explains installation and operation. Finally, a user's manual could explain the installation, operation, and maintenance for a coffeemaker, a kitchen blender, a stereo system with CD player, a tape cassette, or a VCR. Search your home, garage, or workplace for items about which you could write a user's manual.

3. Find examples of instructions or user's manuals for consumer products. These can include instructions for assembling children's toys, refinishing furniture, insulating attics or windows, setting up stereo systems, flushing out a radiator, installing ceiling fans, and so on. Try rummaging around in your kitchen pantry for examples of instructions—look at boxes and cans of food that tell you how to bake brownies or cook spaghetti. Instructions are everywhere.

   Once you find some examples, bring them to class. Then, applying the criteria for good instructions presented in this chapter, determine the success of the examples. If they are successful, explain why and how. If they fail, show where the problems are and rewrite the instructions to improve them.

4. Find examples of instructions or user's manuals written in your work environment. Bring these to class. Using the criteria for instructions presented in this chapter, decide whether the instructions are successful or unsuccessful. If the instructions are good, show how and why. If they are flawed, explain the problem(s). Then rewrite the instructions to improve them.

5. Find an example of a successful instruction or user's manual, either from your work environment or home (a consumer product). To make this instruction even more effective, construct a flowchart that will graphically enhance the writing and make the instruction easier to follow.

6. Find an example of a successful instruction or user's manual, either from your work environment or home (a consumer product). To make this instruction even more effective, draw super comic book pictures to accompany the text. These drawings will graphically enhance the writing, making it easier for your reader to follow the instructions.

7. Good writing demands revision. Following is a flawed instruction. To improve it, rewrite the text, abiding by the criteria for instructions and the rewriting techniques included in this chapter.

Date: November 30, 2000
To: Maintenance Technicians
From: Second Shift Supervisor
Subject: Oven Cleaning

The convection ovens in kiln room 33 need extensive cleaning. This would consist of vacuuming and wiping all walls, door, roofs, and floors. All vents and dampers need to be removed and a tack cloth used to remove loose dust and dirt. Also, all filters need replacing. I am requesting this because when wet parts are placed in the ovens to cure the paint, loose particles of dust and dirt are blown onto the parts, which causes extensive rework. I would like this done twice a week to ensure cleanliness of product.

8. One of the keys to success in writing instructions is audience recognition. One reason flawed instructions fail is lack of audience recognition. For example, an instruction uses high-tech terms but is intended for low-tech readers.

Find an example of an instruction or user's manual geared to a high-tech audience that correctly uses high-tech terms. Then find an example of an instruction geared to low-tech readers that incorrectly uses high-tech words. Next, find an instruction geared to low-tech or lay readers that has correctly defined its terms and explained its technology. Finally, find an example of an instruction that talks down to the reader, using low-tech or lay terms when high-tech words or phrases would have been better.

9. Find examples of instructions or user's manuals that do not provide effective hazard alerts. Using the criteria for effective hazard alerts provided in this chapter, decide where the alerts are flawed. Then improve the instructions or user's manuals by creating effective hazard alerts.

CHAPTER **13** Research

● **Criteria for Writing Research Reports**
When writing from research, recognize audience, use an effective style, and format the research report for ease of access.

● **Process**
Follow a step-by-step process to create successful research documents:

- *Use prewriting techniques.*
- *Use libraries, online sources, and the Internet.*
- *Take notes.*
- *Outline.*
- *Draft your report.*
- *Cite your sources.*

OBJECTIVES

Research skills are important in your school or work environment. You may want to perform research to better understand a technical term or concept; locate a magazine, journal, or newspaper article for your supervisor; or find data on a subject to prepare an oral or written report. Technology is changing so rapidly that you must know how to do research to stay up to date.

You can research information in online catalogs, reference books, indexes for technical subjects, CD-ROMs, and Internet search engines. Reference sources vary and are numerous, so this chapter discusses only research techniques.

When you complete this chapter, you will be able to (a) locate information in the library and/or online, (b) read and analyze sources of information for your report, (c) form an idea for your research report, (d) write a research report, and (e) document your sources of information.

CRITERIA FOR WRITING RESEARCH REPORTS

As with all types of technical writing, writing from research requires that you (a) recognize your audience; (b) use an effective style, appropriate to your reader and purpose; and (c) use effective formatting techniques for reader-friendly ease of access.

AUDIENCE

When writing a research report, you first must recognize the level of your audience. Are your readers high-tech, low-tech, or lay? If you are writing to your boss, consider his or her level of technical expertise. Your boss probably is at least a low-tech reader, but you must determine this based on your own situation. Doing so will help you determine the amount of technical definition necessary for effective communication.

You must decide next whether your reader will understand the purpose of your research report. This will help you determine the amount of detail needed and the tone to take (persuasive or informative). For instance, if your reader has requested the information, you will not need to provide massive amounts of background data explaining the purpose of your research. Your reader has probably helped you determine the scope and purpose. On the other hand, if your research report is unsolicited, your first several paragraphs must clarify your rationale. You'll need to explain why you are writing and what you hope to achieve.

If your report is solicited, you probably know what your reader plans to do with your research. He or she will use it for a briefing, an article to be written for publication, a technical update, and so forth. Thus, your presentation will be informative. If your research report is unsolicited, however, your goal is to persuade your reader to accept your hypotheses substantiated through your research.

EFFECTIVE STYLE

Research reports should be more formal than many other types of technical writing. In a research report, you are compiling information, organizing it, and presenting your findings to your audience using documentation. Because the rules of documentation are structured rigidly (to avoid plagiarism and create uniformity), a research report is also rigidly structured.

The tone need not be stuffy, but you should maintain an objective distance and let the results of your research support your contentions. Again, considering your audience and purpose will help you decide what style is appropriate. For example, if your boss has requested your opinion, then you are correct in providing it subjectively. However, if your audience has asked for the facts—and nothing but the facts—then your writing style should be more objective.

FORMATTING

Reading a research report is not always an easy task. As the writer, you must ensure that your readers encounter no difficulties. One way to achieve this is by using effective formatting. Reader-friendly ease of access is accomplished when you use highlighting techniques such as bullets, numbers, headings, subheadings, and graphics (tables and figures).

In addition, effective formatting includes the following:

- Overall organization (including an introduction, discussion, and conclusion)
- Internal organization (various organizational patterns, such as problem/solution, comparison/contrast, analysis, and/or cause/effect)
- Parenthetical source citations
- Works cited—documentation of sources

Each of these areas is discussed in detail throughout this chapter.

PROCESS

Writing a research report, as with other types of technical correspondence, is easiest when you follow a step-by-step process. Rather than just wandering into a library and hoping that the correct book or periodical will leap off a shelf and into your hands or expecting your online search to reveal useful information immediately, approach your research systematically. Prewrite (to gather your data and determine your objectives), write a draft, and rewrite to ensure that you meet your goals successfully.

PREWRITING RESEARCH TECHNIQUES

1. Select a general topic (or the topic you have been asked to study). Your topic may be a technical term, phrase, innovation, or dilemma. If you're in emergency medical technology, for example, you might want to focus on the current problems with emergency medical service. If you're in telecommunications, you might write about satellite communication. If you work in electronics, select a topic such as robotics. If your field is computer science, you could focus on artificial intelligence.

2. Spot-check sources of information. Check a library or online sources to find material that relates to your subject. A quick review of the *Reader's Guide to Periodical Literature*, catalogs, or the Internet should help you locate some of the available information.

3. Establish a focus. After you have chosen a topic for which you can find available source material, decide what you want to learn about your topic. A focus statement can guide you. In other words, if you are interested in emergency medical technology, you might write a focus statement such as the following:

> I want to research current problems with emergency medical services, including variances in training required, delays in timely response to emergency calls, and limited number of vehicles available.

For telecommunications, you could write:

> I want to research satellite communication maintenance, reliability, and technical innovations.

If you are an electronics technician, you might write:

> I want to discover the uses, impact on employment, and expenses of robotics applications.

Finally, if you are a computer technician, you might write:

> I want to discover the pros and cons regarding artificial intelligence.

With focus statements such as these, you can begin researching your topic, concentrating on articles pertinent to your topic.

4. Research your topic. You may feel overwhelmed by the prospect of research. But there are many sources that, once you know how to use them, will make the act of research less overwhelming.

Books

All books owned by a library are listed in catalogs, usually online. These books are listed under three headings—author, title, and subject. Examples of screens from the three headings are shown in the boxes on pages 324 and 325. In addition, online catalogs include periodical indexes.

Periodicals

Use the following periodical indexes to find articles on your topic.

INDEXES TO GENERAL, POPULAR PERIODICALS

- *Reader's Guide to Periodical Literature.* Covers popular, nontechnical literature from a variety of subject fields.

```
Bicycle racing
    Wheels of fortune [women's team racing] S.
      Hollandsworth. il Women's Sports Fitness 8:
      36-39+ Jl '86.
```

The preceding example tells us that an article on bicycle racing can be found in the July 1986 issue of the magazine *Women's Sports Fitness*, volume 8, pages 36 and following. The article is entitled "Wheels of Fortune," contains illustrations, and is written by S. Hollandsworth. Finally, to clarify the subject matter of the article, "women's team racing" has been added in brackets.

```
                          Author
       TJ     Robillard, Mark J. ─────────────Author
Call Number── 211    Microprocessor based robotics/by ──Title
       .R53    Mark J. Robillard; [illustrated by──Publication
       1983    R.E. Lund].─1st ed.─Indianapolis,    Information
               Ind., USA: H.W. Sams, c1983.
Page Count──    220 p.: ill.; 28 cm. (Intelligent
               machine series; v. 1)

               Bibliography: p. 215.
               Includes index.
               ISBN 0-672-22050-4 (pbk.): $16.95

Subject───────1. Robotics. 2. Microprocessors.
Heading              I. Lund,
               R.E. II. Title. III. Series.
       TJ211.R53 1983     629.8'92
                          dc19
                                83-60160
                          AACR2    MARC
```

```
                          Title
             Microprocessor based robotics
       TJ     Robillard, Mark J.
       211    Microprocessor based robotics/by
       .R53    Mark J. Robillard; [illustrated by
       1983    R.E. Lund].─1st ed.─Indianapolis,
               Ind., USA: H.W. Sams, c1983.
                220 p.: ill.; 28 cm. (Intelligent
               machine series; v. 1)

               Bibliography: p. 215.
               Includes index.
               ISBN 0-672-22050-4 (pbk.): $16.95

               1. Robotics. 2. Microprocessors. I. Lund,
                   R.E. II. Title. III. Series.
       TJ211.R53 1983     629.8'92
                          dc19
                                83-60160
                          AACR2    MARC
```

```
                        Subject
Robotics
  TJ      Robillard, Mark J.
  211      Microprocessor based robotics/by
  .R53     Mark J. Robillard; [illustrated by
  1983     R.E. Lund].—1st ed.—Indianapolis,
           Ind., USA: H.W. Sams, c1983.
            220 p.: ill.; 28 cm. (Intelligent
           machine series; v. 1)

           Bibliography: p. 215.
           Includes index.
           ISBN 0-672-22050-4 (pbk.): $16.95

           1. Robotics. 2. Microprocessors. I. Lund,
           R.E. II. Title. III. Series.
  TJ211.R53 1983         629.8'92
                         dc19
                              83-60160
                         AACR2    MARC
```

```
Schlesinger, Arthur M., 1917-
The challenge of change, il NY Times Mag p20-21
J1 27 '86

                   about

Report on the republic. P.S. Prescott. il por
Newsweek 108:98+ 0 27 '86
```

In the preceding entry, we are told two things. First, an article by Arthur M. Schlesinger can be found in the July 27, 1986, issue of *The New York Times Magazine* on pages 20 and following. The article contains illustrations and is entitled "The Challenge of Change." Next we are told that we can find an article *about* Schlesinger in the October 27, 1986, issue of *Newsweek*, volume 108, on pages 98 and following. The article is written by P. S. Prescott and is titled "Report on the Republic."

INDEXES TO SCHOLARLY AND TECHNICAL JOURNALS

- *Applied Science & Technology Index.* Covers all engineering fields, as well as aeronautics and space sciences, atmospheric sciences, chemistry, computer technology and applications, construction industry, energy resources and research fire prevention, food and the food industry, geology, machinery, mathematics, metallurgy, mineralogy, oceanography, petroleum and gas, physics, plastics, the textile industry and fabrics, transportation, and other industrial and mechanical arts.

- *Business Periodicals Index.* Covers major U.S. publications in marketing, banking and finance, personnel, communications, computer technology, and so on.
- *General Science Index.* Covers the pure sciences, such as biology and chemistry.
- *Nursing and Allied Health Index.* Covers topics from a medical viewpoint.

ELECTRONIC RESOURCES

Today, computers can help you research any topic rapidly and efficiently. Following are several electronic resources that provide bibliographic information, abstracts, and/or full-text articles either online or via CD-ROM:

- *ABI/INFORM.* Provides bibliography, abstracts and some full-text articles about business and management.
- *ERIC (Educational Resources Information Center).* Provides bibliography and abstracts about educational research and resources.
- *MEDLINE.* Provides bibliography and abstracts about medicine and allied health.
- *Newsbank.* Provides bibliography and abstracts about newspaper articles.
- *Nexis/Lexis.* Provides bibliography and full-text articles about news (Nexis) and law (Lexis).
- *PsycINFO.* Provides bibliography and abstracts about psychology and behavioral sciences.
- *SIRS (Social Issues Resources).* Provides bibliography and full-text articles about social issues.

The Internet

Perhaps one of the largest sources of research available today is the World Wide Web. Millions of documents from countless sources are available through the Internet. You can find material on the Internet published either by government agencies, organizations, schools, businesses, or individuals (see Table 13.1, A Sample of Internet Research Sources). And the list of options grows daily. In 1994, 20 newspapers had online Web sites. By 1997, this number had grown to 3,620 online newspapers.

How do you find information online? Use search engines. *Yahoo* and *Excite*, for example, are two Internet search engines that let you search for information from a long list of categories, including the following:

Arts	Government	Politics and Law
Business	Health and Medicine	Recreation
Computers	Hobbies	Science
Education	Money and Investing	Sports
Entertainment	News	Society and Culture

TABLE 13.1	A Sample of Internet Research Sources				
SEARCH ENGINES	ONLINE REFERENCE	ONLINE LIBRARIES	ONLINE NEWSPAPERS	ONLINE MAGAZINES	ONLINE GOVERNMENT SITES
Yahoo	*Webster's Dictionary*	Library of Congress	*New York Times*	*National Geographic*	United Nations
Excite		New York Public Library	CNN	*HotWired*	The White House
Lycos	*Roget's Thesaurus*		*USA Today*	*Atlantic*	
Infoseek	*Britannica Online Encyclopedia*	Cleveland Public Library	*Washington Post*	*The New Republic*	The IRS U.S. Postal Service
AltaVista			*Kansas City Star*	*U.S. News Online*	FedWorld
MetaSearch	*Encyclopedia Smithsonian*	Gutenberg Project	Most city newspapers	*Time* Magazine	Most states' supreme courts, leg- islatures,
MetaCrawler	*The Internet Almanac*	Most city and univer- sity libraries		*Ebony Online*	executive offices, and local gov- ernments
	The Old Farmer's Almanac			*Slate*	
				E-Zine List (index to over 1,000 online magazines)	

To access any of these areas, type a key word, phrase, or name in the appropriate blank space, and press the enter key. The search engine will search the World Wide Web for "hits," documents that match your criteria. One of two things will happen: either the search engine will report "no findings," or it will report that it has found thousands of sites that might contain information on your topic.

In the first instance, "no findings," you'll need to rethink your search strategy. You may need to check your spelling of the key words or find synonyms. For example, if you want to research information about online writing, you could try typing "writing online," "online writing," "electronic writing," "writing electronically," and other similar terms. In the second instance, finding too many hits, you'll need to narrow your search. For example, if you're researching illegalities in baseball, you can't type in "baseball." That's too broad. Instead, try "Shoeless Joe Jackson," "Pete Rose," "crime in baseball," "baseball scandals," and so on.

Researching the Internet presents at least two problems other than finding information. First, is the information you've found trustworthy? Paper-bound newspapers, journals, magazines, and books go through a lengthy publication process involving editing and review by authorities. Not all that's published on the Internet is so professional. Be wary. What you read online needs to be filtered through common sense. Second, remember that although a book, magazine, newspaper, or journal can exist unchanged in print form for years, Web sites change constantly. A site you find today online will not necessarily be the same tomorrow. That's the nature of electronic communication.

In addition to these sources, you can consult the following for help: U.S. government publications, databases, and your reference librarian.

5. Read your researched material and take notes. Once you have researched and located a source (whether it is a book, magazine, journal, or newspaper article), study the material. For a book, use the index and the table of contents to locate your topic. When you have found it, refer to the pages indicated and skim, reading selectively. For shorter documents such as magazine or journal articles or online materials, you can read closely. Reading and rereading the source material is an essential step in understanding your researched information. After you have studied the document thoroughly, go through it page by page and briefly summarize the content.

For a short magazine article, you can make marginal notes on a photocopy. For instance, read a paragraph and then briefly summarize its main point(s) in several words. Such notations are valuable because they are easy to make and provide a clear and concise overview of the article's focus.

You can also take notes on 3″ × 5″ cards. If you do this, be sure to write only one fact, quotation, or paraphrase per card, along with the author's name. This will let you organize the information later according to whatever organizational sequence (chronological narrative, analysis by importance, comparison/contrast, problem/solution, cause/effect, etc.) you prefer. Prepare a bibliography of your sources on 3″ × 5″ cards as well, one source per card.

The following are examples of a bibliography card and a note card with quotation, respectively.

Stephens, Guy M. "To Market, to Market." *Satellite Communications.* November 1987: 15–16.

Future need for C-band

"'We believe there is going to be an important need for C-band capacity into the '90s, into the next century, because it's an ideal way to serve many users, particularly video users,' Koehler said."

Stephens, "To Market," p. 15.

On a *summary* note card, you condense original material by presenting the basic idea in your own words. You can include quotations if you place them in quotation marks, but do not alter the organizational pattern of the original. A summary is shorter than the original.

Future need for C-band

C-band is going to be increasingly important in the '90s because of the increasing number of video owners.

Stephens, "To Market," p. 15.

On a *paraphrase* note card, you restate the original material in your own words without condensing. The paraphrase is essentially the complete version rewritten. A paraphrase is the same length as the original.

Future need for C-band

The author asserts the continued relevance of C-band even in the 1990s and beyond. C-band is the top performer, especially for people who use videos.

Stephens, "To Market," p. 15.

---

6. Isolate the main points. After you complete the analysis of the document (whether you use note cards and/or marginal comments), isolate the main points discussed in the books or periodicals. You will find that of the major points in an article or book, sometimes only three or four ideas will be relevant for your research. Choose the ones discussed repeatedly and/or those which most effectively develop the ideas you want to pursue.

7. Write a statement of purpose. Once you have chosen two to four main ideas from your research, write a purpose statement which expresses the direction of your research. For example, one student wrote the following statement of purpose after performing research on superconductivity:

> The purpose of this report is to reveal the future for this exciting product, which depends on further progress of technology, the development of easily accessible and cost-efficient materials, and industry's need for the final product.

8. Create an outline. After you have written a purpose statement, formulate an outline. An outline will help you organize your paragraphs and ensure that you stay on track as you develop your ideas through quotes and paraphrases. Figures 13.1 and 13.2 are examples of topic and sentence outlines.

I. Sensors used to help robots move
   A. Light
   B. Sound
   C. Touch
II. Touch sensor technology (microswitches)
   A. For gripping
   B. For maintaining contact with the floor
III. Optical sensors (LED/phototransistors)—Like bowling alley foul-line sensors
   A. Less bulky/connected to computer interface
   B. Not just for gripping, but for locating objects by following this sequence:
      1. Scan gripper to locate object
      2. Move gripper arm left and right to center object
      3. Move gripper forward to grasp
      4. Close gripper
   C. Problem—What force to use for gripping?
IV. Force sensors
   A. Spring and microswitch
   B. Optical encoder discs—Microprocessors determine speed of discs to determine force necessary
   C. Integrated circuits with strain gauge and pressure-sensitive paint
   D. Pressure sensors built with conductive foam
V. Conclusion

FIGURE 13.1 Topic Outline

WRITING

You now are ready to write your research report.

1. Review your research. Prior to writing your report, look back over your research sources to make sure that you're satisfied with what you've found. Do you have enough information to develop your points thoroughly? Is the information you've found what you want? If not, it's time to do more research. If you're content with what you've discovered in your research, you can start drafting your text.

2. Organize your report effectively. When you are ready to write, provide an introductory paragraph, discussion (body) paragraphs, a conclusion or recommendation, and your works cited page.

Introduction

Begin with something to arouse your reader's interest. This could include a series of questions, an anecdote, a quote, or data pertinent to your topic. Then use this to lead into your statement of purpose.

I. Because robots must move, they need sensors. These sensors could include light sensors, sound sensors, and touch sensors.
II. Touch sensors can have the following technology:
   A. Microswitches can be used for gripping.
   B. Microswitches can also be used for maintaining contact with the floor. This would keep the robot from falling down stairs, for example.
III. Optical sensors might be better than microswitches.
   A. LED/phototransistors are less bulky than switches.
   B. When connected to a computer interface, optical sensors also can help a robot locate an object as well as grip it.
   C. Here's the sequence followed when using optical sensors:
      1. The robot's grippers scan the object to locate it.
      2. The robot moves its gripper arms left and right to center the object.
      3. The robot moves forward to grip the object.
      4. The robot closes the gripper.
   D. The only problem faced is what force should be used when gripping.
IV. Force sensors can solve this problem.
   A. A combination spring and microswitch can be used to determine the amount of force required.
   B. Optical encoder discs can be used also. A microprocessor determines the speed of the disc to determine the required force.
   C. Integrated circuits with strain gauges and pressure-sensitive paint can be used to determine force.
   D. Another pressure sensor can be built from conductive foam.
V. To conclude, all these methods of tactile sensing comprise a field of inquiry important to robotics.

FIGURE 13.2 Sentence Outline

Discussion

The number of discussion paragraphs will depend on the number of divisions and the amount of detail necessary to develop your ideas. Use quotes and paraphrases to develop your content. Students often ask how much of a research report should be *their* writing, as opposed to researched information. A general rule is to lead into and out of every quotation or paraphrase with your own writing. In other words, (a) make a statement (your sentence), (b) support this generalization with a quotation or paraphrase (referenced material from another source), and (c) provide a follow-up explanation of the referenced material's significance (your sentence).

Conclusion/Recommendations

In a final paragraph, summarize your findings, draw a conclusion about the significance of these discoveries, and recommend future action.

Citing Sources

On a final page, provide an alphabetized list of your research sources. (We discuss this documentation later in this chapter.)

3. Document your sources correctly. Your readers need to know where you found your information and from which sources you are quoting and/or paraphrasing. Therefore, you must document this information. Correct documentation is essential for several reasons:

- You must direct your readers to the books, periodical articles, and online reference sources that you have used in your research report. If your readers want to find these same sources, they depend only on your documentation. If your documentation is incorrect, readers will be confused. Instead, you want your readers to be able to rely on the correctness and validity of your research.

- Do *not* plagiarize. *Plagiarism* is the appropriation (theft!) of some other person's words and ideas without giving proper credit. Writers are often guilty of unintentional plagiarism. This occurs when you incorrectly alter part of a quotation but still give credit to the writer. Your quotation must be *exactly the same* as the original word, sentence, or paragraph. You cannot haphazardly change a word, a punctuation mark, or the ideas conveyed. Even if you have cited your source, an incorrectly altered quotation constitutes plagiarism.

- On the other hand, if you intentionally use another person's words and claim them as your own, omitting quotation marks and source citations, you have committed theft. This is dishonest and could raise questions about your credibility and/or the credibility of your research. Teachers, bosses, and colleagues will have little, if any, respect for a person who purposely takes another person's words or ideas. It is essential, therefore, for you to cite your sources correctly.

To document your research correctly, you must (a) provide parenthetical source citations and (b) supply a works cited page (Modern Language Association) or a references page (American Psychological Association).

PARENTHETICAL SOURCE CITATIONS
The Modern Language Association (MLA) and the American Psychological Association (APA) use a simplified form for source citations. Before 1984, footnotes and endnotes were used in research reports. In certain instances, this form of documentation is still correct. If your boss or instructor requests footnotes or endnotes, you should still use these forms. However, the most modern approach to source citations requires only that we cite the source of our information parenthetically after the quotation or paraphrase.

MLA FORMAT

One Author
After the quotation or paraphrase, parenthetically cite the author's last name and the page number of the information.

> "Viewing the molecular activity required state-of-the-art electron microscopes" (Heinlein 193).

Note that the period follows the parenthesis, not the quotation. And, note that no comma separates the name from the page number and that no lowercase *p* precedes the number.

Two Authors
After the quotation or paraphrase, parenthetically cite the authors' last names and the page number of the information.

> "Though *Gulliver's Travels* preceded *Moll Flanders*, few scholars consider Swift's work to be the first novel" (Crider and Berry 292).

Three or More Authors
Writing a series of names can be cumbersome. To avoid this, if you have a source of information written by three or more authors, parenthetically cite one author's name, followed by *et al.* (Latin for "and others") and the page number.

> "Baseball isn't just a sport; it represents man's ability to meld action with objective—the fusion of physicality and spirituality" (Norwood et al. 93).

Anonymous Works
If your source has no author, parenthetically cite the shortened title and page number.

> "Robots are more accurate and less prone to errors caused by long hours of operation than humans" ("Useful Robots" 81).

APA FORMAT

One Author
If you do not state the author's name or the year of the publication in the lead-in to the quotation, include the author's name, year of publication, and page number in parenthesis, after the quotation.

> "Izzy's stay in Palestine was hardly uneventful" (Cottrell, 1992, p. 118).

(Page numbers are included for quoted material. The writer determines whether page numbers are included for source citations of summaries and paraphrases.)

Two Authors

When you cite a source with two authors, always use both last names with an ampersand (&).

> "Line charts reveal relationships between sets of figures" (Gerson & Gerson, 1992, p. 158).

Three or More Authors

When your citation has more than two authors but fewer than six, use all the last names in the first parenthetical source citation. For subsequent citations, list the first author's name last followed by *et al.*, the year of publication, and for a quotation, the page number.

> "Two-party politics might no longer be the country's norm next century" (Conners et al., 1993, p. 2).

Anonymous Works

When no author's name is listed, include in the source citation the title or part of a long title and the year. Book titles are underlined, and periodical titles are placed in quotation marks.

> Two-party politics might be a thing of the past (*Winning Future Elections*, 1992).

> Many memos and letters can be organized in three paragraphs ("Using Templates," 1994).

WORKS CITED

Parenthetical source citations are an abbreviated form of documentation. In parentheses, you tell your readers only the names of your authors and the page numbers on which the information can be found. Such documentation alone would be insufficient. Your readers would not know the names of the books, the names of the periodicals, or the dates, volumes, or publishing companies. This more thorough information is found on the works cited page or references page, a listing of research sources alphabetized either by author's name or title (if anonymous). This is the last page of your research report.

Your entries should follow MLA or APA standards.

MLA WORKS CITED

A Book with One Author

> Cottrell, Robert C. *Izzy: A Biography of I. F. Stone*. New Brunswick: Rutgers University Press, 1992.

A Book with Two or Three Authors

Tibbets, Charlene, and A. M. Tibbets. *Strategies: A Rhetoric and Reader.* Glenview: Scott, Foresman and Company, 1988.

A Book with Four or More Authors

Nadell, Judith, et al. *The Macmillan Writer.* Boston: Allyn and Bacon, 1997.

A Book with a Corporate Authorship

Corporate Credit Union Network. *A Review of the Credit Union Financial System: History, Structure, and Status and Financial Trends.* Kansas City: U.S. Central, 1986.

A Translated Book

Phelps, Robert, ed. *The Collected Stories of Colette.* Trans. Matthew Ward. New York: Farrar, Straus, Giroux, 1983.

An Entry in a Collection or Anthology

Irving, Washington. "Rip Van Winkle." *Once Upon a Time: The Fairy Tale World of Arthur Rackham.* Ed. Margery Darrell. New York: Viking, 1972. 13–36.

A Signed Article in a Journal

Gerson, Steven M., and Earl Eddings. "Service Learning: Internships . . . with a Conscience." *Missouri English Bulletin* 54 (Fall 1996): 70–75.

A Signed Article in a Magazine

Kroll, Jack. "T. Rex Redux." *Newsweek* 26 May 1997: 74–75.

A Signed Article in a Newspaper

Hoffman, Donald. "Bank Consigned to Vault of Gloom." *The Kansas City Star* 24 Oct. 1988: C1.

An Unsigned Article

"Diogenes Index." *Time* 23 Sep. 1996: 22.

Encyclopedias and Almanacs

"Rocket." *The World Book Encyclopedia.* 1979 ed. Chicago: World Book.

Computer Software

PFS: *Write Sampler*. Computer software. Software Pub. Corp., 1984.

Internet Source

Berst, Jesse. "Berst Alert." *ZDNet* 30 Jan. 1998. 30 Jan. 1998
    <http://www.zdnet.com/anchordesk/story/story_1716.html>.

E-mail

Schneider, Ruth. "Teaching Technical Writing." Personal E-mail. 2 Apr. 1998.

CD-ROM

McWard, James. "Graphics Online." TW/Inform. CD-ROM. New York:
    EduQuest, 1998.

APA References

A Book with One Author

Cottrell, R. C. (1992). *Izzy: A biography of I. F. Stone*. New Brunswick, NJ:
    Rutgers University Press.

A Book with Two Authors

Tibbets, C., & Tibbets, A. M. (1988). *Strategies: A rhetoric and reader*.
    Glenview, IL: Scott, Foresman and Company.

A Book with Three or More Authors

Nadell, J., McNeniman, L., & Langan, J. (1997). *The Macmillan writer*.
    Boston: Allyn & Bacon.

A Book with a Corporate Authorship

Corporate Credit Union Network. (1986). *A review of the credit union
    financial system: History, structure, and status and financial trends*.
    Kansas City, MO: U.S. Central.

A Translated Book

Phelps, R. (Ed.). (1983). *The collected stories of Colette* (M. Ward, Trans.).
    New York: Farrar, Straus & Giroux.

An Entry in a Collection or Anthology

Irving, W. (1972). Rip Van Winkle. In M. Darrell (Ed.), *Once upon a time: The fairy tale world of Arthur Rackham* (pp. 13–36). New York: Viking Press.

A Signed Article in a Journal

Gerson, S. M., & Eddings, E. (1996, Fall). Service learning: Internships . . . with a conscience. *Missouri English Bulletin*, 54, 70–75.

A Signed Article in a Magazine

Kroll, J. (1997, May 26). T. rex redux. *Newsweek*, 74–75.

A Signed Article in a Newspaper

Hoffman, D. (1988, Oct. 24). Bank consigned to vault of gloom. *The Kansas City Star*, p. C1.

An Unsigned Article

Diogenes index. (1996, Sep. 23). *Time*, 22.

Encyclopedias and Almanacs

Rocket. (1979). *The world book encyclopedia*. Chicago: World Book.

Computer Software

PFS: Write Sampler [Computer software]. (1984). Software Pub. Corp.

Internet Source

Berst, J. (1998, Jan. 30). Berst alert. *ZDNet* [on-line]. Retrieved February 1, 1998 from the World Wide Web: http://www.zdnet.com/anchordesk/story/story_1716.html.

CD-ROM

McWard, J. (1998). Graphics on-line [CD-ROM]. TW/Inform. New York: EduQuest.

ALTERNATIVE STYLE SHEETS

Although MLA and APA are popular style sheets, other style sheets are favored in certain disciplines. Refer to these if you are interested or required to do so.

- *U.S. Government Printing Office Style Manual.* Washington, DC: Government Printing Office, 1973.
- *The Chicago Manual of Style,* 14th edition. Chicago: University of Chicago Press, 1994.
- Turabian, Kate L. *A Manual for Writers of Term Papers, Theses, and Dissertations.* Chicago: University of Chicago Press, 1973.

4. Develop your ideas. You've learned how to organize your report (through an introduction, discussion, and conclusion/recommendation) and how to document your sources of research (through parenthetical source citations and a works cited or references page). Writing your research report also requires that you use your research effectively to develop your ideas. Successful use of research demands that you correctly quote, paraphrase, and/or summarize. Summaries are discussed in Chapter 14.

REWRITING

As with all types of writing, drafting the text of your research report is only the second stage of the writing process. To ensure that your report is effective, revise your draft as follows:

1. *Add new detail for clarity and/or persuasiveness.* Too often, students and employees assume that they have developed their content thoroughly when, in fact, their assertions are general and vague. This is especially evident in research reports. You might provide a quotation to prove a point, but is this documentation sufficient? Have you truly developed your assertions? If an idea within your report seems thinly presented, either add another quotation, paraphrase, or summary for additional support or explain the significance of the researched information.

2. *Delete dead words and phrases and researched information that does not support your ideas effectively.* Good writing in a work environment is economical writing. Thus, as always, your goal is to communicate clearly and concisely. Delete words that serve no purpose, maintaining a low fog index. In addition, review your draft for clarity of focus. The goal of a research report is not to use whatever researched information you've found wherever it seems valid. Instead, you want to use quotations, paraphrases, and summaries only when they help develop your statement of purpose. If your research doesn't support your thesis, it's counterproductive and should be eliminated. In the rewriting stage, delete any documented research that is tangential or irrelevant.

3. *Simplify your words for easy understanding.* The goal of technical writing is to communicate, not to confuse. Write to be understood. Don't say *grain-consuming animal units* if you mean *chickens*. Don't call the October 1987 stock market crash a *fourth-quarter equity retreat.* Don't describe an airplane crash as *uncontrolled contact with the ground.*

4. *Move information within your report to ensure effective organization.* How have you organized your report? Did you use a problem/solution format?

Did you use comparison/contrast or cause/effect? Is your report organized as a chronological narrative or by importance? Whichever method you've used, you want to be consistent. To ensure consistency, rewrite by moving any information that is misplaced.

5. *Reformat your text for reader-friendly ease of access.* Look at any technical journal. You'll notice that the writers have guided their readers through the text by using headings and subheadings. You'll also notice that many journals use graphics (pie charts, bar graphs, line drawings, flow charts, etc.) to clarify the writer's assertions. You should do the same. To help your readers follow your train of thought, reformat any blocks of wall-to-wall words. Add headings, subheadings, itemized lists, white space, and graphics.

6. *Correct any errors.* This represents your greatest challenge in writing a research report. You not only must be concerned with grammar and mechanics, as you are when writing a memo, letter, or report, but also with accurate quoting, paraphrasing, summarizing, parenthetical source citations, and works cited.

When revising, pay special attention to these concerns. If you quote, paraphrase, or summarize incorrectly, you run the risk of plagiarizing. If you fail to provide correct parenthetical source citations and/or works cited, you will make it impossible for your readers to find these same sources of information in their research or to check the accuracy of your data. Research demands accuracy and reliability.

CHAPTER HIGHLIGHTS

1. You can research a topic either in a library or online at your computer.
2. You need to consider the audience's level of technical knowledge when you write a research paper.
3. Reader-friendly highlighting techniques help your audience access information in your research report.
4. Narrowing a topic can help you find sources of information.
5. A focus statement lets you determine the direction of your document.
6. Use discrimination when you research on the Internet.
7. Careful source citations help you avoid plagiarism.
8. The Modern Language Association (MLA) and the American Psychological Association (APA) are two widely used style sheets for citing sources.

ACTIVITIES

1. Correctly format and alphabetize a works cited page that contains the following entries:
   - An anonymously written magazine article
   - A magazine article signed by two authors

- A journal article signed by one author
- A book with three or more authors
- A book with an editor
- A signed newspaper editorial
- An online document
- A CD-ROM document
- An e-mail message

2. Summarize in one sentence any paragraph from this textbook. Provide a parenthetical source citation and works cited information.

3. Read a one- or two-page article from a magazine, journal, or online source. Then practice note taking. Writing in the margins or between paragraphs, briefly note the key point(s) made in each paragraph. (These notes can be limited to one or two words.)

4. Using a one- or two-page article from a magazine or journal, practice note taking on 3″ × 5″ cards. To do so, first write the correct works cited information on one card. Then, on separate cards, take notes about approximately four key ideas discussed in the article. Write only one note per card; give the card a title for future reference; provide either quotations or paraphrases; and then write the author's name, the article's title, and the page number on the bottom of the card.

5. Select a technical topic from your major field or your job and write a research report. You might want to consider a controversy in your area of interest (such as the greenhouse effect, hazardous waste management, or computer viruses) or the impact of a technical innovation (such as micromachines or the Internet).

The Summary

OBJECTIVES

You might be required to write a summary if, for example, your boss is planning to give a speech at a civic meeting (Better Business Bureau, Rotary Club, or Businesswomen's Association) and needs some up-to-date information for his or her presentation. Bosses often are too busy to perform research, wade through massive amounts of researched data, or attend a conference at which new information might be presented. Therefore, your boss asks you to research the topic and provide the information in a shortened version. He or she wants you to provide a summary of your findings. Similarly, your boss might need to give a briefing (discussed in Chapter 15) to upper-level management. Again, time is a problem. How can your boss get the appropriate information and digest it rapidly? The answer is for *you* to read the articles, attend the conference or meeting, and then summarize your research. At other times, you will be asked to summarize your own writing. For example, if you are writing a long report, you might include an "executive summary," an overview of the report's key points (we discuss this in Chapter 16).

As a class assignment, your teacher might want you to write a summary of a formal research report. The summary, though shorter than a research report, would still require that you practice research skills, such as reading, note taking, and writing paraphrases.

To write a summary, you'll need to study the research material. Then, in condensed form (a summary is no more than 5 to 15 percent of the length of the original source), you'll report on the author's main points. Preparing a summary puts to good use your research, analyzing, and writing skills.

The following will help you to write a summary:

- Criteria for writing summaries
- Process to follow
- Process log with examples

CRITERIA FOR WRITING SUMMARIES

A well-constructed summary, though much shorter than the original material being summarized, highlights the author's important points. A summary is like a *Reader's Digest* approach to writing. Although the summary will not cover every fact in the original, after reading the summary you should have a clear overview of the original's main ideas. There are several criteria for accomplishing this.

OVERALL ORGANIZATION

As with any good writing, a summary contains an introduction, a discussion, and a conclusion.

Introduction

Begin with a topic sentence. This sentence will present the primary focus of the original source and list the two or three major points to be discussed. You must also tell your reader what source you are summarizing. You can accomplish this in one of three ways. You can either list the author's name and article title in the topic sentence; preface your summary with a works cited notation providing the author's name, title, publication date, and page numbers; or follow your topic sentence with a footnote and give the works cited information at the bottom of the page.

NAME/TITLE IN TOPIC SENTENCE
In "Robotics: Tactile Sensing" (*Radio Electronics*, August 1986: 71–72),
Mark J. Robillard states that since robots move to do their jobs, they need to be equipped with an assortment of tactile sensors.

WORKS CITED PREFACING TOPIC SENTENCE
Robillard, Mark J. "Robotics: Tactile Sensing." *Radio Electronics*. August 1986: 71–72.

Because robots must move around to do their jobs, they need to be equipped with an assortment of tactile sensors.

Topic Sentence Followed by Explanatory Footnote
Because robots move around to do their jobs, they need to be equipped with an assortment of tactile sensors.*

_____
_____
_____
_____
_____
_____
_____
_____
_____

*This paragraph summarizes Mark J. Robillard's "Robotics: Tactile Sensing," *Radio Electronics*, Aug. 1986: 71–72.

Discussion

In this section, briefly summarize the main points covered in the original material. To convey the author's ideas, you can paraphrase, using your own words to restate the author's point of view.

Conclusion

To conclude your summary, you can either reiterate the focus statement, reminding the reader of the author's key ideas; highlight the author's conclusions regarding his or her topic; or state the author's recommendations for future activity.

INTERNAL ORGANIZATION

How will you organize the body of your discussion? Since a summary is meant to be objective, you should present not only what the author says but also how he or she organizes the information. For example, if the author has developed his or her ideas according to a problem/solution format, your summary's discussion should also be organized as a problem/solution. This would give your audience the author's content and method of presentation. Similarly, if the author's article is organized according to cause/effect, comparison/contrast, or analysis (classification/division), this would determine how you would organize your summary.

DEVELOPMENT

To develop your summary, you'll need to focus on the following:

- *Most important points*. Since a summary is a shortened version of the original, you can't include all that the author says. Thus, you should include only the two or three key ideas within the article. Omit irrelevant details, examples, explanations, or descriptions.

- *Major conclusions reached.* Once you've summarized the author's key ideas, then state how these points are significant. Show their value or impact.
- *Recommendations.* Finally, after summarizing the author's major points and conclusions, you'll want to tell your audience if the author recommends a future course of action to solve a problem or to avoid potential problems.

STYLE

The summary, like all technical writing, must be clear, concise, accurate, and accessible. Therefore, you'll want to abide by the fog index. Watch out for long words and sentences. Avoid technical jargon. Most importantly, be sure that your summary truly reflects the author's content. Your summary must be an unbiased presentation of what the author states and include none of your opinions.

LENGTH

As mentioned earlier, the summary will be approximately 5 to 15 percent the length of the original material. To achieve this desired length, omit references to the author (after the initial reference in the works cited or topic sentence). You'll also probably need to omit some types of material:

- Past histories
- Definitions
- Complex technical concepts
- Statistics
- Tables and/or figures
- Tangential information (such as anecdotes and minor refutations)
- Lengthy examples
- Biographical information

AUDIENCE RECOGNITION

You must consider your audience in deciding whether to omit or include information. Although you usually would omit definitions from a summary, this depends on your audience. Technical writing is useless if your reader does not understand your content. Therefore, determine whether your audience is high tech, low tech, or lay.

GRAMMAR AND MECHANICS

As always, flawed grammar and mechanics will destroy your credibility. Not only will your reader think less of your writing and research skills, but also errors in grammar and mechanics might threaten the integrity of your summary. Your summary will be inaccurate and, therefore, invalid.

PROCESS

Writing your summary will be easiest if you follow the same step-by-step process detailed throughout this text.

PREWRITING

To gather your data and determine your objectives, do the following:

1. Locate your periodical article, book chapter, or report. If you are summarizing a meeting or seminar, take notes. If your boss or instructor gives you your data, one major obstacle has been overcome. If, however, you need to visit a library to research your topic, refer to Chapter 13 for helpful hints on research skills.

2. Write down your works cited or references information. Once you've found your research material, *immediately* write down the author's name, the article's title, the name of the periodical or book in which this source was found, the date of publication, and the page numbers. Doing so will save you frustration later. For example, if you lose your copy of the article and have not documented the source of your information, you'll have to begin your research all over again. Instead, spend a few minutes at the beginning of your assignment documenting your sources. This will ensure that you do not spend several harried hours the night before the summary is due frantically scouring the library for the missing article. (Chapter 13 provides the correct works cited or references format.)

3. Read the article to acquire a general understanding of its content.

4. Reread the article, this time taking notes. Determine exactly what the author's thesis is and what main points are discussed. You can take notes in one of two ways. Either you can underline important points, or you can make marginal or interlinear notes on a photocopy. For a summary of an article or chapter in a book, we suggest that you write notes directly on the photocopy. To do so, read and reread a paragraph. After you've read and understood the paragraph, write a one- or two-word notation in the margin which sums up that paragraph's main idea.

WRITING

Once you've gathered your data and determined your author's main ideas, you're ready to write a rough draft of your summary.

1. Review your prewriting. Look back over your marginal notes and your underlining. Have you omitted any significant points? If you have, now is the time to include them. Have you included any ideas which are insignificant or tangential? If so, delete these to ensure that your summary is the appropriate length.

2. Write a rough draft, using the sufficing technique. Once you've reviewed the data you've gathered, *quickly* draft the text of your summary. Follow the criteria for writing an effective summary discussed earlier in this chapter, including

- A topic sentence clearly stating the author's main idea. Also provide the works cited information here, in a title, or in an explanatory footnote.
- Organization paralleling the author's method of organization (comparison/contrast, argument/persuasion, cause/effect, analysis, chronology, spatial, importance, etc.).
- Clear transitional words and phrases.

- Development through paraphrases. Restate the author's ideas in your own words.

- A conclusion in which you reiterate the main points discussed, state the significance of the author's findings, and/or recommend a future course of action.

REWRITING

Once you've written a rough draft of your summary, it's time to revise. Follow the six-step revision checklist to ensure that your finished product will be acceptable to your readers.

1. Add detail for accuracy. When looking over your draft, be sure that you've covered all of the author's major points. If you haven't, now is the time to add any omissions. Be sure that your summary accurately covers the author's primary assertions.

2. Delete unnecessary information, biased comments, and/or dead words and phrases. Deleting is especially important in a summary. Because a summary must be no more than 5 to 15 percent of the length of the original source, your major challenge is brevity. Therefore, review your draft to see if you have included any of the author's points that are *not* essential. Such inessential information may include side issues that are interesting but not mandatory for your reader's understanding.

Another type of information you'll want to delete will be complex technical theories. Although such data might be valuable, a summary is not the vehicle for conveying this kind of information. If you have included such theories, delete them. These deletions serve two purposes. First, your summary will be stronger since it will focus only on key ideas and not on tangential arguments. Second, the summary will be briefer due to your deletions.

Delete biased comments. If you inadvertently have included any of your attitudes toward the topic, remove these biases. A summary should not present your ideas; it should reflect only what the author says.

Finally, as with all good technical writing, conciseness is a virtue. Long words and sentences are not appreciated. Reread your draft and delete any unnecessary words and phrases. Strive for an average of fifteen-word sentences and one- or two-syllable words. We're not suggesting that every multisyllabic word is incorrect, but keep them to a minimum. If an author has written about telecommunications or microbiology, for example, you shouldn't simplify these words. You should, however, avoid long words that aren't needed and long sentences caused by wordy phrases.

3. Simplify "impressive" words and complex technical terms. Good technical writing doesn't force the reader to look up words in a dictionary. Your goal in a summary is to present complex data in a brief and easily understandable package. This requires that you avoid difficult words. In a summary, *difficult* means two things. First, as with all good technical writing, you want to use words which readers understand immediately. Don't write *supersede* when you mean *replace*. Don't write *remit* when you mean *pay*. Second, you'll also need to simplify technical terms. To do so, you can define them parenthetically or merely replace the technical term with its definition.

4. Move information within the summary to parallel the author's organization. Does your summary reflect the order in which the author has presented his or her ideas? Your summary must not only tell the reader what the author has said but also how the author has presented these ideas. Make sure that your summary does this. If your summary fails to adhere to the author's organization, move information around. Cut and paste. Maintain the appropriate comparison/contrast, argument/persuasion, or cause/effect sequence.

5. Correct errors. Reread your draft and look for grammatical and mechanical errors. Don't undermine your credibility by misspelling words or incorrectly punctuating sentences. In addition, accurate information is mandatory. You might be writing this summary for your boss, who is giving a speech at a civic meeting or briefing upper-level management. Your boss is trusting you to provide accurate information. Therefore, to ensure that your boss is not embarrassed by errors, review your text against the original source. Make sure that all the information is correct.

6. Avoid sexist language. Sexist language is always inappropriate.

As one of our regular rewriting tips, we ask you to *enhance* the text by adding pronouns and positive words. For summaries, however, such enhancements would be incorrect. Since the summary must only restate what the author has written and be devoid of your attitude, total objectivity is necessary.

Use the Checklist on page 348 to help you write an effective summary.

PROCESS LOG

Following is one student's successful summarization of an article, including the student's rough draft, revisions, and finished copy.

PREWRITING

First, the student found an article, wrote down the required works cited information, read the article, and then took marginal notes and underlined key points.

WRITING

The student wrote a rough draft from the notes and underlining (Figure 14.1).

REWRITING

Student peer evaluators suggested revisions (Figure 14.2).

The student made these changes and prepared a finished version of the summary (Figure 14.3).

SUMMARY CHECKLIST

✔ Does your summary provide the works cited information for the article that you're summarizing?

✔ Is the works cited information correct?

✔ Does your summary begin with an introduction clearly stating the author's primary focus?

✔ Does your summary's discussion section explain the author's *primary* contentions and omit secondary side issues?

✔ In the discussion section, do you explain the author's contentions through pertinent facts and figures while avoiding lengthy technicalities?

✔ Is your content accurate? That is, are the facts that you've provided in the summary exactly the same as those the author provided to substantiate his or her point of view?

✔ Have you organized your discussion section according to the author's method of organization?

✔ Did you use transitional words and phrases?

✔ Have you omitted direct quotations in the summary, depending instead on paraphrases?

✔ Does your conclusion either reiterate the author's primary contentions, reveal the author's value judgment, or state the author's recommendations for future action?

✔ Is your summary completely objective, avoiding any of your own attitudes?

✔ Have you used an effective technical writing style, avoiding long sentences and long words?

✔ Are your grammar and mechanics correct?

✔ Have you avoided sexist language and multicultural biases?

CHAPTER HIGHLIGHTS

1. A summary is a compressed version of a much longer document or speech.
2. You can summarize chapters, books, speeches, reports, and material from the Internet.
3. A well-written summary is about one-tenth the length of the original.
4. Include an introduction, a discussion, and a conclusion to give your summary a coherent design.
5. Include source citations when appropriate.
6. The writing process helps you create an effective summary.

In "Robotics" (Radio Electronics, August 1986), Mark J. Robillard states, "Most robots must move around to accomplish their tasks" (71). The information that is needed to accomplish these tasks is gathered through an assortment of sensors—touch sensors, sound sensors, and light sensors. Depending on what type of tactile information is obtained through these sensors, objectives of the robot can be met. A robot's gripper could crush an object or not exert enough force to hold on to an object if it doesn't have sensors to determine the amount of force needed. Microswitches are used in the form of touch sensors. They are the simplest form of sensors used for this purpose. To eliminate weight and space used for switches, LED/phototransistor pairs can be used. "If you've ever been bowling, that setup should look familiar" (71). The phototransistor is then interfaced to a computer or other type of controller. The pairs of sensors provide more than just a there/not there signal. "The amount of light that is reflected provides an indication of how close the object is" (72). This approach is patented by "Heath's Hero 2000" (72) and uses optical encoder disks. Integrated circuits that use "strain guages and pressure sensitive pain" (72) are yet another way to detect the amount of force applied to an object.

All of these different methods of allowing a robot to interface with the real world "comprise a field of inquiry that is as large as robotics itself" (72).

FIGURE 14.1 Rough Draft

Could you separate Bibliography from your — *underline periodical titles*

topic sentence? In "Robotics" (Radio Electronics, August 1986), Mark J. Robillard

states, "Most robots must move around to accomplish their tasks" (71). *Avoid quotes*

paraphrase instead The information that is needed to accomplish these tasks is gathered

through an assortment of sensors—touch sensors, sound sensors, and

Repetitious light sensors. Depending on what type of tactile information is

obtained through these sensors, objectives of the robot can be met.

Do you need this? A robot's gripper could crush an object or not exert enough force to

hold on to an object if it doesn't have sensors to determine the

amount of force needed. Microswitches are used in the form of touch

sensors. They are the simplest form of sensors used for this purpose. To

Combine these sentences eliminate weight and space used for switches, LED/phototransistor pairs

can be used. "If you've ever been bowling, that setup should look famil- *Don't quote*

iar" (71). The phototransistor is then interfaced to a computer or other

type of controller. The pairs of sensors provide more than just a —— *Combine these sentences*

Don't quote there/not there signal. "The amount of light that is reflected provides

an indication of how close the object is" (72). This approach is patented *Do you need this?*

by "Heath's Hero 2000" (72) and uses optical encoder disks. Integrated

Don't quote circuits that use "strain guages and pressure sensitive pain" (72) are yet

another way to detect the amount of force applied to an object.

—— *Write just one ¶*

All of these different methods of allowing a robot to interface with

the real world "comprise a field of inquiry that is as large as robotics

itself" (72). *Don't quote*

FIGURE 14.2 Rough Draft with Editing

Robillard, Mark J. "Robotics." *Radio Electronics* August 1986: 71–72.

Robots, which must move to do their jobs, require an assortment of tactile sensors. These sensors help the robots locate items and use the appropriate amount of force for gripping an object. Microswitches, in the form of touch sensors, are the simplest sensing devices. These microswitches can be used to grip an object. In addition, microswitches can help the robot maintain contact with the ground to avoid falling down stairs. LED/photo-transistors, similar to those used on bowling alley foul lines, are less bulky than microswitches. The phototransistor, when inter-faced with a computer, provides more than a there/not there sig-nal. The light reflected indicates the exact placement of the object. The above sensors, however, have difficulty gauging the appropriate force required for gripping. This problem could be solved by using optical encoder disks and integrated circuits to determine the appropriate force. All of the above tactile sensing devices constitute an important part of successful robotics.

FIGURE 14.3 Revised Summary

ACTIVITIES

1. Locate an article that interests you (one within your field of expertise, your degree program, and/or an area which you would like to pursue). Study this article and summarize it according to the criteria provided in this chapter.

2. Locate an article that interests you. After reading it, take marginal notes (one-to three-word notations per paragraph) highlighting the article's key points.

3. Once you've read an article and made marginal notes, write either a topic or a sentence outline (discussed in Chapter 13).

4. Read three to five articles. Then determine what method of organization the authors have used. Have the authors used analysis? Others might use division, focusing on parts of a whole, comparison/contrast, argument/per-suasion, cause/effect, and so on.

5. Many textbooks begin or end chapters with summaries. Find such a sum-mary in one of your textbooks. Then read the accompanying chapter. Is the summary effective? If so, why? If not, why not? If the summary is ineffec-tive, how would you rewrite it?

6. After attending a lecture, meeting, or conference, summarize its content. Provide the speaker's name, the location of the presentation, and the date of presentation for the source citation.

7. Using the Internet, research 10 companies in your major field. Write a summary of each home page and connecting links.

Reports

OBJECTIVES

At one time or another, you'll be asked to write a report. Reports can vary in length. Generally, a shorter report (approximately one to five pages) will be formatted differently than a longer report (more than five pages long). This chapter focuses on the design of shorter reports. Chapter 16 discusses the design of longer reports, using a proposal as an example. Your reports will satisfy one or all of the following needs:

- Supply a record of work accomplished
- Record and clarify complex information for future reference
- Present information to a large number of people
- Record problems encountered
- Document schedules, timetables, and milestones
- Recommend future action
- Document current status
- Record procedures

The most common types of reports include the following:

1. Accident/incident reports. What happened, how did it happen, when did it happen, why did it happen, who was involved?

2. *Feasibility reports.* Can we do it, should we do it?

3. *Inventory reports.* What's in storage, what's been sold, what needs to be ordered?

4. *Staff utilization reports.* Is labor sufficient and efficiently used?

5. *Progress/activity reports* (weekly, monthly, quarterly, annually). What's our status?

6. *Travel reports.* Where did I go, what did I learn, whom did I meet, and so on?

7. *Lab reports.* How did we do it?

8. *Performance appraisal reports.* How's an employee doing on the job?

9. *Study reports.* What's wrong?

10. *Justification reports.* Here's why we need the material (or will pursue this action) on this date.

Although there are many different types of reports and individual companies have unique demands and requirements, certain traits are basic to all report writing.

CRITERIA FOR WRITING REPORTS

All reports share certain generic similarities in format, development, and style.

ORGANIZATION

Every report should contain four basic units: heading, introduction, discussion, and conclusion/recommendations.

Heading

The heading includes the date on which the report is written, the name(s) of the people to whom the report is written, the name(s) of the people from whom the report is sent, and the subject of the report (as discussed in Chapter 5, the subject line should contain a *topic* and a *focus*).

DATE: August 13, 2000
TO: Shelley Stine
FROM: Julie Jones
SUBJECT: REPORT ON TRIP TO SOUTHWEST REGIONAL
           CONFERENCE ON ENGLISH (FORT WORTH, TX)

Introduction

The introduction supplies an overview of the report. It can include three optional subdivisions:

• Purpose—a topic sentence(s) explaining why you are submitting the report (rationale, justification, objectives) and exactly what the report's subject matter is.

- Personnel—names of others involved in the reporting activity.
- Dates—what period of time the report covers.

Objectives: I attended the National Electronic Packaging Conference in Anaheim, CA, to review innovations in vapor phase soldering.

Dates: September 26–30, 2000

Personnel: Susan Lisk and Larry Rochelle

Some businesspeople omit the introductory comments in writing reports and begin with the discussion. They believe that introductions are unnecessary because the readers know why the reports are written and who is involved.

These assumptions are false for several reasons. First, it is false to assume that readers will know why you're writing the report, when the activities occurred, and who was involved. Perhaps if you are writing only to your immediate supervisor, there's no reason for introductory overviews. However, even in this situation you might have an unanticipated reader because

- Immediate supervisors change—they are promoted, fired, or retired or go to work for another company.
- Immediate supervisors aren't always available—they're sick for the day, on vacation, or off site for some reason.

Second, avoiding introductory overviews assumes that your readers will remember the report's subject matter. This is false because reports are written not just for the present, when the topic is current, but for the future, when the topic is past history. Reports go on file—and return at a later date. At that later date,

- You won't remember the particulars of the reported subject matter.
- Your colleagues, many of whom weren't present when the report was originally written, won't be familiar with the subject.
- You might have outside, lay readers who need additional detail to understand the report.

An introduction—which seemingly states the obvious—is needed to satisfy multiple readers, readers other than those initially familiar with the subject matter, and future readers who are unaware of the original report.

Discussion

The discussion of the report summarizes your activities and the problems you encountered. This is the largest section of the report and involves development, organization, and style (more on these later).

Conclusion/Recommendations

The conclusion allows you to sum up, to relate what you've learned, or to state what decisions you have made regarding the activities reported. The recommendations allow you to suggest future action, to state what you believe you and/or your company should do next.

The conference was beneficial. Not only did it teach me how the computer can save us time and money, but also I received hands-on training. Because the computer can assist our billing and inventory control, let's buy and install three terminals in bookkeeping before our next quarter.

DEVELOPMENT

Now that you know what subdivisions are traditional in reports, your next question is, "What do I say in each section? How do I develop my ideas?"

First, answer the reporter's questions.

1. *Whom* did you meet or contact, who was your liaison, who was involved in the accident, who was on your technical team, etc.?
2. *When* did the documented activities occur (dates of travel, milestones, incidents, etc.)?
3. *Why* are you writing the report and/or why were you involved in the activity (rationale, justification, objectives)? Or, for a lab report, for example, why did the electrode, compound, equipment, or material act as it did?
4. *Where* did the activity take place?
5. *What* were the steps in the procedure, what conclusions have you reached, or what are your recommendations?

Second, when providing the foregoing information, *quantify!* Don't hedge or be vague or imprecise. Specify to the best of your abilities with photographic detail.

The following justification is an example of vague, imprecise writing.

> Installation of the machinery is needed to replace a piece of equipment deemed unsatisfactory by an Equipment Engineering review.

Which machine are we purchasing? Which piece of equipment will it replace? Why is the equipment unsatisfactory (too old, too expensive, too slow)? When does it need to be replaced? Where does it need to be installed? Why is the installation important? A department supervisor will not be happy with the preceding report. Instead, supervisors need information *quantified*, as follows:

> The <u>exposure table</u> needs to be installed by <u>9/00</u> so that we can <u>manufacture printed wiring products with fine line paths and spacing (down to .0005 inches).</u> The table will replace the <u>outdated printer</u> in <u>Dept. 76.</u> Failure to install the table <u>will slow the production schedule by 27%.</u>

Note that the underlined words and phrases provide detail by quantifying.

STYLE

Style includes conciseness, simplicity, and highlighting techniques. As already discussed, you achieve conciseness by eliminating wordy phrases. Say *consider* rather than *take into consideration*; say *now* rather than *at this present time*. You achieve simplicity by avoiding old-fashioned words: *utilize* becomes *use*, *initiate* becomes *begin*, *supersedes* becomes *replaces*.

The value of highlighting has already been shown in this chapter. The parts of reports reviewed earlier use headings (Introduction, Discussion, Conclusion/Recommendation). Graphics can also be used to help communicate content, as evident in the following example. A recent demographic study of Kansas City predicted growth patterns for Johnson County (a large county south of Kansas City):

> Johnson County is expected to add 157,605 persons to its 1980 population of 270,269 by the year 2010. That population jump would be accompanied by a near doubling of the 96,925 households the county had in 1980. The addition of 131,026 jobs also is forecast for Johnson County by 2010, more than doubling its employment opportunity.

This report is difficult to access readily. We are overloaded with too much data. Luckily, the report provided a table (Table 15.1) for easier access to the data. Through highlighting techniques (tables, white space, headings), the demographic forecast is made accessible at a glance.

TABLE 15.1 Johnson County Predicted Growth by 2010			
	POPULATION	HOUSEHOLDS	EMPLOYMENT
1980	270,269	96,925	127,836
2010	427,874	192,123	258,862
%change	+58.3%	+98.2%	+102%

TYPES OF REPORTS

All reports include a heading (date, to, from, subject), an introduction, a discussion, and conclusion/recommendations. However, different types of short reports customize these generic components to meet specific needs. Let's look at the criteria for five common types of reports: trip reports, progress reports, lab reports, feasibility reports, and incident reports.

TRIP REPORTS

When you leave your work site to go to a conference, analyze problems in another work environment, give presentations, or make sales calls, you must report on these work-related travels. Your supervisors not only require that you document your expenses and time while off site, but they also want to be kept up to date on your work activities. Following is an overview of what you'll include in an effective trip report.

1. **Heading**

Date

To

From

Subject (topic + focus)

2. **Introduction (overview, background)**

Purpose: In the purpose section, document the date(s) and destination of your travel. Then comment on your objectives or rationale. What motivated the trip, what did you plan to achieve, what were your goals, why were you involved in job-related travel?

You might also want to include these following optional subheadings:

Personnel: With whom did you travel?

Authorization: Who recommended or suggested that you leave your work site for job-related travel?

3. **Discussion (body, findings, agenda)**

Using subheadings, document your activities. This can include a review of your observations, contacts, seminars attended, or difficulties encountered.

4. **Conclusion/recommendations**

Conclusion. What did you accomplish—what did you learn, whom did you meet, what sales did you make, what of benefit to yourself, colleagues, and/or your company occurred?

Recommendations. What do you suggest next? Should the company continue on the present course (status quo) or should changes be made in personnel or in the approach to a particular situation? Would you suggest that other colleagues attend this conference in the future, or was the job-related travel not effective? In your opinion, what action should the company take?

Figure 15.1 presents an example of a trip report.

PROGRESS REPORTS

Your supervisors want to know what you're doing at work. They want to know what progress you're making on a project, whether you're on schedule, what difficulties you might have encountered, and/or what your plans are for the next reporting period. Because of this, supervisors ask you to write progress (or activity or status) reports—daily, weekly, monthly, quarterly, or annually. The following are components of an effective progress report.

1. **Heading**

Date

To

From

Subject: Include the topic about which you are reporting and the reporting interval (date).

DATE: February 26, 2000
TO: Pat Berry
FROM: Debbie Rulo
SUBJECT: TRIP REPORT—RENTON WEST SEMINAR ON
           ELECTRONIC PACKAGING

INTRODUCTION

On Tuesday, February 23, 2000, I attended the Renton West
National Electronic Packaging Seminar, held in Ruidoso, NM. My
goal was to acquire hands-on training and to learn new techniques for electronic packaging for our Telemetry Department.
Also in attendance from our department were Richard DiBono
and Bill Cole.

DISCUSSION

Richard, Bill, and I attended the following seminars:

• *Production Automation*
This two-hour workshop was presented by Dr. Wang Ng, a noted
scientist from Southwest Texas State University. During Dr. Ng's
presentation, we reviewed four foam-encapsulation automated
techniques for electronic packaging and received hands-on training. Dr. Ng worked individually with each seminar participant.

• *Vapor Phase Soldering*
The hour-long presentation was facilitated by Garth Nelson, a
manufacturing supervisor from Spark Welding and Soldering, Inc.
(Colorado Springs, CO). Mr. Nelson showed a slide presentation
on techniques for processing double-sided chip components.

• *Electronics for Extreme Temperatures*
This hour-long presentation was led by Randy Towner and
Leanna Wilson, professors at the University of Nevada, Las Vegas.
Their scientific data on packaging under temperature extremes
was accompanied by a workbook and a multimedia presentation.

• *Robotics and You*
Denise Pakula, Canyon Electronics, Tempe, AR, spoke
about her company's robotics packaging equipment. This

FIGURE 15.1 Trip Report

Page 2
Debbie Rulo
February 26, 2000

half-hour presentation focused on the various machines
Canyon manufactures and their diverse capabilities.

CONCLUSION/RECOMMENDATION

Every presentation we attended was beneficial. However, the following information will clarify which workshop(s) would profit
our company the most:

1. Dr. Ng's program was the most useful and informative. His
   interactive presentation skills were superb, including hands-
   on activities, small-group discussions, and individual
   instruction. Richard, Bill, and I suggest that you invite Dr.
   Ng to our site for further consultation and training.

2. Vapor phase soldering is costly. We could not pay back our
   investment within this quarter. In addition, Garth Nelson's
   training techniques are outdated—training videos are not as
   interactive as our personnel request in their annual training
   evaluations.

3. The Towner/Wilson scientific data on packaging under
   temperature extremes would benefit our supervisors, who
   are interested in up-to-date information. However, our new
   hires would be overwhelmed by the data. If we considered
   inviting these professors in for consultation, we would
   have to assess audience carefully.

4. Canyon Electronics' half-hour presentation was geared
   more to sales than to instruction. Ms. Pakula's focus was on
   new models of robotics equipment. Our current equipment
   is satisfactory. If the Purchasing Department is looking for
   new vendors, we could suggest that they contact Canyon.

FIGURE 15.1 Trip Report (Cont.)

Date: May 31, 2000
To: Joanna Freeman
From: Lupe Salinas
Subject: JANUARY PROGRESS REPORT ON SALES CALLS

2. **Introduction (overview, background)**

Objectives. These can include the following:

• Why are you working on this project (what's the rationale)?
• What problems motivated the project?
• What do you hope to achieve?
• Who initiated the activity?

Personnel. With whom are you working on this project (i.e., work team, liaison, contacts)?

Previous activity. If this is the second, third, fourth, etc. report in a series, remind your readers what work has already been accomplished. Bring them up to date with background data or a reference to previous reports.

3. **Discussion (findings, body, agenda)**

Work accomplished. Using subheadings, itemize your work accomplished either through a chronological list or a discussion organized by importance.

Work remaining. Tell your reader what work you plan to accomplish next. List these activities, if possible, for easy access.

A visual aid, such as a Gantt chart or a pie chart, fits well after these two sections. The chart will graphically depict both work accomplished and work remaining. (See Chapter 9 for discussions of Gantt and pie charts.)

Problems encountered. Inform your reader(s) of any difficulties encountered (late shipments, delays, poor weather, labor shortages) not only to justify your possibly being behind schedule but also to show the readers where you'll need help to complete the project.

4. **Conclusion/Recommendations**

Conclusion. Sum up what you've achieved during this reporting period and provide your target completion date.

Recommendations. If problems were presented in the discussion, you can recommend changes in scheduling, personnel, budget, or materials which will help you meet your deadlines.

Figure 15.2 presents an example of a progress report.

LAB REPORTS

Professionals in electronics, engineering, medical fields, the computer industry, and other technologies often rank the ability to communicate as highly as they

DATE: April 2, 2000
TO: Buddy Ramos
FROM: Pat Smith
SUBJECT: FIRST QUARTERLY REPORT—PROJECT 80
            CONSTRUCTION

INTRODUCTION

In response to your December 10, 1999, request, following is our first quarterly report on Project 80 Construction (Downtown Airport). Department 93 is in the start-up phase of our company's 2000 build plans for the downtown airport and surrounding site enhancements. These construction plans include the following:

1. Airport construction—terminals, runways, feeder roads, observation tower, parking lots, maintenance facilities.

2. Site enhancements—northwest and southeast collecting ponds, landscaping, berms, and signage.

DISCUSSION

<u>Work Accomplished</u>
In this first quarter, we've completed the following:

1. *Subcontractors:* Toby Summers and Karen Kuykendahl worked with our primary subcontractors (Apex Engineering and Knoblauch and Sons Architects). Toby and Karen arranged site visitations and confirmed construction schedules. This work was completed January 12, 2000.

2. *Permits:* Once site visitations were held and work schedules agreed upon, Karen and Toby acquired building permits from the city. They accomplished this task on January 20, 2000.

3. *Core Samples:* Core sample screening has been completed by Department 86 with a pass/fail ratio of 76.4 percent pass to 23.6 percent fail. This meets our goal of 75 percent. Sample screening was completed January 30, 2000.

FIGURE 15.2 Progress Report

Page 2
Pat Smith
April 2, 2000

4. *Shipments:* Timely concrete, asphalt, and steel beam shipments this quarter have provided us a 30-day lead on scheduled parts provisions. Materials arrived February 8, 2000.

5. *EPA Approval:* EPA agents have approved our construction plans. We are within guidelines for emission controls, pollution, and habitat endangerment concerns. Sand cranes and pelicans nest near our building site. We have agreed to leave the north plat (40 acres) untouched as a wildlife sanctuary. This will cut into our parking plans. However, since the community will profit, we are pleased to make this concession. EPA approval occurred on February 15, 2000.

Work Remaining
To complete our project, we need to accomplish the following:

1. *Advertising:* Our advertising department is working on brochures, radio and television spots, and highway signs. Advertising's goal is to make the construction of a downtown airport a community point of pride and civic celebration.

2. *Signage:* With new roads being constructed for entrance and exit, our transportation department is working on street signage to help the public navigate our new roads.

In addition, transportation is working with advertising on signage designs for the downtown airport's two entrances. These signs will juxtapose the city's symbol (a flying pelican) with an airplane taking off. The goal is to create a logo that simultaneously promotes the preservation of wildlife and suggests progress and community growth.

FIGURE 15.2 Progress Report (Cont.)

Page 3
Pat Smith
April 2, 2000

3. *Landscaping:* We are working with Anderson Brothers Turf and Surf to landscape the airport, roads, and two ponds. Our architectural design team, led by Fredelle Schneider, is selecting and ordering plants, as well as directing a planting schedule. Anderson Brothers also is in charge of the berms and pond dredging. Fredelle will be our contact person for this project.

4. *Construction:* The entire airport must be built. Thus, construction comprises the largest remaining task.

The Gantt chart in Figure 1 clarifies our status at this time.

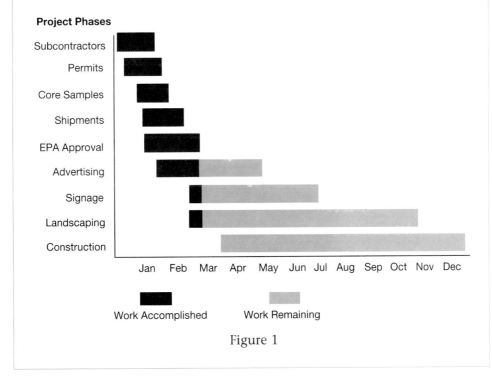

Figure 1

FIGURE 15.2 Progress Report (Cont.)

Page 4
Pat Smith
April 2, 2000

Problems Encountered
Core samples are acceptable throughout most of our construction site. However, the area set aside for the northwest pond had a heavy rock concentration. We believed this would cause no problem. Unfortunately, when Anderson Brothers began dredging, they hit rock, which had to be removed with explosives. Since this northwest pond is near the sand crane and pelican nesting sites, EPA requested that we wait until the birds were resettled. The extensive rock removal and wait for wildlife resettlement has slowed our progress. We are behind schedule on this phase. This schedule delay and increased rock removal will affect our budget.

CONCLUSION/RECOMMENDATION

Though we have just begun this project, we still are approximately 15 percent of the way toward our goal. We anticipate a successful completion, especially since deliveries have been timely.

Only the delays at the northwest pond site present a problem. We are behind schedule and over cost. With additional personnel to speed the rock removal and increased funds, we can meet our target dates. Darlene Laughlin, our city council liaison, is the person to see about corporate investors, city funds, and big-ticket endowments. With your help, and Darlene's cooperation, we should meet our build schedules.

FIGURE 15.2 Progress Report (Cont.)

do technical skills. Conclusions derived from a technical procedure are worthless if they reside in a vacuum. The knowledge you acquire from a lab experiment *must* be communicated to your colleagues and supervisors so they can benefit from your discoveries. That's the purpose of a lab report—to document your findings. You write a lab report after you've performed a laboratory test to share with your readers

- Why the test was performed.
- How the test was performed.
- What the test results were.
- What follow-up action (if any) is required.

The following are components of a successful lab report.

1. **Heading**

Date

To

From

Subject (topic + focus)

2. **Introduction (overview, background, purpose)**

Why is this report being written? To answer this question, provide any or all of the following:

- The rationale (What problem motivated this report?)
- The objectives (What does this report hope to prove?)
- Authorization (Under whose authority is this report being written?)

3. **Discussion (body, methodology)**

How was the test performed? To answer this question, provide the following:

- Apparatus (What equipment, approach, or theory have you used to perform your test?)
- Procedure (What steps—chronologically organized—did you follow in performing the test?)

4. **Conclusion/recommendations**

Conclusion. The conclusion of a lab report presents your findings. Now that you've performed the laboratory experiment, what have you learned or discovered or uncovered? How do you interpret your findings? What are the implications?

Recommendations. What follow-up action (if any) should be taken?

You might want to use graphics to supplement your lab report. Schematics and wiring diagrams are important in a lab report to clarify your activities, as shown in Figure 15.3.

Date: July 18, 2000
To: Dr. Jones
From: Sam Ascendio, Lab Technician
Subject: LAB REPORT ON THE ACCURACY OF DECIBEL
             VOLTAGE GAIN (A) MEASUREMENTS

INTRODUCTION

Purpose
Technical Services has noted inaccuracies in recent measurements.
In response to their request, this report will present results of tested
A (gain in decibels) of our ABC voltage divider circuit. Measured A
will be compared to calculated A. This will determine the accuracy
of the measuring device.

DISCUSSION

Apparatus
   • Audio generator
   • Decade resistance box
   • 1/2 W resistor: four 470 ohm, two 1 kilohm, 100 kilohm
   • AC millivoltmeter

Procedure
   1. Figure A shows a voltage divider. For each value of R (resis-
      tances) in Table 1, voltage gain was calculated (table attached).
   2. An audio generator was adjusted to give a reading of 0 dB
      for input voltage.
   3. Output voltage was measured on the dB scale. This reading
      is the measured A and is recorded in Table 1.
   4. Step 3 was repeated for each value of R listed in Table 1.
   5. Figure B shows three cascaded voltage dividers. A1 = v2/v1,
      A2 = v3/v2, and A3 = v4/v3. These voltage gains added
      together give the total voltage, as recorded in Table 2.
   6. The circuit in Figure B was connected.
   7. Input voltage was set at 0 dB on the 1-V range of the AC
      millivoltmeter.
   8. Values V2, V3, and V4 were read and recorded in Table 3.

FIGURE 15.3 Lab Report

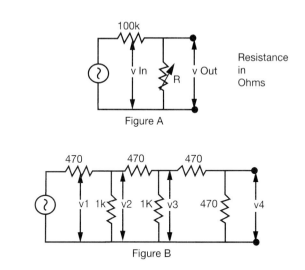

Figure A

Figure B

TABLE 1

R	Calculated A	Measured A
240 kilohm	-3.02 dB	-3.21 dB
100 kilohm	-6.02 dB	-6.13 dB
46 kilohm	-10.00 dB	-9.71 dB
11 kilohm	-20.00 dB	-20.00 dB
1 kilohm	-40.00 dB	-40.00 dB

TABLE 2		TABLE 3	
A1	.500	V1	0 dB
A2	.500	V2	-5.97 dB
A3	.500	V3	-11.97 dB
A	.125	V4	-18.06 dB

CONCLUSION/RECOMMENDATIONS

Accuracy between measured and calculated A in Table 1 was between .01 and .1 dB. This is acceptable. Accuracy between measured A in Table 3 and calculated voltage gain in Table 2 was between .01 and .1 dB—also very accurate. These tests show minimal error. No further action should be taken.

FIGURE 15.3 Lab Report (Cont.)

Occasionally, your company plans a project but is uncertain whether the project is feasible. For example, your company might be considering the purchase of equipment but is concerned that the machinery will be too expensive, the wrong size for your facilities, or incapable of performing the desired tasks. Perhaps your company wants to expand and is considering new locations. The decision makers, however, are uncertain which locations would be best for the expansion. Maybe your company wants to introduce a new product to the marketplace, but your CEO wants to be sure that a customer base exists before funds are allocated.

One way a company determines the viability of a project is to perform a feasibility study and then write a feasibility report documenting the findings. The following are components of an effective feasibility report.

1. **Heading**

Date

To

From

Subject (topic + focus)

Subject: FEASIBILITY REPORT ON XYZ PROJECT
(focus)   (topic)

---

2. **Introduction (overview, background)**

Objectives. Under this subheading, you can answer any of the following questions:

- What is the purpose of this feasibility report? Until you answer this question, your reader doesn't know. As mentioned earlier in this chapter, it's false to assume prior knowledge on the part of your audience. One of your responsibilities is to provide background data. To answer the question regarding the report's purpose, you should provide a clear and concise statement of intent.

- What problems motivated this study? To clarify for your readers the purposes behind the study, *briefly* explain either

   —what problems cause doubt about the feasibility of the project (i.e., is there a market, is there a piece of equipment available which would meet the company's needs, is land available for expansion?)

   —what problems led to the proposed project (i.e., current equipment is too costly or time-consuming, current facilities are too limited for expansion, current net income is limited by an insufficient market)

- Who initiated the feasibility study? List the name(s) of the manager(s) or supervisor(s) who requested this report.

Personnel. Document the names of your project team members, your liaison between your company and other companies involved, and/or your contacts at these other companies.

3. Discussion (body, findings)

Under this subheading, provide accessible and objective documentation.

Criteria. State the criteria upon which your recommendation will be based. Criteria are established so you have a logical foundation for comparison of personnel, products, vendors, costs, options, schedules, and so on.

Analysis. In this section, compare your findings against the criteria. In objectively written paragraphs, develop the points being considered. You might want to use a visual such as a table to organize the criteria and to provide easy access.

4. Conclusion/recommendations

Conclusion. In this section, you go beyond the mere facts as evident in the discussion section: you state the significance of your findings. Draw a conclusion from what you've found in your study. For example, state that "Tim is the best candidate for Director of Personnel" or "Site 3 is the superior choice for our new location."

Recommendations. Once you've drawn your conclusions, the next step is to recommend a course of action. What do you suggest that your company do next? Which piece of equipment should be purchased, where should the company locate its expansion, or is there a sufficient market for the product?

Figure 15.4 presents an example of a feasibility report.

INCIDENT REPORTS

If a problem occurs within your work environment that requires investigation and suggested solutions, you might be asked to prepare an *incident report* (also called a *trouble report* or *accident report*).

Engineering environments requiring maintenance reports rarely provide employees with easy-to-fill-in forms. To write an incident report when you have not been given a printed form, include the following components:

1. Heading

Date
To
From
Subject (topic + focus)

Subject: REPORT ON CHILLED WATER LEAKS IN D/823
      (focus)   (topic)

DATE: August 13, 2000
TO: Pat Hobby
FROM: Nick Adams
SUBJECT: RECOMMENDED PURCHASE OF THREE AAA VIDEO
           CASSETTE RECORDERS (VCRs)

INTRODUCTION

Purpose
   The purpose of this report is to recommend the purchase of three
   new VCRs.

Problem
   We need these VCRs in the Management Development Training
   Department for the following reasons:

   1. Currently, our two ABC VCRs (purchased in 1997) break down
      approximately twice per week. When this occurs, they are unus-
      able for ten hours per week and cost the department $150 each
      for weekly repairs.
   2. Our department has acquired four new trainers. We offer 25 percent
      more management development courses this year than we did last
      year. Enrollment has increased 115 percent. These factors, coupled
      with the increased use of video presentations, require working VCRs.

Personnel
   Our vendor contacts are as follows:

   ABC Electeck   XYZ Inc.   AAA Electronic
   Steve Ross   Jay Rochlin   Karen Allen
   415–628–9704   415–489–7286   415–459–3872

DISCUSSION

Criteria
   The following criteria were considered when I compared the three
   top-selling VCRs:

   1. Maintenance—We need to purchase a VCR which comes with a
      frequent (either quarterly or biannual) service agreement (at no
      extra charge).

FIGURE 15.4 Feasibility Report

Page 2
Nick Adams
August 13, 2000

2. Service Personnel—The service people should be trained by the company that sells the VCR.
3. Warranty—The warranty should be for at least one year.
4. Cost—The VCR should cost no more than $1,000.

Analysis of Three VCRs
The following table compares the top three VCRs we researched:

Table 1 Criteria Comparison			
CRITERIA	AAA	ABC	XYZ
Maintenance Package	Quarterly	Biannual	No Options
Service Personnel	AAA Certified	Subcontracted	No Options
Warranties	1 Year Parts and Labor	90 Days Parts and Labor	1 Year Parts and Labor
Cost	$950.00	$899.99	$1,500.00

CONCLUSION

Although the ABC and XYZ VCRs are good pieces of equipment, neither meets all our criteria. In particular, their service personnel are not trained by the VCR manufacturers. Also, the XYZ VCR, our current model, is the most expensive and offers the fewest service options.

RECOMMENDATION

Given the combination of cost, service contracts, and warranties, AAA is our best choice for purchase. With three new AAA VCRs, we will be able to offer more efficient and effective seminars to a greater number of company personnel.

FIGURE 15.4 Feasibility Report (Cont.)

2. Introduction

Purpose. In this section, document when, where, and why you were called to perform maintenance. What motivated your visit to the scene of the problem?

3. Discussion (body, findings, agenda, work accomplished)

Using subheadings or itemization, quantify what you saw (the problems motivating the activity) and what you did to solve the problem.

4. Conclusion

Explain what caused the problem.

5. Recommendations

Relate what could be done in the future to avoid similar problems.

Figure 15.5 presents an example of an incident report.

PROCESS

Now that you know the criteria for reports in general and for specific types of reports (trip reports, progress reports, lab reports, feasibility reports, and incident reports), the next step is to construct these reports. How do you begin? As always, *prewrite*, *write*, and *rewrite*.

PREWRITING

We've presented several techniques for prewriting—reporter's questions, clustering/mind mapping, flowcharting, and brainstorming/listing. An additional technique is called *branching*.

As with flowcharting and mind mapping, branching allows you to depict information graphically so you can not only gather data but also visualize it. This type of prewriting benefits both left-brain and right-brain people—those who are linear (outline oriented) as well as those who are more graphically attuned. Figure 15.6 shows an example of branching.

Branching is ideally suited for short reports. You can focus on whether the primary subject is a trip report, progress report, lab report, feasibility report, or incident report in the *main idea*. In addition, the *subordinate points* easily correspond to the introduction, discussion, and conclusion/recommendations. Finally, you can develop your ideas more specifically in the *subheadings*. Figure 15.7 shows an example of branching for a trip report.

You can see in Figure 15.7 how branching meshes with reporter's questions. Branching allows you to sketch out your organization and visualize your content. The reporter's questions help you make that content factual, precise, specific, and quantified.

WRITING

Once you've sketched an outline for your report using branching, the next step is to write the text. To write your report, do the following:

Date: October 16, 2000
To: Tom Warner
From: Carlos Sandia
Subject: REPORT ON CHILLED WATER LEAKS IN D/823

INTRODUCTION

On October 15, 2000, a flood was reported in D/823. I was called in to repair this leak. Following is a report on my findings and maintenance activities.

DISCUSSION

Agenda

8:15 p.m.   The flood was reported.
8:25 p.m.   I arrived at D/823 and discovered that the water level in the expansion tank at HVAC unit #253R-01 was 4 feet above normal and the tank valve was open.
8:30 p.m.   A second flood was reported at HVAC unit #937-01 in the same department.
8:35 p.m.   I shut off the open tank valves.
9:00 p.m.   The expansion tank levels and pressure returned to normal.
9:30 p.m.   All leaks were secure.

CONCLUSION

The chilled water leaks were caused by a malfunctioning level switch on the chilled water expansion tanks in the west boiler rooms. This caused the water level in the tanks to rise, which increased the system pressure to 150 lb.

RECOMMENDATIONS

The level control switch will be repaired. An alarm system will be installed by November 1, 2000.

FIGURE 15.5 Incident Report

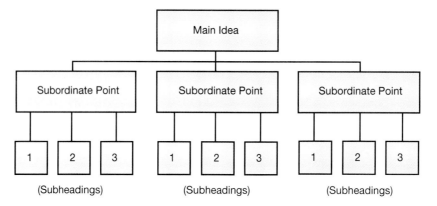

FIGURE 15.6 Branching

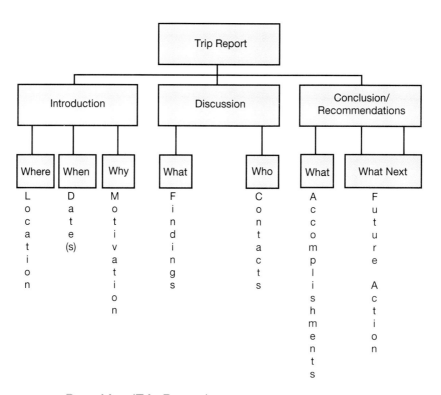

FIGURE 15.7 Branching (Trip Report)

1. Reread your prewriting. Review what you've sketched in branching. Determine whether you've covered all important information. If you believe you've omitted any significant points, add them for clarity. If you've included any irrelevant ideas, omit them for conciseness.

2. Assess your audience. Your decisions regarding points to omit or include will depend on your audience. If your audience is familiar with your subject, you might be able to omit background data. However, if you have multiple audiences or an audience new to the situation, you'll have to include more data than you might have assumed necessary in your prewriting.

3. Draft the text. Focusing on your major headings (introduction, discussion, conclusion/recommendations), use the sufficing technique mentioned in earlier chapters. Just get your ideas down on paper in a rough draft without worrying about grammar.

4. Organize your content. You can organize the discussion portion of your report using any one of these four modes: *chronology, importance, comparison/contrast*, and *problem/solution*. Which of these modes you use depends on your subject matter.

For example, if the subject of your trip report is a price check of sales items at two stores, comparison/contrast would be appropriate, as in Figure 15.8.

If the subject of your incident report is a site evaluation, chronology might work in the report's discussion, as in the following example:

	DISCUSSION
8:00 A.M.	I arrived at the site and met with the supervisor to discuss procedures.
9:00 A.M.	We checked the water tower for possible storm damage. Only 10 shingles were missing.
10:00 A.M.	We checked the irrigation channel. It was severely damaged, the wall cracked in six places and water seeping through its barriers. Surrounding orchards will be flooded.
11:00 A.M.	We checked fruit bins. No water had entered.
1:00 P.M.	We checked the freezer units. The storm had disrupted electricity for four hours. All contents were destroyed.
2:00 P.M.	The supervisor and I returned to his office to evaluate our findings.

As discussed in earlier chapters, chronology is an easy method of organization to use and to follow. However, it is not always your most successful choice. Chronology inadvertently buries key data. For instance, in the preceding report findings, the most important discoveries occur at 10:00 A.M. and at 1:00 P.M.,

To: Meagan Clem
From: Mary Jane Post
Date: January 12, 2000
Subject: PRICE CHECK REPORT—HANDY SANDY HARDWARE

INTRODUCTION

On Thursday, January 8, 2000, I compared our sale prices on plumbing items with those of Handy Sandy Hardware, 1000 W. 29th St., Newtown, WI.

DISCUSSION

Item	Hughes's Sale Price	Handy Sandy's Sale Price
1/2" copper tubing	$4.89 (10')	$4.99 (10')
3/4" copper tubing	10.50 (10')	9.99 (10')
1/2" CPVC pipe	2.39 (10')	2.99 (10')
3/4" CPVC pipe	3.49 (10')	5.99 (10')
1 1/2" PVC pipe	3.99 (10')	4.99 (10')
Acme 800 faucet	40.95	39.95
Acme 700 faucet	22.95	25.95

CONCLUSION

On five of the seven items (71.4%), Hughes's had the lower price.

RECOMMENDATIONS

We should continue to compare our prices to Handy Sandy's. We should also try to lower any prices which exceed theirs.

FIGURE 15.8 Trip Report Organized by Comparison/Contrast

hidden in the middle of the list. To avoid making your readers guess where the important information is, use the third method of organization—importance—in which you list the most important point first, lesser points later, as in the following example:

> DISCUSSION
> 1. Freezer units: Electricity was out for four hours. All contents were destroyed.
> 2. Irrigation channel: The wall was cracked in six places. Seepage was significant and sure to affect the orchards.
> 3. Water tower: Minor damage to shingles.
> 4. Fruit bins: No damage.

If you're writing a progress report to document a problem and suggest a solution, use problem/solution to organize your data, as follows:

> DISCUSSION
> *Problem*—Production schedules on our M23 and B19 are behind three weeks due to machinery failures. Our numerical control device no longer maintains tolerance. This is forcing us to rework equipment, which is costing us $200 per reworked piece.
> *Solution*—We must purchase a new numerical control device. The best option is an Xrox 1234. This machine is guaranteed for five years. Any problems with tolerance during this period are covered. Xrox either will correct the errors on site or provide us a loaner until the equipment is fixed. To avoid further production delays, we must purchase this machine by 1/18/00.

REWRITING

After a gestation period in which you let the report sit so you can become more objective about your writing, retrieve the report and rewrite. Perfect your text by using the following rewriting techniques:

1. Add detail for clarity. Have you answered the reporter's questions as thoroughly as needed? Don't assume your readers will know the whys and wherefores. Spell it all out, exactly. You often have multiple readers, many of whom do not know your motivations or objectives. They need clarity.

2. Delete dead words and phrases for conciseness. For example, in your recommendations, don't say, "Acme's opinion is that apparently the proposed ideas will not successfully supersede those already implemented." Instead, simply write, "Recommendations: No further action is required."

3. Simplify old-fashioned words and phrases. What does *pursuant* mean? And how about *issuance of this report?* The ultimate goal of technical writing is to communicate, not to confuse.

4. Move information within your discussion for emphasis. Make sure that you've either maintained chronology (if you're documenting an agenda) or used importance to focus on your main point. If you've confused the two, cut and paste—move information around to achieve your desired goals.

5. *Reformat your text for accessibility.* Highlight your key points with underlining or boldface. Use graphics to assist your readers. Don't overload your readers with massive blocks of impenetrable text.

6. *Enhance your text for style.* Be sure to quantify when needed, personalize with pronouns for audience involvement, and be positive. Stress words like *benefit*, *successful*, *achieved*, and *value* rather than bowing to negative words like *can't*, *failed*, *confused*, or *mistaken*.

7. *Proofread and correct the report for grammatical and contextual accuracy.*

8. *Avoid sexist language.* *Foreman* should be *supervisor*, *chairman* should be *chairperson* or *chair*, and *Mr. Swarth*, *Mr. James*, and *Sue* must be *Mr. Swarth*, *Mr. James*, and *Ms. Aarons*.

Use the following checklist to help you in writing your short report.

REPORT CHECKLIST

✔ *Does your subject line contain a topic and a focus?* If you write only "Subject: TRIP REPORT" or "Subject: FEASIBILITY REPORT," you have not communicated thoroughly to your reader. Such a subject line merely presents the focus of your correspondence. But what's the topic? To provide both topic and focus, you need to write "Subject: TRIP REPORT ON SOLVENT TRAINING COURSE, ARCO CORPORATION—3/15/00" or "Subject: FEASIBILITY REPORT ON COMPANY EXPANSION TO BOLKER BLVD."

✔ *Does the introduction explain the purpose of the report, document the personnel involved, and/or state when and where the activities occurred?*

✔ *When you write the discussion section of the report, do you quantify what occurred?* In this section, you must clarify precisely. Supply accurate dates, times, calculations, and problems encountered.

✔ *Is the discussion accessible?* To create reader-friendly ease of access, use highlighting techniques, such as headings, boldface, underlining, and itemization. You also might want to use graphics, such as pie charts, bar charts, or tables.

✔ *Have you selected an appropriate method of organization in your discussion?* You can use chronology, importance, comparison/contrast, and/or problem/solution to document your findings.

✔ *Does your conclusion present a value judgment regarding the findings presented in the discussion?* The discussion states the facts; the conclusion decides what these facts mean.

✔ *In your recommendations, do you tell your reader(s) what to do next or what you consider to be the appropriate course of action?*

✔ *Have you maintained a low fog index for readability?*

✔ *Have you effectively recognized your audience's level of understanding (high tech, low tech, lay, management, subordinate, colleague) and written accordingly?*

✔ *Is your report accurate?* Correct grammar and calculations make a difference. If you've made errors in spelling, punctuation, grammar, or mathematics, you will look unprofessional.

PROCESS LOG

Let's look at how one student used the writing process (prewriting, writing, and rewriting) to construct her progress report.

PREWRITING

First, the student used branching so she could visually gather data and determine objectives (Figure 15.9).

WRITING

Next, the student wrote a quick draft without worrying about correctness. She focused on the main units of the information she discovered in prewriting (Figure 15.10).

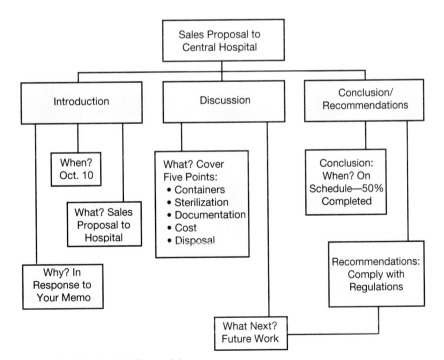

FIGURE 15.9 Student's Branching

November 6, 2000

To: Carolyn Jensen
From: Shuan Wang.
Subject: PROGRESS REPORT.

Purpose: This is a progress report on the status of a sales proposal you requested in your memo of October 10, 2000. The objective of this proposal is the sale of a total program for infectious waste control and disposal to a Central Hospital. The proposal will cover the following five areas.

1. Containers
2. Steam sterilization
3. Cost savings
4. Landfill operating
5. Computerized documentation

Work Completed: The following is a list of items that are finished on the project.

Containers
I have finished a description of the specially designed containers with disposable biohazard bag.

Steam sterilization
I have set a instructions for using the Biological Detector, model BD 12130 as a reliable indicator of sterilization of infectious wastes.

Cost savings
Central Hospital sent me some information on installing a pathological incinerator as well as the constant personel and maintenance costs to operate them. I ran this information through our computer program "Save-Save". The results show that a substantial savings will be realized by the use of our services.

Future Work: I have an appointment to visit Central Hospital on November 7, 2000. At that time I will make on-site evaluations of present waste handling practices at Central Hospital. After carefully studing this evaluations, I will report to you on needed changes to comply with governmental guidelines.

Conclusion: The project is proceeding on schedule. Approximately 50 percent of work is done (see graph). I don't see any problem at this time and should be able to meet the target date.

FIGURE 15.10 Student's Rough Draft

REWRITING

After drafting this report, the student submitted her copy to a peer review group, which helped her revise her text (Figure 15.11).

November 6, 2000

To: Carolyn Jensen
From: Shuan Wang   *Needs topic*
Subject: PROGRESS REPORT×   *& focus*

Reformat-   *— Redundant*
underline   Purpose: This is a progress report on the status of a sales proposal
& use all   you requested in your memo of October 10, 2000. The objective   *wordy?*
CAPS   of this proposal is the sale of a total program for infectious waste
             control and disposal to a Central Hospital. The proposal will
             cover the following five areas. *weak word*

   1. Containers
   2. Steam sterilization
   3. Cost savings
   →4. Landfill operating *— Needs new title*
   5. Computerized documentation
Reformat   *wordy*

Work Completed: The following is a list of items that are finished
on the project.

Containers
I have finished a description of the specially designed containers
with disposable biohazard bag. *when?*

Steam sterilization   *Huh?*
I have set a instructions for using the Biological Detector, model
BD 12130 as a reliable indicator of sterilization of infectious *when*
wastes. *wordy*

Cost savings   *vague*
Central Hospital sent me some information on installing a pathological incinerator as well as the constant personnel and maintenance costs to operate them. I ran this information through our
computer program "Save-Save." The results show that a substantial
savings will be realized by the use of our services.   *vague*

FIGURE 15.11 Peer Group Revisions

Reformat

Future Work: I have an appointment to visit Central Hospital on November 7, 2000. At that time I will make on-site evaluations of present waste handling practices at Central Hospital. After carefully studing this evaluations, I will report to you on needed changes to comply with governmental guidelines.

Reformat

Conclusion: The project is proceeding on schedule. Approximately 50 percent of work is done (see graph). I don't see any problem at this time and should be able to meet the target date.

FIGURE 15.11 Peer Group Revisions (Cont.)

After discussing the suggested changes with her peer review group, the student revised her draft and submitted her finished copy (Figure 15.12).

INTEROFFICE CORRESPONDENCE

DATE: November 6, 2000
TO: Carolyn Jensen
FROM: Shuan Wang
SUBJECT: PROGRESS REPORT ON CENTRAL HOSPITAL
            SALES PROPOSAL

INTRODUCTION

Purpose
In response to your October 10, 2000, request, following is a
report on the status of our Central Hospital sales proposal. This
proposal will present Central Hospital our total program for
infectious waste control and disposal. The proposal will cover
these five topics:

- Containers
- Steam sterilization
- Cost savings
- On-site evaluations
- Landfill disposal

DISCUSSION

Work Completed
1. Containers—On October 5, I finished a description of the
   specially designed containers with biohazard bags.
2. Steam sterilization—On November 4, I wrote instructions
   for using our Biological Detector, model BG 12130. This
   detector measures infectious waste sterilization levels.
3. Cost savings—Central Hospital sent me their cost charts for
   pathological incinerator expenses and maintenance costs. I
   used our computer program "Save-Save" to evaluate these
   costs. The program results show that our services can save
   Central Hospital $15,000 per year on annual incinerator
   costs, after a two-year break-even period.

FIGURE 15.12 Final Copy

Page 2
Shuan Wang
November 6, 2000

Work Remaining
1. On-site evaluations—When I visit Central Hospital on November 11, 2000, I will evaluate their present waste handling system. After studying these evaluations, I will report changes necessary for government compliance.
2. Landfill disposal—I will contact the Environmental Protection Agency on November 13, 2000, to receive authorization for disposing sterilized waste at our sanitary landfill.

CONCLUSION

The project is proceeding on schedule. Approximately 55 percent of our work is completed. I see no problems at this time. We should meet our December 31, 2000, target date.

FIGURE 15.12 Final Copy (Cont.)

CHAPTER HIGHLIGHTS

1. Reports are used to document many different occurrences on the job.
2. Use headings, such as "Introduction," "Discussion," and "Conclusion/recommendations," when designing your report.
3. Trip reports document work-related travel.
4. Progress reports recount work accomplished and work remaining on a project.
5. Lab reports document the findings from a lab analysis.
6. Feasibility reports are used to determine the viability of a proposed project.
7. Branching is a visual prewriting technique that will help you write effective reports.

ACTIVITIES

1. Write a progress report. The subject of this report can involve a project or activity at work. Or, if you haven't been involved in job-related projects, write about the progress you're making in this class or another course you're taking. Write about the progress you're making on a home improvement project (refinishing a basement, constructing a deck, painting and papering a room). Write about the progress you're making on a hobby (rebuilding an antique car, constructing a computer, or making model trains, etc.). Whatever your topic, first prewrite (using branching), then write a draft, and, finally, postwrite, revising the text. Abide by all the criteria presented in this chapter regarding progress reports.

2. Write a lab report. The subject of this report can involve a test you're running at work or in one of your classes. Whatever topic you select, follow the three stages in the writing process to construct your report: prewrite (using branching), write a draft, and then revise the draft in postwriting. Use the criteria regarding lab reports presented in this chapter to help you write the report.

3. Write a feasibility report. You can draw your topic either from your work environment or home. For example, if you and your colleagues were considering the purchase of new equipment, the implementation of a new procedure, expansion to a new location, or the marketing of a new product, you could study this idea and then write a feasibility report on your findings. If nothing at work lends itself to this topic, then consider plans at home. For example, are you and your family planning a vacation, the purchase of a new home or car, the renovation of your basement, or a new business venture? If so, study this situation. Research car and home options, study the market for a new business, and/or get bids for the renovation. Then write a feasibility report to your family documenting your findings. Whether your topic comes from business or home, gather your data in prewriting (using branching), draft your text in writing, and then revise in rewriting. Follow the criteria for feasibility reports provided in this chapter to help you write the report.

4. Write an incident report. You can select a topic either from work or home. If you have encountered a problem at work, write an incident report documenting the problem and providing your solutions to the incident. If nothing has happened at work lending itself to this topic, then look at home. Has your car broken down, did the water heater break, did you or any members of your family have an accident of any sort, did your dog or cat knock over the vase your mother-in-law gave you for Christmas? Consider such possibilities, and then write an incident report documenting the incident. Follow the criteria for incident reports provided in this chapter, and use writing process techniques (prewriting, writing, and rewriting).

5. Find examples of lab reports, trip reports, progress reports, feasibility reports, and/or incident reports. You can find these at your company or visit another job site. Bring these reports to class. Then, using the criteria presented in this chapter, form small groups and discuss whether the reports are successful or unsuccessful. If they are good, specify how and why. If they are flawed, discuss what's wrong, prescribe solutions, and rewrite the reports to improve them.

6. Look around your company or visit another job site and find examples of short reports not discussed in this chapter. Bring these reports to class. Then, using the criteria for short reports presented in this chapter, form small groups and discuss how these unique reports differ from or abide by the criteria we've presented. Also discuss whether the reports you've found are successful or unsuccessful. If they are good, discuss why and how. If they are flawed, talk about how they could be improved and rewrite them.

7. Interview an employee and a supervisor in your profession or the field of your choice. Ask these people why they write reports, what their reports are supposed to accomplish, who their audiences are, what length reports are preferred in their company, how many reports they write in a week or month, and so on. Use your imagination regarding the questions you ask. Then, once you've received your answers, form small groups and convey your findings to your classmates. To do so, either present an oral presentation or—better yet—write a group report!

8. As a class, take a field trip. Visit a publishing firm, see a play, go to a museum, hear a guest speaker on campus, or interview a professional technical writer, for example. Then, in small groups, write a trip report about your observations. Once the reports are written, compare and contrast the writing. Which group of students (or which student) has written the best report? To make this judgment, specify why the report is successful. Doing so will provide for the class a model of good writing, as agreed on by class consensus. Class members can replicate this report for future success. In addition, discuss how less successful reports fell short of the desired objectives. This will help show students what to avoid in future report-writing activities.

9. Revision is the key to good writing. An example of a seriously flawed progress report that needs revising is shown in Figure 15.13. To improve this report, form small groups and first decide what's missing according to

DT1234 Test Equipment Project

Date: February 17, 2000
Software Engineer: Sylvia Light
Section Supervisor: Sam Tintereto

Responsibilities:
    Test Executives
    Test Routines
    Calibration
    Enhancements
    Troubleshooting Software

Test Executives:	Complete	Incomplete
Entry	X	
Troubleshooting		X
Calibration		X
Enhancements		X

Test Routines:	Complete	Incomplete
ProLog		X
PreSet		X
Flight		X
Switch		X
Contact		X
Clear		X

Peripherals:	Complete	Incomplete
6000		X
5335		X
1001		X
4250		X

Note: All calibrations, enhancements, and troubleshooting software
are incomplete.

FIGURE 15.13 Flawed Report

this chapter's criteria for good progress reports. Then use the rewriting techniques presented in this chapter to rewrite and revise the report.

Case Studies

In small groups, read the following case studies and write a feasibility report.

1. You manage an engineering department at Acme Aerospace. Your current department supervisor is retiring. Thus, you must recommend the promotion of a new supervisor to the company's executive officer, Kelly Adams. You know that Acme seeks to promote individuals who have the following traits:

   - Familiarity with modern management techniques and concerns, such as TQM (total quality management), teamwork, global economics, and the management of hazardous materials.
   - An ability to work well with colleagues (subordinates, lateral peers, and management).
   - Thorough knowledge of one's areas of expertise.

   You have the following candidates for promotion. Using the information provided about each and the criteria for feasibility reports discussed in this chapter, write your report recommending your choice for a new supervisor.

   A. *Pat Jefferson.* Pat has worked for Acme for 12 years. Pat, in fact, has worked up to a position as a lead engineer by having started as an assembler, then working in test equipment, quality control, and environmental safety and health (ESH). As an engineer in ESH, Pat was primarily in charge of hazardous waste disposal. Pat's experience is lengthy, although Pat has only taken two years of college coursework and one class in management techniques. Pat is well liked by all colleagues and is considered to be a team player.

   B. *Kim Kennedy.* Kim is a relatively new employee at Acme, having worked for the company for two years. Kim was hired directly out of college after earning an M.B.A. degree from the Mountaintop College School of Management. As such, Kim is extremely familiar with today's management climate and modern management techniques. Kim's undergraduate degree was a B.S. in business with a minor in engineering. Currently, Kim works in the engineering department as a departmental liaison, communicating the engineering department's concerns to Acme's other departments. Kim has developed a reputation as an excellent co-worker who is well liked by all levels of employees.

   C. *Chris Clinton.* Chris has a B.S. degree in engineering from Poloma College and an M.B.A. degree from Weatherford University. Prior to working for Acme, Chris served on the IEEE (Institute of Electrical and Electronic Engineering) Commission for Management Innovation, specializing in global concerns and Total Quality Management. In 1991, Chris was hired by Acme and since then has worked in various capacities. Chris is now lead engineer in the engineering Department. Chris has earned high scores on every yearly evaluation, especially regarding knowledge of engineering. Whenever you have needed assistance with new management techniques, Chris has been a valued resource. Chris's only negative points on evaluations have resulted from difficulties with colleagues, some of whom regard Chris as haughty.

Whom will you recommend for supervisor? Write your feasibility report stating your decision.

2. You are the accountant at AAA Computing, a retail store specializing in computer hardware. Your boss states that all new computers sold should be accompanied by an optional service contract. This service will be held by an outside vendor. Your job is to research several vendors and write a feasibility report recommending the best choice for your company. To do so, you know your boss will emphasize the following criteria:

- Years in business/expertise—to be sure that customers receive quality service, the vendor should have a good track record and be familiar with computer hardware innovations.
- Quick response/turnaround—since many of your customers depend on their computers for daily business operations, the vendor should provide on-site service for minor repairs and 24-hour turnaround service on major repairs.
- Cost-effective pricing—the less the vendor charges, the more you can mark up the service cost. Thus, AAA can receive more profit.

The following vendors have proposed their services. Using the information provided about each company and the criteria for feasibility reports provided in this chapter, write your report recommending a vendor for AAA's service contracts.

A. *QuickBit.* This company has 16 months experience in computer hardware service. QuickBit, although without a lengthy track record, is co-owned by three graduates of the Silicon Valley Institute of Technology's renowned B.A. program in computer information services. The three individuals are very knowledgeable about computers, having received superior instruction and hands-on training using today's most up-to-date technology. Because QuickBit is small, it can provide immediate, 24-hour on-call service. Furthermore, QuickBit owns a twelve-wheeler truck fully stocked with parts. Therefore, all service can be handled on site. QuickBit charges $50 per hour plus parts.

B. *ROM on the Run.* This company has been in business for ten years. They employ 50 servicepeople, have a lengthy track record, and have provided service for ARC Telecommunications, Capital Bank, Helping Hand Hospital, and the State Penitentiary at Round Rock. Because of their years in business and the fact that all their employees have at least two-year certificates from vocational colleges, they charge $85.50 per hour plus parts. ROM on the Run promises on-site service within 24 hours on all service requests.

C. *You Bet Your Bytes (YBYB).* YBYB has been in business for four years, employs ten servicepeople (all of whom have at least a B.S. degree in computer systems), and specializes in retail outlets. YBYB charges $70 per hour plus parts. They advertise that they respond within thirty minutes for on-site service calls. Furthermore, they advertise that most parts are carried on their service trucks. Any repairs requiring unavailable parts can be made within 24 hours if the service call is received by 3:30 p.m. Calls after that time require two working days.

Which company do you recommend? Write the feasibility report stating your choice.

3. Your company, Telecommunications R Us (TRU), has experienced a 45 percent increase in business, a 37 percent increase in warehoused stock, and a 23 percent increase in employees. You need more room. Your executive officer, Polina Gertsberg, has asked you to research existing options. To do so, you know you must consider these criteria:

- *Ample space for further expansion.* Gertsberg suggests that TRU could experience further growth upwards of 150 percent. You need to consider room for parking, warehouse space, additional offices, and a cafeteria—approximately 20,000 sq. ft. total.

- *Cost.* Twenty million dollars should be the top figure, with a preferred payback of five years at 10 percent.

- *Location.* Most of your employees and customers live within 15 miles of your current location. This has worked well for deliveries and employee satisfaction. A new location within this 15-mile radius is preferred.

- *Aesthetics.* Ergonomics suggest that a beautiful site improves employee morale and increases productivity.

After research, you've found three possible sites. Based on the following information and on the criteria for feasibility reports discussed in this chapter, write your report recommending a new office site.

A. *Site 1 (11717 Grandview).* This four-story site, located 12 miles from your current site, offers three floors of finished space equaling 18,000 sq. ft. The fourth floor is an unfinished shell equaling an additional 3,000 sq. ft. As is, the building will sell for $19 million. If the current owner finishes the fourth floor, the addition would cost $4 million more. For the building as is, the owner asks for payment in five years at 12 percent interest. If the fourth floor is finished by the owner, payment is requested in seven years at 10 percent. The building has ample parking space but no cafeteria, although a building next door has available food services. Site 1 is nestled in a beautifully wooded area with hiking trails and picnic facilities.

B. *Site 2 (808 W. Blue Valley).* This one-story building offers 21,000 sq. ft. that includes 100 existing offices, a warehouse capable of holding eighty storage bins that measure 20 ft. tall × 60 yds. long × 8 ft. wide, and a full-service cafeteria. Because the complex is one story, it takes up 90 percent of the lot, leaving only 10 percent for parking. Additional parking is located across an eight-lane highway which can be crossed via a footbridge. The building, located 18 miles from your current site, has an asking price of $22 million at 8.75 percent interest for five years. Site 2 has a cornfield to its east, the highway to its west, a small lake to its north where flocks of geese nest, and a strip mall to its south.

C. *Site 3 (1202 Red Bridge Avenue).* This site is 27 miles from your current location. It has three stories offering 23,000 sq. ft., a large warehouse with four-bay loading dock, and a cafeteria with ample seating and vending machines for food and drink. Because this site is located near a heavily industrialized area, the asking price is $15 million at 7.5 percent interest for five years.

Which site do you recommend? Write your feasibility report stating your choice.

16 Proposals

OBJECTIVES

In the preceding chapter, we discussed various types of reports, including trip reports, progress reports, lab reports, incident reports, and feasibility reports. The majority of these reports will be limited to no more than five pages. A report on your job-related travel would rarely require multipage documentation. Similarly, most progress reports address only daily, weekly, or monthly activities. The same applies to most lab reports. Since the subject matter will be limited, the report will be short.

However, in some instances, your subject matter might be so complex that a shorter report will not suffice. For example, your company asks you to write a report proposing the purchase of a new facility. You will have to write a longer report—an *internal proposal* for your company's management. Or, perhaps your company is considering offering a new service or manufacturing a new product. Your responsibility is to write—an *external proposal* selling the benefits of this new corporate offering to a prospective client. External proposals are also written in response to *RFPs (requests for proposals)*. Often, companies, city councils, and state or federal agencies need to procure services from other corporations. A city, for example, might need extensive road repairs. To receive bids and analyses of services, the city will write an RFP, specifying the scope of their needs. Competing companies will respond to this RFP with an external proposal.

In each of these instances, you ask your readers to make significant commitments regarding employees, schedules, equipment, training, facilities, and finances. Only a long report, possibly complete with research, will convey your content sufficiently and successfully. (We discussed research writing in Chapter 13.)

To help you write your proposals effectively, this chapter provides the following: criteria for proposals, a sequential process for writing proposals, and a sample report.

CRITERIA FOR PROPOSALS

To guide your readers through a proposal, you'll need to provide the following:

- Title page
- Cover letter
- Table of contents
- List of illustrations
- Abstract (or summary)
- Introduction
- Discussion (the body of the proposal)
- Conclusion/recommendation
- Glossary
- Works cited page (if you're documenting research; this is discussed in Chapter 13)
- Appendix

TITLE PAGE

The title page serves several purposes. On the most simple level, a title page acts as a dust cover or jacket keeping the actual report clean and neat. More important, the title page tells your reader the

- Title of the proposal (thereby providing clarity of intent)
- Name of the company, writer, or writers submitting the proposal
- Date on which the proposal was completed

If the external proposal is being mailed outside your company to a client, you also might include on the title page the audience to whom the report is addressed. If the internal proposal is being submitted within your company to peers, subordinates, and/or supervisors, you might want to include a routing list of individuals who must sign off and/or approve the proposal.

Following are two sample title pages. Figure 16.1 is for an internal proposal; Figure 16.2 is for an external proposal.

PROPOSED CABLE TRANSMISSION NETWORK
FROM CHEYENNE, WY
TO
HARTFORD, CT

Prepared by: _____ Date: _____
                    Pete Niosi
                    Network Planner

Reviewed by: _____ Date: _____
                    Leah Workman
                    Manager, Capital Planning

Recommended by: _____ Date: _____
                    Greg Foss
                    Manager, Facilities

Recommended by: _____ Date: _____
                    Shirley Chandley
                    Director, Implementation Planning

Approved by: _____ Date: _____
                    Ralph Houston
                    Vice President, Network Planning

FIGURE 16.1 Title Page for an Internal Proposal

COVER LETTER

Your cover letter prefaces the proposal and provides the reader an overview of what is to follow. It tells the reader

- Why you are writing
- What you are writing about (the subject of this proposal)
- What exactly of importance is within the proposal
- What you plan to do next as a follow-up
- When the action should occur
- Why that date is important

Each of these points is discussed in greater detail in Chapter 6.

TABLE OF CONTENTS

Proposals are read by many different readers, each of whom will have a special area of interest. For example, the managers who read your proposals will be interested

PROPOSAL TO MAINTAIN COMPUTER EQUIPMENT

For
Acme Products, Inc.
2121 New Tech Avenue
Bangor, ME

Submitted by
Thomas Brasher
Engineering Technician

August 13, 2000

FIGURE 16.2 Title Page for an External Proposal

in cost concerns, time frames, and personnel requirements. Technicians, in contrast, will be interested in technical descriptions and instructions. Not every reader will read each section of your proposal.

Your responsibility is to help these different readers find the sections of the proposal which interest them. One way to accomplish this is through a table of contents. The table of contents should be a complete and accurate listing of the main *and* minor topics covered in the proposal. In other words, you don't want just a brief and sketchy outline of major headings. This could lead to page gaps; your readers would be unable to find key ideas of interest. In the table of contents below, we can see that the proposal section contains approximately eight pages of data. What's covered in those eight pages? Anything of value? We don't know. The same applies to the appendix, which covers four pages. What's in this section?

FLAWED TABLE OF CONTENTS
Table of Contents

In contrast, an effective table of contents fleshes out this detail so your readers know exactly what's covered in each section. By providing a thorough table of contents, you'll save your readers time and help them find the information they want and need. Figure 16.3 is an example of a successful table of contents.

In the examples, note that the actual pagination (page 1) begins with the introduction section. Page 1 begins with your main text, not the front matter. Instead, information prior to the introduction is numbered with lowercase Roman numerals

Table of Contents

FIGURE 16.3 Successful Table of Contents

(i, ii, iii, etc.). Thus, the title page is page i, and the cover letter is page ii. However, you never print the numbers on these two pages. Therefore, the first page with a printed number is the table of contents. This is page iii, with the lowercase Roman numeral printed at the foot of the page and centered.

LIST OF ILLUSTRATIONS

If your proposal contains several tables and/or figures, you'll need to provide a list of illustrations. This list can be included below your table of contents, if there is room on the page, or on a separate page. As with the table of contents, your list of illustrations must be clear and informative. Don't waste your time and your reader's time by providing a poor list of illustrations like the one below.

FLAWED LIST OF ILLUSTRATIONS	
List of Illustrations	
Fig. 1	2
Fig. 2	4
Fig. 3	5
Fig. 4	5
Fig. 5	9
Table 1	3
Table 2	6

The example above provides your reader with very little information. All the reader can ascertain from this list is that you've used some figures and tables. However, the reader will have no idea what purpose each illustration serves. Instead of supplying such a vague list, you should accompany the table and figure numbers with descriptive titles, as follows:

SUCCESSFUL LIST OF ILLUSTRATIONS		
List of Illustrations		
Figure 1.	Revenues Compared to Expenses	2
Figure 2.	Average Diesel Fuel Prices Since 1978	4
Figure 3.	Mainshaft Gear Outside Face	5
Figure 4.	Mainshaft Gear Inside Face	5
Figure 5.	Acme Personnel Organization Chart	9
Table 1.	Mechanism Specifications	3
Table 2.	Costs: Expenditures, Savings, Profits	6

ABSTRACT (OR SUMMARY)

As mentioned earlier, a number of different readers will be interested in your proposal. One group of those readers will be management—supervisors, managers, and highly placed executives. How do these readers' needs differ from others? Because these readers are busy with management concerns and might have little technical knowledge, they need your help in two ways: they need information quickly, and they need it presented in low-tech terminology. You can achieve both these objectives through an abstract or summary.

The abstract is a brief overview of the proposal's key points geared toward a low-tech reader. If the intended audience is composed of upper-level management, this unit might be called an executive summary. To accomplish the required brevity, you should limit your abstract to approximately three to ten sentences. These sentences can be presented as one paragraph or as smaller units of information separated by headings. Each proposal you write will focus on unique ideas. Therefore, the content of your abstracts will differ. Nonetheless, abstracts should focus on the following: (a) the *problem* necessitating your proposal, (b) your suggested *solution*, and (c) the *benefits* derived when your proposed suggestions are implemented. These three points work for external as well as internal proposals.

For example, let's say you're asked to write an internal proposal suggesting a course of action (limiting excessive personnel, increasing your company's workforce, improving your corporation's physical facilities, etc). First, your abstract should specify the problem requiring your planned action. Next, you should mention the action you're planning to implement. This leads to a brief overview of how your plan would solve the problem, thus benefiting your company.

If you were writing an external proposal to sell a client a new product or service, you would still focus on problem, solution, and benefit. The abstract would remind the readers of their company's problem, state that your company's new product or service could alleviate this problem, and then emphasize the benefits derived.

In each case, you not only want to be brief, focusing on the most important issues, but you should avoid high-tech terminology and concepts. The purpose of the abstract is to provide your readers with an easy-to-understand summary of the entire proposal's focus. Your executives want the bottom line, and they want it quickly. They don't want to waste time deciphering your high-tech hieroglyphics. Therefore, either avoid all high-tech terminology completely or define your terms parenthetically.

The following is an example of a brief low-tech abstract from an internal proposal.

ABSTRACT

Due to deregulation and the recent economic recession, we must reduce our workforce by 12 percent.

Our plan for doing so involves

- Freezing new hires
- Promoting early retirement
- Reassigning second-shift supervisors to our Desoto plant
- Temporarily laying off third-shift line technicians

Achieving the above will allow us to maintain production during the current economic difficulties.

INTRODUCTION

Your introduction should include two primary sections: (a) purpose, and (b) problem.

Purpose

In one to three sentences, tell your readers the purpose of your proposal. This purpose statement informs your readers *why* you are writing and/or *what* you hope to achieve. This statement repeats your abstract to a certain extent. However, it's not redundant; it's a reiteration. Although numerous people read your report, not all of them read each line or section of it. They skip and skim.

The purpose statement, in addition to the abstract, is another way to ensure that your readers understand your intent. It either reminds them of what they've just read in the abstract or informs them for the first time if they skipped over the abstract. Your purpose statement is synonymous with a paragraph's topic sentence, an essay's thesis, the first sentence in a letter, or the introductory paragraph in a shorter report.

The following is an effective purpose statement.

> Purpose: The purpose of this report is to propose the immediate installation of the 102473 Numerical Control Optical Scanner. This installation will ensure continued quality checks and allow us to meet agency specifications.

Problem

Whereas the purpose statement should be limited to one to three sentences for clarity and conciseness, your discussion of the problem must be much more detailed.

For example, if you're writing an internal proposal to add a new facility, your company's current work space must be too limited. You've got a problem which must be solved. If you're writing an external proposal to sell a new piece of equipment, your prospective client must need better equipment. Your proposal will solve the client's problem.

Your introduction's focus on the problem, which could average one to two pages, is important for two reasons. First, it highlights the importance of your proposal. It emphasizes for your readers the proposal's priority. In this problem section, you persuade your readers that a problem truly exists and needs immediate attention.

Second, by clearly stating the problem, you also reveal your knowledge of the situation. The problem section reveals your expertise. Thus, after reading this section of the introduction, your audience should recognize the severity of the problem and trust you to solve it.

One way to help your readers understand the problem is through the use of highlighting techniques, especially headings and subheadings.

Figure 16.4 provides a sample introduction stating purpose and problem.

1.0 INTRODUCTION

1.1 Purpose

This is a proposal for a storm sewer survey for Yakima, Washington. First, the survey will identify storm sewers needing repair and renovation. Then it will recommend public works projects that would control residential basement flooding in Yakima.

1.2 Problem

1.2.1. *Increased Flooding*

Residential basement flooding in Yakima has been increasing. Fourteen basements were reported flooded in 1999, whereas eighty-three residents reported flooded basements in 2000.

1.2.2. *Property Damage*

Basement flooding in Yakima results in thousands of dollars in property damage. The following are commonly reported as damaged property:
   a. Washers
   b. Dryers
   c. Freezers
   d. Furniture
   e. Furnaces
Major appliances cannot be repaired after water damage. Flooding can also result in expensive foundation repairs.

1.2.3. *Indirect Costs*

Flooding in Yakima is receiving increased publicity. Flood areas, including Yakima, have been identified in newspapers and on local newscasts. Until flooding problems have been corrected, potential residents and businesses may be reluctant to locate in Yakima.

1.2.4. *Special-Interest Groups*

Citizens over 55 years old represent 40 percent of the Yakima, WA, population. In city council meetings, senior citizens with limited incomes expressed their distress over property damage. Residents are unable to obtain federal flood insurance and must bear the financial burden of replacing flood-damaged personal and real property. Senior citizens (and other Yakima residents) look to city officials to resolve this financial dilemma.

FIGURE 16.4 Proposal Introduction

DISCUSSION

The discussion section of your proposal constitutes its body. In this section, you sell your product, service, or suggested solution. As such, the discussion section represents the major portion of the proposal, perhaps 85 percent of the text.

What will you focus on in this section? Since every proposal will differ, we can't tell you exactly what to include. However, your discussion can contain any or all of the following:

- Analyses
  —Existing situation
  —Solutions
  —Benefits
- Technical descriptions of mechanisms, tools, facilities, and/or products
- Technical instructions
- Options
  —Approaches/methodologies
  —Purchase options
- Managerial chains of command (organizational charts)
- Biographical sketches of personnel
- Corporate and/or employee credentials
  —Years in business
  —Satisfied clients
  —Certifications
  —Previous accomplishments
- Schedules
  —Implementation schedules
  —Reporting intervals
  —Maintenance schedules
  —Delivery schedules
  —Completion dates
  —Payment schedules
  —Projected milestones (forecasts)
- Cost charts

You will have to decide which of these sections will be geared toward high-tech readers, low-tech readers, or a lay audience. Once this decision is made, you'll write accordingly, defining terms as needed. However, one way to handle multiple audience levels is through a glossary (we'll discuss this later in this chapter).

In addition to audience recognition, you should also enhance your discussion with figures and tables for clarity, conciseness, and cosmetic appeal.

CONCLUSION/RECOMMENDATION

As for shorter reports, you must sum up your proposal, providing your readers with a sense of closure. The conclusion can restate the problem, your solutions, and the benefits to be derived. In doing so, remember to quantify. Be specific—state percentages and amounts.

Your recommendation will suggest the next course of action. Specify when this action will or should occur and why that date is important.

The conclusion/recommendation section can be made accessible through highlighting techniques, including headings, subheadings, underlining, boldface, itemization, and white space.

Your conclusion/recommendation, like your abstract, will be read primarily by executives. Thus, write to a low-tech reader.

Figure 16.5 is an example of a successfully written conclusion/recommendation from an internal proposal.

GLOSSARY

Because you will have numerous readers with multiple levels of expertise, you must be concerned about your use of high-tech language (abbreviations, acronyms, and terms). Although some of your readers will understand your terminology, others won't. However, if you define your terms each time you use

---

3.0 CONCLUSION

Our line capability between San Marcos and LaGrange is insufficient. Presently, we are 23 percent under our desired goal. Using the vacated fiber cables will not solve this problem because the current configuration does not meet our standards. Upgrading the current configuration will improve our capacity by only 9 percent and still present us the risk of service outages.

4.0 RECOMMENDATION

We suggest laying new fiber cables for the following reasons. They will
- Provide 63 percent more capacity than the current system
- Reduce the risk of service outages
- Allow for forecasted demands when current capacity is exceeded
- Meet standard configurations

If these new cables are laid by September 1, 2000, we will predate state tariff plans to be implemented by the new fiscal year.

FIGURE 16.5 Proposal Conclusion/Recommendation

them, two problems will occur: you will insult high-tech readers, and you will delay your audience as they read your text. To avoid these pitfalls, use a glossary.

A glossary is an alphabetized list of high-tech terminology placed after your conclusion/recommendation. When your first high-tech, unfamiliar abbreviation, acronym, or term is used, follow it with an asterisk (*). Then, at the bottom of the page, in a footnote, write

*This and subsequent terms followed by an asterisk are defined in the glossary beginning on page ___.

Subsequent high-tech terms should be followed by an asterisk, but you don't have to place the footnote at the bottom of each page.

A glossary is invaluable. Readers who are unfamiliar with your terminology can turn to the glossary and read your definitions. Those readers who understand your word usage can continue to read without stopping for unneeded information.

Figure 16.6 is a sample glossary.

WORKS CITED (OR REFERENCES)

If you've used research to write your proposal, you will need to include a works cited page. This page documents the sources (books, periodicals, interviews, computer software, etc.) you've researched and quoted or paraphrased. For more about conducting research, see Chapter 13.

GLOSSARY

BSCE:	Bachelor of Science, Civil Engineering
Drainage Studies:	The study of moving surface or surplus water
Gb:	Gigabyte
Interface:	Communication with other agencies, entities, and systems to discuss subjects of common interest
POP:	Place of purchase
Smoke Test:	Test using underground smoke bombs to give visual, above-ground signs of sewer pipe leaks
Video Scanner:	A portable video camera which examines the inside surface of sewer pipe
Water Parting:	A boundary line separating the drainage districts of two streams
Watershed Area:	An area bounded by a water parting and draining into a particular watercourse

FIGURE 16.6 Glossary

APPENDIX

A final, optional component is an appendix. Appendices allow you to include any additional information (survey results, tables, figures, previous report findings, relevant letters and/or memos, etc.) that you have not built into your proposal's main text.

The contents of your appendix should not be of primary importance. Any truly important information should be incorporated within the proposal's main text. Valuable data (proof, substantiation, or information which clarifies a point) should appear in the text where it is easily accessible. Information provided within an appendix is buried, simply because of its placement at the end of the report. You don't want to bury key ideas. An appendix is a perfect place to file nonessential data that provides documentation for future reference.

Before you submit your proposal, be sure you've included all the necessary information by consulting the Proposal Checklist below.

PROCESS

Proposals, which include descriptions, instructions, cost analyses, scheduling assessments, and personnel considerations, are more demanding than other kinds of technical correspondence. Therefore, writing according to a process approach is even more important than for other kinds of writing discussed in this textbook.

PROPOSAL CHECKLIST

Have you included the following in your proposal?

✔ *Title page* (listing title, audience, author or authors, and date)

✔ *Cover letter* (stating why you're writing and what you're writing about; what exactly you're providing the reader(s); what's next—follow-up action)

✔ *Table of contents* (listing all major headings, subheadings, and page numbers)

✔ *List of illustrations* (listing all figures and tables, including their numbers and titles, and page numbers)

✔ *Abstract* (stating in low-tech terms the problem, solution, and benefits)

✔ *Introduction* (providing a statement of purpose and a lengthy analysis of the problem)

✔ *Discussion* (solving the readers' problem by discussing topics such as procedures, specifications, timetables, materials/equipment, personnel, credentials, facilities, options, and costs)

✔ *Conclusion* (restating the benefits and recommendation for action)

✔ *Glossary* (defining terminology)

✔ *Appendix* (optional additional information)

For your proposal, you'll have more data to gather, more information to organize, and more text to revise. To help you tackle these tasks,

- Prewrite,
- Write, and then
- Rewrite.

PREWRITING

Throughout this textbook, we have provided numerous prewriting techniques geared toward helping you gather data and determine your objectives. Any or all of these can be used while prewriting your proposal. For example, you might want to use

- *Listing/brainstorming* to outline the key components of your proposal or for your technical description, if you plan to include one.
- *Reporter's questions (who, what, when, where, why, and how)* to help you gather data for any of the sections in your proposal.
- *Flowcharting* to organize procedures and/or schedules.
- *Branching* and *mind mapping* to organize your managerial sections (organizational charts for chains of command, personnel responsibilities, etc.).

In addition to these prewriting techniques, you might need to perform research prior to writing. For instance, let's say your engineering firm is submitting a proposal to an out-of-state corporation. Prior to bidding on the job, you'll need to find out what kinds of certifications or licenses this state requires. To do so, you'll have to research the state's requirements and read state laws regarding construction certifications and licenses. (We discuss techniques for doing research in Chapter 13.)

Perhaps you'll need to survey residents, clients, or personnel to get their ideas regarding proposed changes. If so, here's what you should do.

Surveys

Before you can take a survey, you must decide what questions to ask. Brainstorming/listing would be an excellent way to gather such data. Follow this procedure:

1. Rapidly jot down whatever questions come to mind regarding your topic.
2. Don't editorialize at this point. Don't try to organize the list; don't delete any questions that emerge.
3. Once you have a list of possible topics, review the list.
   a. Add any omissions (keep in mind your reporters' questions during this stage).
   b. Delete any redundancies or irrelevant ideas.
   c. Organize the list according to some rational order.

Your next step is to format the list into a survey questionnaire, as in Figure 16.7.

STUDENT AND STAFF QUESTIONNAIRE
FOR
PROPOSED COLLEGE DAY-CARE CENTER

1. Are you male or female? _____

2. Are you married? _____

3. Are you in a single- or dual-income family? _____

4. Are you a student or a staff member? _____

5. How many children do you have? _____

6. What are their ages? _____ _____ _____ _____

7. Would you be interested in having a day-care center at this college? _____

8. How much would you be willing to pay per hour for child care at this college? _____

9. Do you think a day-care center would improve attendance at this college? _____

If student:

10. Are you enrolled full time or part time at this college? _____

11. Do you attend mornings, afternoons, evenings, or a combination of the above? (Please specify.) _____

12. Would you be willing to work in this day-care center part time? _____

If staff:

13. What hours do you work at this college (primarily mornings, afternoons, evenings, or a combination of the above)? (Please specify.) _____

14. What hours would you need to use the day-care center?
_____

Optional:

Name: _____
Address: _____
Phone: _____
Thank you for your assistance.

FIGURE 16.7 Questionnaire

WRITING

After gathering your data and organizing your thoughts through prewriting, your next step is to draft your proposal.

1. Review your prewriting. Double-check your listing, brainstorming, mind mapping, flowcharting, reporter's questions, interviews, research, and/or surveys. Is all the necessary information there? Do you have what you need? If not, now is the time to add new detail or research your topic further. In contrast, perhaps you've gathered more information than necessary. Maybe some of your data is irrelevant, contradictory, or misleading. Delete this information. Focus your attention on the most important details.

2. Organize the data. Each of your proposal's sections will require a different organizational pattern. Following are several possible approaches.

- Abstract: problem/solution/benefit
- Introduction: cause/effect (The problem unit is the cause of your writing; the purpose statement represents the effect.)
- Main text: This unit will demand many different methods of organization, including
  —Analysis (cost charts, approaches, managerial chains of command, personnel biographies, etc.)
  —Chronology (procedures, scheduling)
  —Spatial (descriptions)
  —Comparison/contrast (options—approaches, personnel, products)
- Conclusion/recommendation: analysis and importance (Organize your recommendations by importance to highlight priority or justify need.)

3. Write using sufficing techniques. When you draft your text, don't worry about correct grammar, highlighting techniques, or graphics. Just get the information down as rapidly as you can. You can revise the draft during rewriting, the third stage of the writing process.

4. Format your writing according to the criteria for effective proposals.

REWRITING

After you've written a rough draft of your proposal, the next step is to revise it—fine tune, hone, sculpt, and polish your draft.

1. Add detail for clarity. In addition to rereading your rough draft and adding a missing *who, what, when, where, why,* and *how* where necessary, add your graphics. Go back to each section of your proposal and determine where you could use any of the following:

- Tables. Your cost section lends itself to tables.
- Figures. Your introduction's problem analysis and any of the main text sections could profit from the following figures:

—Line charts (excellent for showing upward and downward movement over a period of time. A line chart could be used to show how a company's profits have decreased, for example.)

—Bar charts (effective for comparisons. Through a bar or grouped bar chart, you could reveal visually how one product, service, or approach is superior to another.)

—Pie charts (excellent for showing percentages. A pie chart could help you show either the amount of time spent or amount of money allocated for an activity.)

—Line drawings (effective for technical descriptions)

—Photographs (effective for technical descriptions)

—Flowcharts (a successful way to help readers understand procedures)

—Organizational charts (excellent for giving an overview of managerial chains of command).

We discuss each of these types of graphics in detail in Chapter 9.

Another important addition to your proposal in this rewriting stage is a glossary. Now that you've written your rough draft, go back over each page to decide which abbreviations, acronyms, and high-tech terms must be placed in your glossary. After each of these high-tech words or phrases, add an asterisk. After the first abbreviation, acronym, or high-tech word or phrase, add a footnote stating that subsequent words or phrases followed by an asterisk will be defined in the glossary. Then write your glossary and add it to your proposal.

2. Delete dead words and phrases for conciseness. Because proposals are long, you've already made your reader uncomfortable. People don't like wading through massive pages of text. Anything you can do to help your reader through this task will be appreciated. Avoid making the proposal longer than necessary. (Chapter 3 gives detailed explanations for achieving conciseness.)

3. Simplify old-fashioned words. A common adage in technical writing is, "Write the way you speak." In other words, try to be more conversational in your writing. We aren't suggesting that you use slang or colloquialisms. But a word like *supersede* should be simplified: instead, write *replace*. Check your text to see whether you've used words that are unnecessarily confusing.

Deleting and simplifying, when used together, will help you lower your fog index. This is especially important in your abstract and conclusion/recommendation sections, which are geared toward low-tech management. Review these sections of your proposal to determine if you have written at the appropriate level. If you have used terminology which is too high tech, simplify.

4. Move information. Each section of your proposal will use a different organizational method. Your abstract, for example, should be organized according to a problem/solution/benefit approach. Your introduction will be organized according to cause (the problem) and effect (the purpose). The organization of your main text will vary from section to section. You might analyze according to importance, set up schedules and procedures chronologically, describe spatially, and so forth.

In rewriting, revise your proposal to ensure that each section maintains the appropriate organizational pattern. To do so, move information around (cut and paste).

5. *Reformat for reader-friendly ease of access.* If you give your readers wall-to-wall words, they will doze off while attempting to wade through your proposal. To avoid this, revise your proposal by reformatting. Indent to create white space. Add headings and subheadings. Itemize ideas, boldface key points, and underline important words or phrases. Using graphics (tables and figures) will also help you avoid long, overwhelming blocks of text. By reformatting your proposal, you will make the text inviting and ease reader access.

6. *Enhance the tone of your proposal.* Although you want to keep your proposal professional, remember that people write to people. Your reader is a human being, not a machine. Therefore, to achieve a sense of humanity, enhance the tone of your text by using pronouns and positive, motivational words. These are important in your statements of benefit and recommended courses of action. *Sell* your ideas.

7. *Correct errors.* If you're proposing a sale, you must provide accurate figures and information. Proposals, for example, are legally binding. If you state a fee in the proposal or write about schedules, your prospective client will hold you responsible; you must live up to those fees or schedules. Make sure that your figures and/or data are accurate.

Reread each of your numbers, recalculate your figures, and double-check your sources of information. Correct errors prior to submitting your report. Failure to do so could be catastrophic for your company or your client.

Of course, you must also check for typographical, mechanical, and grammatical errors. If you submit a proposal containing such errors, you will look unprofessional and undermine your company's credibility.

8. *Avoid sexist language.* As noted throughout this textbook, both men and women read your writing. Don't talk about *foremen, manpower,* or *men and girls.* Change *foremen* to *supervisors, manpower* to *workforce* or *personnel,* and *men and girls* to *men and women.* Don't address your cover letter to *gentlemen.* Either find out the name of your readers or omit the salutation according to the simplified letter style discussed in Chapter 6.

SAMPLE PROPOSAL

On pages 413–425 a sample proposal (Figure 16.8) can be used as a model for your writing.

CHAPTER HIGHLIGHTS

1. You might have multiple readers for a proposal. Consider your audiences' needs. To communicate with different levels of readers, include abstracts or executive summaries, glossaries, and parenthetical definitions.

2. A proposal could include the following:

- title page
- cover letter
- list of illustrations
- abstract (or executive summary)
- introduction
- discussion
- conclusion
- recommendation
- glossary
- works cited
- appendix

3. Subheadings will make your proposal more accessible.

4. You could use a questionnaire to generate content for your proposal.

ACTIVITIES

1. Write an external proposal. To do so, create a product or a service and sell it through a long report. Your product can be an improved radon detection unit, a new fiber optic cable, safety glasses for construction work, bar codes for pricing or inventory control, an improved four-wheel steering system, computer graphics for an advertising agency, etc. Your service may involve dog grooming, automobile servicing, computer maintenance, home construction (refinishing basements, building decks, room additions, etc.), freelance technical writing, at-home occupational therapy, or telemarketing. The topic is your choice. Draw from your job experience, college course work, or hobbies. To write this proposal, follow the process provided in this chapter—prewrite, write a draft, and then rewrite to revise.

2. Write an internal proposal. You can select a topic from either work or school. For example, your company or department is considering a new venture. Research the prospect by reading relevant information. Interview involved participants and/or survey a large group of people. Once you've gathered your data, document your findings and propose to management the next course of action. If you choose a topic from school, you could propose a day-care center, on-campus bus service, improved computer facilities, tutoring services, coed dormitories, pass/fail options, and so on. Research your topic by reading relevant information and/or by interviewing/surveying students, faculty, staff, and administration. Once you've gathered your data, document your findings and recommend a course of action. In each instance, be sure to prewrite, write, and rewrite.

3. Find a previously written proposal (at work or one already submitted by a prior student) and improve it according to the criteria presented in this chapter. Add or change any or all of the report's components, including the cover letter, table of contents, list of illustrations, abstract, introduction, main text, conclusion/recommendation, glossary, etc.

4. Find a previously written proposal (at work or one already submitted by a prior student) and improve it by adding graphics where needed. You can use tables and/or figures to clarify a point and to make the report more cosmetically appealing.

Case Study

The technical writing department at Bellaire Educational Supplies/Technologies (BEST) needs new computer equipment. Currently, the department has outdated hardware, outdated word processing software, an outdated printer, and limited graphics capabilities. Specifically, the department is using computers with 12″ black-and-white monitors, hard drives with only 256 kilobytes of memory, and one, 10 megabyte hard disk drive. The word processing package used is *WordPro 3.0*, a version created in 1989. Since then, *WordPro* has been updated four times; the latest version is 6.5. The department printer is a black-and-white Amniprint dot matrix machine. The current word processing package has no clip art. To create art, the department must go off site to a part-time graphic artist who charges $35 an hour, so the department uses very few graphics.

Because of these problems, the company's user manuals, reports, and sales brochures are being poorly reviewed by customers. Further, BEST has no Web site for product advertisement and/or company recognition. The bottom line: BEST is falling behind the curve, and profits are off 27 percent from last year.

As technical writing department manager, you have consulted with your five staff members (Earl Eddings, Shannon Conner, Mike Thurmand, Amber Badger, and Lauren Hensley) to correct these problems. As a team, you have decided the company needs to purchase new equipment:

- *Six new personal computers.* Each computer must have a 17″ color monitor, 32 megabytes of memory, a 2-gigabyte hard drive, a 12-speed CD-ROM drive, and a VGA graphics card.
- *Two laser printers.* These must have a print speed of 24 pages per minute, resolution of 600×600 dots per inch, and 4 megabytes of memory, expandable to 132 megabytes.
- *Word processing software. WordPro 6.5* with these capabilities: voice-activated annotations, typing, and correcting; automatic footnoting and endnoting; envelope labeling; grammar and spell checking; thesaurus; help options; automatic index generating; 50 or more scalable fonts.
- *Graphics software.* For professional-quality newsletters and brochures, BEST needs the capability for quick demonstrations and design tips; a layout checker with at least 10 online vies; 20 true type fonts, each scalable; a

table creation toolbar; 2,000 clip art images; and a logo creator with 50 border design options.

- *Scanner.* To increase your graphics potential, BEST also needs a flatbed scanner with these specifications: 300 dots per inch image resolution; 155 pages per minute gray-scale scanning capability; 8-bit, 256 gray-scale color support; and approximately 8 1/2″ × 12″ bed size.

Using the criteria provided in this chapter, write a proposal to BEST's CEO, Jim McWard. In this proposal, explain the problem, discuss the solution to this problem, and then highlight the benefits derived once the solution has been implemented. These benefits will include increased productivity, better public relations, increased profits, and less employee stress. Develop these points thoroughly, and provide Mr. McWard the names of vendors for the required hardware/software. To find these vendors, you could search the Internet.

Orson Medical Supplies
"Improving the Quality of Life"
12345 College Blvd.
Overland Park, OR 90091
(976) 988–2000

December 8, 2000

Dr. Richard Davis, Director
La Habra Retirement Center
220 Cypress
La Habra, CA 90631

Dear Dr. Davis:

Submitted for your review is our proposal regarding the Electronic
Demand Cannula oxygen-saving system. This document is in
response to your September 25, 2000, letter and our subsequent
discussions.

Within our report, you will find the following supporting materials
geared toward your requests:

- Product specifications . . . . pages 4–5
- Operating instructions . . . . page 6
- Qualifications and experience . . . . page 7
- Cost . . . . page 8
- Warranty . . . . page 10

Thank you for your interest in our product. We look forward to
serving you and will call within two weeks to finalize arrangements.

Sincerely,

Robert Maxwell

Robert Maxwell
Marketing Director

Enclosure: Proposal

FIGURE 16.8 Proposal

PROPOSAL
FOR
REDUCING OXYGEN EXPENSES

Prepared for
Dr. Richard Davis, Director
La Habra Retirement Center

by
Robert Maxwell
Marketing Director
Orson Medical Supplies

December 8, 2000

FIGURE 16.8 Proposal (Cont.)

TABLE OF CONTENTS

LIST OF ILLUSTRATIONS

FIGURE 16.8 Proposal (Cont.)

ABSTRACT

Expenses for medical oxygen have increased steadily for several years. Now the federal government is reducing the amount of coverage that Medicare allows for prescription oxygen.

These cost increases can be reduced through the use of our new Electronic Demand Cannula (EDC). The EDC delivers oxygen to the patient only when the patient inhales. Oxygen does not flow during the exhalation phase. Therefore, oxygen is conserved.

This oxygen-saving feature can reduce your oxygen expenses by as much as 50 percent. Patients who use portable oxygen supplies can enjoy prolonged intervals between refilling, thus providing more freedom and mobility.

iv

FIGURE 16.8 Proposal (Cont.)

1.0 INTRODUCTION

1.1 PURPOSE

This is a proposal to sell the new Electronic Demand Cannula (EDC)* to the La Habra Retirement Center, La Habra, California. This bid to sell offers you a special discount when you purchase our EDCs in the quantities suggested in this proposal.

1.2 PROBLEMS

1.2.1 *High Costs*
Since 1995, the price of medical-grade oxygen has skyrocketed. It cost $10 per 1,000 cubic feet (cu ft) in 1995. Today, medical-grade oxygen costs $26 per 1,000 cu ft. In fact, you can expect next year's oxygen expenses to double the amount you spent this year.

1.2.2 *Governmental/Insurance Involvement*
Many factors have contributed to this soaring cost, including demand, product liability, and inflation. However, two factors contributed the most. First, legislation reduced the amount that Medicare pays for prescription oxygen. Second, few insurance companies offer programs covering long-term prescription oxygen. Therefore, you, or your patients, must pay the additional expenses.

1.2.3 *Decreased Quality of Service*
Since prescription oxygen has risen in cost so dramatically, few medical service companies can produce affordable EDCs and stay competitive. Since 1996, according to *Medical Digest Bulletin*, 80 percent of medical service vendors have gone out of business. Your ability to receive quality service at an affordable price has diminished.

*This and subsequent terms marked by an asterisk (*) are defined in the glossary.

FIGURE 16.8 Proposal (Cont.)

2.0 DISCUSSION

2.1 IMPLEMENTATION OF ELECTRONIC DEMAND CANNULA

Since the price of oxygen will not go down, you must try to use less while obtaining the same clinical benefits.

Orson Medical Supplies, a leader in oxygen-administering technology, proposes the implementation of our new Electronic Demand Cannula* (EDC). Using state-of-the-art electronics, the EDC senses the patient's inspiratory effort.* When a breath is detected, the EDC dispenses oxygen through the patient's cannula.* The patient receives oxygen only when he or she needs it.

Continuously flowing cannulas waste gas during exhalation and rest. Clinical studies have proved that 50 percent of the oxygen used by cannula patients is wasted during that phase. These same tests also revealed that blood oxygen saturation* does not significantly vary between continuous and intermittent flow cannulas. The patient receives the same benefit from less oxygen. Table 1 explains this in greater detail.

Table 1
BLOOD OXYGEN SATURATION
USING THE ELECTRONIC DEMAND CANNULA
versus
CONTINUOUS FLOW CANNULAS

Prescribed Flowrate	Breaths per Minute (bpm)	Blood Oxygen Saturation %	
1pm*	bpm*	Intermittent	Continuous
0.5 1pm	12	96%	98%
1 1pm	12	98%	99%
2 1pm	12	99%	100%
3 1pm	12	100%	100%
4 1pm	12	100%	100%

We have included a technical description of the EDC to help explain how this system will benefit oxygen cannula users.

2

FIGURE 16.8 Proposal (Cont.)

2.2 TECHNICAL DESCRIPTION

The EDC (Electronic Demand Cannula) is an oxygen-administering device which is designed to conserve oxygen. The EDC is composed of six main parts: oxygen inlet connector, visual display indicators (LEDs),* power switch, patient connector, AC adapter connector, and high-impact plastic case (see Figures 1 and 2).

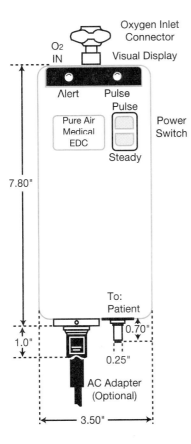

Figure 1 EDC Front View

OXYGEN INLET CONNECTOR: The oxygen inlet connector is a DISS No. 1240 (Diameter Index Safety System) and is made of chrome-plated brass.

VISUAL DISPLAY: Two light-emitting diodes (LED) provide visual indications of important functions. Alarm functions are monitored by a red LED, Motorola No. R32454. An indication of each delivered breath is given by the pulse display, which is a yellow LED, Motorola No. Y32454.

POWER SWITCH: The power switch is an ALCO No. A72-3, DPSI slide switch. The dimensions are .5" x .30": button height is .20". Electrical Specifications: Dry contact rating is 1 amp, contract resistance is 20 milliohms, and the life expectancy is 100,000 actuations.

3

FIGURE 16.8 Proposal (Cont.)

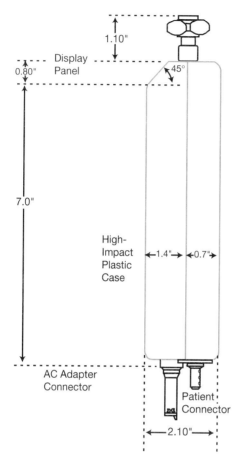

PATIENT CONNECTOR: Attachment of the patient cannula system is made at the patient connector, which is located at the bottom of the case. The white nylon connector, Air Logic No. F-3120-85, is a 10-32, UNF male threaded, straight barbed connector for 1/8" ID flexible tubing.

AC ADAPTER CONNECTOR: An optional AC adapter* and battery charger assembly, part number PA-32, plugs into the AC adapter connector, which is located at the bottom left-hand side of the case. The connector is a male, D-subminiature, 12-pin flush insert supplied by Dupont Connector Systems. Their part number is DCS: 68237009.

HIGH-IMPACT PLASTIC CASE: The case housing is made from an impact-resistant, flame-retardant, oxygen-compatible ABS plastic*.

Figure 2 EDC Side View

2.3 OPERATING INSTRUCTIONS

The EDC (Electronic Demand Cannula) is an oxygen-saving and -administering device (see Figures 1 and 2). By following these five easy steps, you will be able to enjoy the benefits of intermittent demand oxygen.

WARNING: Federal law prohibits the sale or use of this device without the order of a physician.

4

FIGURE 16.8 Proposal (Cont.)

1. Attach your oxygen supply to the Oxygen Inlet Connector located at the top of the case.
2. Move the Pulse-Steady Switch to the Pulse position to begin intermittent demand flow.
3. Connect your nasal cannula to the Patient Outlet Connector located at the bottom of the case.
4. Adjust your oxygen supply to the oxygen flow prescribed by your physician.
5. Put your nasal cannula on and breathe normally. The pulse light will turn on when a breath is delivered.

You are now ready to conserve oxygen by as much as 50 percent. Should you have the need to go back to continuous flow, just push the Pulse-Steady Switch to the "steady" position.

2.4 QUALIFICATIONS AND EXPERIENCE

Orson Medical Supplies has been an international leader in the field of respiratory therapy since 1947. Pure Air introduced the first IPPB* respirator (Intermittent Positive Pressure Breathing) on the market. In 1960, responding to the needs of doctors and therapists, we produced the first life support volume ventilator, the VV-1. The VV-1 became the industry standard by which all other ventilators were measured.

In 1981, Orson Medical Supplies introduced the first computer-controlled life support system, the VV-2. Technology developed for this product has found application in other areas as well. Recently, we introduced one such product, the Electronic Demand Cannula.

Orson Medical Supplies is located in Overland Park, Oregon. The main manufacturing and engineering facility employs 450 people. Regional sales and service branch offices are located throughout the United States.

2.5 PERSONNEL

Each of our engineering facilities is staffed by trained technicians ready to answer your questions. The following individuals have been assigned to La Habra Retirement Center:

5

FIGURE 16.8 Proposal (Cont.)

Randy Draper
Randy (BS, Electrical Engineering, South Central Texas University, 1988) has worked at Orson Medical Supplies since 1989. In 11 years at Orson, Randy has been promoted from service technician to supervisor. Randy specializes in developing new medical equipment. He has supervised the development teams which worked on the X29 respirator, the Z284-00 ventilator, and the Omega R-449 the sphygmomanometer. Randy was lead development specialist for the Pure Air EDC.

Randy will be in charge of your account. Please contact him directly regarding any questions you might have about the EDC.

Ruth Bressette
Ruth (BS, Mechanical Engineering, Pittsburgh State University, 1992) has worked at Orson Medical Supplies since 1993. She has risen in our company from service technician to manager of troubleshooting/maintenance. Ruth has received the highest-level certification (Master Technician) offered by the IEEE for service on every piece of equipment developed, manufactured, and sold by Orson.

Ruth will be the manager of your Orson equipment maintenance and troubleshooting crew. Her responsibility is to ensure that your equipment is kept in outstanding working condition. She will schedule maintenance checks and promptly assign technicians to troubleshoot potential malfunctions.

Douglas Loeb
Doug (AA, Electrical Engineering Technology, Plainview Community College, 1986) is one of our most accomplished troubleshooters. Having worked at Orson for 14 years, Doug is commended annually for his speed, accuracy, and skill. Your equipment is in good hands with Doug. He will be your primary troubleshooter and maintenance person.

2.6 COST

Orson Medical Supplies is pleased to offer our Pure Air EDC at cost-effective pricing. Table 2 explains the benefits you'll derive when purchasing in quantity.

6

FIGURE 16.8 Proposal (Cont.)

Table 2 LIST PRICE VS. DISCOUNT PRICE FOR EDC

LIST PRICE	QUANTITY WARRANTY*	EXTENDED *PER EDC*	TOTAL COST
$350	1–9 units	$75	$425
DISCOUNT PRICE			
$310	10–24 units	$60	$370
$300	25+ units	$50	$350

As you can see, Orson is happy to offer you substantial savings when you purchase our Pure Air EDC in volume. At these prices, and assuming normal use, the oxygen cost savings will exceed your initial investment in less than one year, as shown in Figure 3.

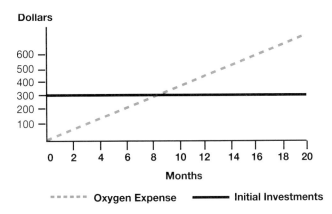

Figure 3 Return on Investment

2.7 WARRANTY

Orson Medical Supplies warrants this product to be free of manufacturing defects for a one-year period after the original date of consumer purchase. This warranty does not include damage done to the product due to accident, misuse, improper installation or operation, or unauthorized repair. This warranty also does

*Orson's extended warranty is discussed in Section 2.8.

7

FIGURE 16.8 Proposal (Cont.)

not include replacement of parts due to normal wear. If the product becomes defective within the warranty period, we will replace or repair it free of charge.

This warranty gives you special legal rights. You may also have other rights which vary from state to state. Some states do not allow exclusions or limitations of incidental damage, so the above limitations may not apply to you.

2.8 EXTENDED WARRANTY

In addition to the coverage provided in our unconditional warranty, you might want to take advantage of our extended warranty package. For the prices provided in Table 2, Orson Medical Supplies will extend the warranty to cover a three-year period after the original date of consumer purchase. This three-year extended warranty not only covers manufacturing defects, but also Orson will replace worn parts free of charge.

The extended warranty does not cover damage to the product resulting from accident, misuse, improper installation or operation, or unauthorized repair.

For answers to any of your questions regarding repair, replacement, warranty, or extended warranty, please call 1–800–555–ORSN, or write to Manager, Customer Relations, Orson Medical Supplies, 12345 College Blvd., Overland Park, OR 90091.

3.0 CONCLUSION

3.1 MAJOR CONCERN

Prescription oxygen expenses are escalating while government support has been reduced. The cost increase to the patient and the health care facility will be enormous.

3.2 RECOMMENDATION

To offset the inevitable rise of oxygen expenses, we recommend the use of the Electronic Demand Cannula.

8

FIGURE 16.8 Proposal (Cont.)

4.0 GLOSSARY

ABS plastic:	Acrylonitrile butadiene styrene. A durable and long-lasting plastic.
AC adapter:	A remote power supply used to convert alternating current to direct current.
Blood oxygen saturation:	The partial pressure of oxygen in alveolar blood recorded in percent.
Cannula:	A small tube inserted into the nose, specifically for administering oxygen.
Electronic Demand Cannula (EDC):	An electronically controlled device that dispenses oxygen only when triggered by an inspiratory effort.
Inspiratory effort:	The act of inhaling.
Light-emitting diode (LED):	A solid-state semiconductor device that produces light when current flows in the forward direction.

LIST OF SYMBOLS

BPM:	Breaths per minute
DISS:	Diameter index safety system
DPDT:	Double pole, double throw
EDC:	Electronic Demand Cannula
IPPB:	Intermittent positive pressure breathing
LPM:	Liters per minute

9

FIGURE 16.8 Proposal (Cont.)

CHAPTER 17

Oral Presentations

OBJECTIVES

Many confident and successful people are frightened by the idea of giving an oral presentation. This chapter suggests ways to prepare and to present information verbally so the experience will be rewarding rather than frightening.

Too often college students fail to realize the importance of speech classes, or they may be in a curriculum which does not require oral communication classes. However, after they graduate and begin their careers, they quickly realize that oral communication skills are important.

You may be called on to make oral presentations at work for any number of reasons.

- Your boss needs your help in preparing his or her presentation. You conduct research, interview appropriate sources, and prepare reports. When you've concluded your research, you might be asked to present your findings orally.

- Your company is making numerous cuts in its workforce. As a supervisor, you have to present an oral briefing justifying the preservation of your staff.

- A customer needs work done; you and several co-workers have studied the problem and have prepared an oral presentation proposing alternative solutions.

- Your company asks you to represent it at a conference by giving a speech.
- At a departmental meeting, you are asked to report orally on the work you and subordinates have completed and to explain future activities.
- You are asked to give a sales speech promoting, explaining, and selling a service or product to a potential customer.
- Your company asks you to represent it at a city council meeting. You will give an oral presentation explaining your company's desired course of action or justifying activities already performed.
- Your company asks you to visit a civic club meeting and to provide an oral presentation to maintain good corporate/community relations.

In addition to the preceding instances, you must communicate daily with your peers, your supervisors, your subordinates, and the public.

Think about the number of times you have spoken to someone today or this week. Each time you communicated orally, you reflected something about yourself and your company. The goal of effective oral communication, of course, is to ensure that your verbal skills make a good impression.

TYPES OF ORAL PRESENTATIONS

Four types of oral presentations include the following:

1. The impromptu speech
2. The memorized speech
3. The manuscript speech
4. The extemporaneous speech

THE IMPROMPTU SPEECH

You are asked a question in a meeting. In this instance, you will have to organize your thoughts and answer the question quickly. You will have no time to prepare. For example, you might be asked to explain the progress of a project, your company's stand on a civic issue, how your business will meet revised construction schedules, or how your club will finance a proposed activity. To answer these questions, you must give an *impromptu speech*.

THE MEMORIZED SPEECH

The least effective type of oral presentation for technical communication is the *memorized speech*. This is a well-prepared speech which has been committed to memory. Although such preparation might make you feel less anxious, too often these speeches sound mechanical and impersonal. They are stiff and formal, and they allow no speaker-audience interaction. Moreover, technical data, statistics, and definitions do not lend themselves easily to memorization. You might think you've memorized your technical information, but the odds are against it.

THE MANUSCRIPT SPEECH

In a *manuscript speech*, you read from a carefully prepared manuscript. The entire speech is written on paper. This may lessen your speaker anxiety and help you accurately communicate your technical data, but such a speech can become monotonous, wooden, and boring to your audience. Imagine what college would be like if all your professors read their lectures without writing on the board, creating short class projects, asking and answering questions, using the overhead projector, or ever making eye contact. Your audience will not be able to maintain interest in a speech read from a manuscript. However, if you are presenting purely statistical information, you will need to read from a manuscript.

THE EXTEMPORANEOUS SPEECH

Extemporaneous speeches are probably the best and most widely used method of oral communication. You carefully prepare your speech by conducting necessary research, and then you create a detailed outline. However, you do not write the speech in a complete manuscript form. Instead, you study your outline and present the speech from it or from 3″ × 5″ cards on which you have written important points. This type of speech helps you avoid dull and mechanical presentations, allows you to interact with your audience, and still ensures that you correctly present complex technical data.

CRITERIA FOR EFFECTIVE ORAL PRESENTATIONS

To prepare for a successful oral presentation, follow these guidelines:

OVERALL ORGANIZATION

As with any well-written communication, a speech consists of an introduction, a discussion, and a conclusion.

Introduction

The introduction should arouse and capture your audience's interest. This is the point in the speech where you are drawing in your listener, hoping to create enthusiasm and a positive impression.

To create a positive impression, you should address the audience politely by saying "Good Morning" or "Good Afternoon." You can welcome participants to the conference or seminar; you can thank them for inviting you to speak.

To arouse audience interest, you can begin with

- An anecdote (a short, appropriate story)
- Facts and figures
- A suitable joke
- A question
- A quotation
- A startling comment

The important factor is to stimulate your listener's interest. Once you have grasped their interest, then state the subject of your speech, just as you would provide a thesis statement for an essay or topic sentence for a paragraph.

Discussion

In the discussion section of your speech, provide details to support your thesis statement. These can be presented in any order, but there are several traditional ways in which you can organize your data: (a) comparison/contrast, (b) problem/solution, (c) argument/refutation/persuasion, (d) least to most important, and (e) chronological.

You also want to maintain coherence throughout your speech. To do so,

- Restate your topic often. Constant restating of the topic is required because listeners have difficulty retaining spoken ideas. Whereas a reader can refer to a previously discussed point simply by turning back a page or two, listeners do not have this option. Furthermore, a listener is easily distracted from your speech by external noises, room temperature, uncomfortable chairs, or movement inside and outside the meeting site. Restating your topic helps your reader maintain focus.

- Use transitional words and phrases to help your listener follow your train of thought. Some appropriate transitional terms are *first*, *second*, *next*, *therefore*, *furthermore*, *another reason for*, and so on.

- Use clear topic sentences to state each new point.

Conclusion

Conclude your speech by (a) reiterating the main points discussed and (b) recommending a future course of action.

DEVELOPMENT

To maintain credibility, develop your discussion with thorough and accurate research (discussed further in the process section of this chapter).

STYLE

As with good writing, effective oral communication demands clarity, conciseness, and audience recognition. To achieve clarity, stick to your point. Don't allow yourself to digress about side issues or irrelevancies. Maintain your focus. Concise oral presentations depend on the same skills evident in concise writing—word and sentence length. Trim your sentences of excess words (fifteen words per sentence is still your average, preferred length). Stick to monosyllabic words versus multisyllabic ones. Finally, speak to your audience's level of understanding. Either avoid highly technical terms or define them.

DELIVERY

Effective oral communicators interact with and establish a dynamic relationship with their audiences. The most thorough research will be wasted if you are unable to create rapport and sustain your audience's interest.

Eye Contact

Don't keep your eyes glued to your notes. Look into your listeners' eyes (or at least glance over the heads of those sitting in the back row). Your audience probably has been in your position at one time and feels more sympathy for you than you might guess. They're not out to "get" you. They're in the audience to listen and to learn.

Rate

Since your audience wants to listen and learn, you must speak at a rate geared toward those two goals. Determine your normal rate of speech, and cut it in half. *Slow down*. Do not speak in a monotone, however. To keep your audience alert, vary your rate of speech. Speed up your delivery when discussing less interesting facts. Slow down for more interesting points. Match your rate of speaking to the content of your speech, just as actors vary their speech rate to reflect emotion and changes in content.

Enunciation

Don't mumble. Speak each syllable of every word clearly and distinctly. No one will ask you to repeat yourself, so speak more clearly than you might in a more conversational setting. Slowing down your delivery will help you enunciate clearly.

Pitch

When we speak, our voices create high and low sounds. In your presentation, capitalize on this fact. Vary your pitch by using even more high and low sounds than you do in your normal, day-to-day conversations. Modulate to stress certain key words or major points in your oral presentation.

Pauses

As mentioned earlier, you want to speak slowly. One way to achieve a successful pace is to pause within the oral presentation. Pause to ask for and to answer questions, to allow ideas to sink in, and to use visual aids or give the audience handouts. These pauses won't lengthen your speech; they'll only improve it. In fact, a well-prepared speech will allow for pauses and will have budgeted time effectively. Know in advance if your speech is to be five minutes, ten minutes, or an hour long. Then plan your speech according to the time restraints, building pauses into your presentation. Practice the speech beforehand so you can determine where to pause and how often.

Emphasis

You can't underline or boldface comments made in oral presentations. However, just as in written communication, you will want to emphasize key ideas. Your body language, pitch, gestures, and enunciation will enable you to highlight words, phrases, or even entire sentences.

APPEARANCE

When you speak in front of people, they see you as well as hear you. Therefore, you don't want the way you look to interfere with what you're saying. With this in mind, avoid wearing clothes or jewelry which might distract the audience. Studies show that dark clothes are best to wear when making oral presentations. You might be representing your company or trying to make a good impression for yourself when you speak, so dress appropriately.

BODY LANGUAGE AND GESTURES

During an oral presentation, nonverbal communication can be as important as verbal communication. Your appearance is important, but in addition, the way you present your speech through your movements and tone of voice will affect your listeners. If you are enthusiastic about your topic, your listeners will respond enthusiastically. If you are bored or ambivalent, your tone and mannerisms will reflect your attitude. This will be communicated to your audience. To communicate effectively, then, be aware of your body language and your gestures.

One aspect of body language already discussed is eye contact. In addition, you must be concerned with how you stand and how you control your hands and arms. If, for example, you spend the entire speech standing in one place as if rooted to the spot, your audience will get bored. On the other hand, if you jump around the room like an atom whirling through a superconductor, you'll make your listeners dizzy. Find a happy medium. Move around the dais casually, speaking to different sides of the room.

Furthermore, watch your hands. If you're nervous and your hands shake, hold onto a chair, table, or lectern (something immobile), or place your hands in your pockets. Again, however, don't just maintain one posture. Vary your hand placements. The same applies to your arms. You don't want to wave them wildly and frighten your audience, but neither do you want to fold them stiffly across your chest. This body language communicates a negative attitude.

Many aspects of body language overlap with gestures (hand and arm movements, for example). One additional point relates to effective hand movements. As mentioned earlier, you need to help your listeners follow your train of thought. Through hand motions, you can emphasize ideas and provide transitions. For instance, you could put one finger up for a first point, two fingers up for a second point, and so forth. Be inventive but controlled.

VISUAL AIDS

Most speakers find that illustrations help them communicate more effectively. You must be the judge of whether visual aids will enhance your speech. Avoid using them if you think they'll distract from your presentation or if you lack confidence in your ability to create them and integrate them effectively.

Table 17.1 lists advantages, disadvantages, and helpful hints for using visual aids. For all types of visual aids, practice using them *before* you give your speech.

TABLE 17.1	Visual Aids: Advantages and Disadvantages		
TYPE	ADVANTAGES	DISADVANTAGES	HELPFUL HINTS
Chalkboards	Are inexpensive Help audiences take notes Allow you to emphasize a point Allow audiences to focus on a statement Help you be spontaneous Break up monotonous speeches	Make a mess Can be noisy Make you turn your back to the audience Can be hard to see Can be hard to read if your handwriting is poor	Clean the board well. Have extra chalk. Stand to the side as you write. Print in large letters. Write slowly. Avoid talking with your back to the audience. Don't erase too soon.
Chalkless Boards	Same as above except for expense	Are expensive Require unique, erasable pens Can stain your clothes Can be hard to erase if left on the board too long	Use blue or red ink. Cap your pens to avoid drying out. Use pens made especially for these boards. Erase soon after use.
Flip Charts	Can be prepared in advance Are neat and clean Can be reused Are inexpensive Are portable Help you avoid a non-stop presentation Allow for spontaneity Help audiences take notes Allow you to emphasize ideas Encourage audience participation Allow easy reference by turning back to prior pages Allow highlighting through different color	Are limited by small size Require an easel Require neat handwriting Won't work well with large groups Can run out of paper Markers can run out of ink	Have two pads of paper. Have numerous markers. Use different colors. Print in large neat letters. Turn pages when through with an idea so audiences won't be distracted. Don't write on the back of pages.

TYPE	ADVANTAGES	DISADVANTAGES	HELPFUL HINTS
Overhead Transparencies	Are inexpensive Can be used with lights on Can be prepared in advance Can be reused Can be used for large audiences Allow you to return to a prior point Allow you to face audiences	Require an overhead projector Can be hard to focus Require an electrical outlet Can become scratched and smudged Can be too small for viewing Can malfunction (bulbs burn out)	Use larger print. Protect the transparencies with separating sheets of paper. Frame the transparencies for better handling. Turn the overhead off to avoid distractions. Focus the overhead before your speech. Keep spare bulbs. Don't write on transparencies. Face the audience.
Slides	Are portable Are easy to protect Can be used for large groups Can be prepared in advance Are entertaining and colorful Allow for later reference	Are immobile Require dark rooms Deny speaker/audience interaction (eye contact) Are somewhat expensive Are challenging to create Require screen, machinery, and electrical outlets Can get out of order Don't allow audience to take notes Machinery can malfunction	Use a pointer. Use remote control. Check order of presentation. Check working condition of machinery.
Videotapes	Allow instant replay Can be freeze-framed for emphasis Can be economically duplicated Can be rented or leased inexpensively Are entertaining	Require costly equipment (monitor and recorder) Are difficult to move Can malfunction Require dark rooms Require compatible equipment Deny easy note taking Deny speaker-audience interaction	Practice operating the equipment. Avoid long tapes.

TYPE	ADVANTAGES	DISADVANTAGES	HELPFUL HINTS
Films	Are easy to use Are entertaining Have many to choose from Can be used for large groups	Require dark rooms Require equipment and outlets Don't allow note taking Deny speaker-audience interaction Malfunction Become dated	Use up-to-date films. Avoid long films. Provide discussion time. Use to supplement the speech, not replace it. Practice with the equipment.
Computerized Presentations	Are entertaining Offer flexibility, allowing you to move from topic to topic randomly Can be customized and updated Can be used for large groups Allow speaker-audience interaction	Require equipment and outlets Require dark rooms Malfunction	Practice with the equipment. Bring spare cables. Be prepared with a backup plan if the system crashes.
PowerPoint Presentations	Are entertaining Keep an audience interested Provide an outline for audiences to follow Can be duplicated in handouts Can be prepared in advance Can be reused Can be used for large audiences Allow you to return to a prior point Are easy to project Can include animation	Required dark rooms Require computer equipment (unless made into transparencies) Equipment can malfunction Can detract from the speaker Can be too small for viewing Can distance the audience from the presenter Can contribute to stilted delivery	Have correct computer equipment (cables, monitors, CPUs, etc.). Make backup transparencies. Practice your presentation. Have handouts available in case of computer malfunction.

PROCESS

As with writing memos, letters, and reports, approach your oral presentation systematically. Follow the three-part step-by-step process to organize your speech effectively.

- Prewrite
- Write
- Make your presentation

PREWRITING

Prewriting is the work you do to get started on your speech. Similar to prewriting for written communication, when you prewrite a speech, you (a) consider audience, (b) determine objectives, and (c) gather data.

Consider Audience

Before giving your speech, ask yourself the following questions:

- Are you speaking laterally to peers?
- Are you speaking to supervisors?
- Are you speaking to subordinates?
- Are you speaking to lay customers?
- Is your audience high tech or low tech?
- Is your audience composed of multiple levels of expertise and authority?
- Does your audience know you?
- Is your audience interested in the subject matter?
- Will your listeners be receptive to the topic, or will they be hostile?

Answering these questions will help you decide how technical you can be, when you need to define terms, and what tone you should take. You won't communicate effectively if your audience fails to understand your terminology or if your tone is offensive. Your only reason for speaking is to communicate to your audience. Tailor your content and style accordingly.

Determine Objectives

Determining your reason for speaking is as important as recognizing your audience. Is the purpose of your speech to *inform*? Or do you hope to *persuade*, *explain procedures*, or *motivate*?

INFORM

If your speech is to inform, all you want to do is update your listeners. Such a speech could be about new taxation laws affecting listeners' pay, new management hirings, or budget restraints affecting capital equipment purchases. Speeches which inform don't require any action on the part of your audience. Your listeners won't change the tax laws, alter hiring practices, or increase the budget. The informative speech merely keeps the audience up to date on business changes.

In such a speech, you'll want to clarify *when* changes will occur, *who* will be affected, and *how* the changes will affect your audience. Leave time for questions and answers to clarify any confusing points. However, in an informative speech, little audience involvement is necessary.

PERSUADE, EXPLAIN PROCEDURES, AND/OR MOTIVATE

In contrast, speeches that persuade, explain procedures, and/or motivate demand audience involvement. You're not just informing; you're asking your listeners to act. For instance, you might be speaking about the need to hold more regular and constructive quality circle meetings (persuasion). You might be telling your audience how to perform a series of steps for computer operation (explaining procedures). You might be telling your audience that they can achieve a higher level of productivity through mutual trust and cooperation (motivation). In each instance, you want your audience to leave the speech ready to implement your suggestions.

Gather Data

The best delivery by the most polished professional will lack credibility if the speaker has nothing to say. Therefore, you must study and research your topic thoroughly before you write about it.

Chapter 13 discusses how to collect information in a library and online. Businesspeople also rely on numerous other sources for data, including

- Interviews
- Questionnaires and surveys
- Conversations in meetings
- Company monthly, quarterly, and year-end reports

Gathering data from company reports requires that you read and document your information (discussed in Chapter 13). Gathering data through interviews, questionnaires, or conversations, however, requires assistance from other people. To do so,

- Ask politely for their assistance.
- Explain your reasons for seeking the interview and information.
- Set up convenient interview dates.
- Come prepared.
  - —Research the subject matter so you'll be prepared to ask appropriate questions.
  - —Write your interview questions before the meeting to save time.
  - —Be sure you have whatever paper, writing utensils, or recording devices you'll need for the interview.
- Thank them politely for their help, and/or make another date for follow-up discussions.

WRITING

Once you obtain information, your next step is to write a draft for your speech. As mentioned earlier in this chapter, avoid committing the entire speech to paper. Instead, organize your data and then outline the details.

You can begin writing by using either a presentation plan, an outline, and/or note cards.

Presentation Plan

Figure 17.1 (on p. 439) is an example of a presentation plan.

Outline

You may want to write a more detailed outline focusing on your speech's major units of discussion and supporting documentation. The *skeleton speech outline* (Figure 17.2, p. 440) and *student-written outline* (Figure 17.3, pp. 441-442) provide you with templates for your speech presentation.

Note Cards

If you decide that presenting the speech from the outline will not work for you, then you can write highlights of the speech on 3″ × 5″ cards. One caution—avoid writing complete sentences or filling the cards side to side. Just write short notes (phrases or key words) which will aid your memory.

Once your speech is outlined and note cards are prepared, practice, incorporating all your visual aids. If you were told to present a twenty-minute speech, that is how long you should talk. To speak longer than requested will infringe on another speaker's time, on question-and-answer time, or on coffee-break time. Practicing will help you meet any time constraints.

PRESENTATION

Use the following speaker's checklist to determine whether your presentation will be successful.

SPEAKER'S CHECKLIST

✔ Is my choice of topic relevant and appropriate?

✔ Is my beginning polite and upbeat?

✔ Have I used an attention-getting device?

✔ Is my focus statement clear?

✔ Do I have an effective overall organization?

✔ Are my major points evident?

✔ Do I remind my audience of my topic at the beginning of the speech, in the middle, and at the end?

✔ Do I move smoothly from point to point using transitional words and phrases?

✓ Have I documented my findings using effective research?

✓ Do I define any unfamiliar terms?

✓ Do I use my note cards or outline well?

✓ Are my visual aids presented effectively?

✓ Is my conclusion effective?

✓ Do I enunciate clearly?

✓ Is my grammar correct?

✓ Is my body language effective?

✓ Do I maintain eye contact?

✓ Are my volume, tone, and pitch good?

✓ Is my pace effective (not too slow or too fast)?

✓ Am I confident and sincere?

POST-SPEECH ACTIVITIES

After your speech, be prepared for a question-and-answer session. Politely invite your audience to participate by saying, "If you have any questions, I am happy to answer them."

When an audience member asks a question, make sure everyone hears it. If not, you can repeat the question. If you don't understand the question, ask the audience member to repeat it and clarify.

If you have no answer, tell the audience. You will only mar the effect of your presentation if you stumble over an answer and appear to be unsure of yourself. Credibility is essential to your success; faking an answer will not ensure credibility.

If a member of the audience is rude or has disliked your speech, you can say that you will be happy to discuss the matter after the presentation. Don't let a successful speech be marred by one disgruntled listener.

CHAPTER HIGHLIGHTS

1. When you have no time to prepare an oral presentation, you must make an impromptu speech.

2. Memorized speeches are the least effective, because they are often too stiff and formal.

3. Reading a manuscript speech can be ineffective.

4. The extemporaneous speech is the most successful type of oral presentation.

5. Writing key points on 3" × 5" note cards will help you prepare for the speech.

Directions: Use this plan to help you clarify your thinking about the topic.

Topic: _____

Objectives: What do you want your audience to believe or do as a result of your presentation?

Development: What main points are you going to develop in your presentation?

1.

2.

3.

Organization Will this presentation lend itself to a particular organizational format, such as comparison/contrast, chronology, or analysis?

Visuals: What visuals will you use?

FIGURE 17.1 Presentation Plan

Title: _____

Purpose: _____

   I. Introduction _____

      A. Attention getter: _____

         _____

      B. Focus statement: _____

         _____

   II. Body _____

      A. First main point: _____

         1. Documentation/subpoint: _____

            a. Documentation/subpoint: _____

               _____

            b. Documentation/subpoint: _____

               _____

         2. Documentation/subpoint: _____

            _____

         3. Documentation/subpoint: _____

      B. Second main point: _____

         1. Documentation/subpoint: _____

            _____

         2. Documentation/subpoint: _____

            a. Documentation/subpoint: _____

               _____

            b. Documentation/subpoint: _____

               _____

         3. Documentation/subpoint: _____

      C. Third main point: _____

         1. Documentation/subpoint: _____

          2. Documentation/subpoint: _____

          3. Documentation/subpoint: _____

            a. Documentation/subpoint: _____

            b. Documentation/subpoint: _____

               _____

  III. Conclusion

      A. Summary of main points: _____

         _____

      B. Recommended future course of action: _____

         _____

FIGURE 17.2 Skeleton Outline

Title: How Leather Is Made
Purpose: To Explain Procedure
I. Introduction
    A. Attention getter: How many of you own leather products? How many of you know how leather is made? Working at Houston's Leathery, I am often asked to explain the procedure for making leather.
    B. Focus statement: Leather comes from animal skin preserved by a process called tanning. Tanning requires the application of chromium salt, a chemical solution. The process of leather tanning is divided into three phases: pretanning, tanning, and dyeing and finishing.
II. Body
    A. Phase 1: Pretanning requires six steps.
        1. The skin is cured so it will not decompose. Curing can be accomplished in either of two ways.
            a. Salt can be spread on the underside of the skin and kept in packs for approximately a week.
            b. The skin can be soaked in a lime solution for several days.
        2. The cured skins are sent to a tannery where they are trimmed and the best parts are kept for leather.
        3. The excess salt from the skins is removed by soaking them in water.
        4. The skins are run through a machine in which revolving blades remove excess fat.
        5. The skins are placed in vats of lime solution for several days.
        6. Pretanning is complete when the skins are pickled by placing them in salt and chemical solutions.
    B. Phase 2: The tanning stage requires three steps.
        1. The tanning process begins by soaking the skins in revolving drums containing tanning agents.
        2. The skins are put through a wringer to remove excess moisture and chemicals.
        3. The skins are split to proper thickness and measured by a leather gauge.
    C. Phase 3: Dyeing and finishing are the last phase of making leather.
        1. Dyeing requires four steps.
            a. A machine smooths and stretches the skin while removing any remaining moisture.
            b. The skin is dried in an oven.

FIGURE 17.3 Student-Written Sentence Outline

> c. The leather is buffed and sanded to achieve a smooth-grained texture.
> d. Color is added by drum dyeing or spray dyeing.
>   i. Drum dyeing penetrates both sides of the leather.
>   ii. Spray dyeing covers only one side and only small blemishes.
> 2. There are five basic types of finishes.
>   a. Finished leather is created by spraying a pigment onto the outer surface of the leather. This finish is the most durable because it resists water, soil, and stains.
>   b. Aniline finish is achieved by soaking the leather in aniline dye. This type of finish lets you see the actual grain of the leather, but it leaves the leather unprotected.
>   c. Semi-aniline finish is a combination of the above two finishes. An aniline dye is applied and an opaque pigment is sprayed to achieve an even color.
>   d. Suede finish is achieved when the leather is split and buffed to a silky nap.
>   e. Shearling finish is derived from sheep's wool. The wool is left on the skin when tanned and then used as the lining for the garment.
> III. Conclusion
>   A. Summation: I now have explained the three phases of the tanning process and the five types of finishes.
>   B. Recommendations: Not only should you have a better understanding of how leather is made, but also you will be able to select the finish that will best suit your needs.

FIGURE 17.3 Student-Written Sentence Outline (Cont.)

6. Begin your speech in an interesting way: ask a question, use an anecdote, or cite data to involve your audience.

7. Restate your topic often to help your audience follow you more easily.

8. Adequate research ensures a factual delivery.

9. Looking at your listeners and speaking to individuals will keep your audience interested.

10. Speak slowly.

11. Choose your wardrobe carefully.

12. Use visual aids to enhance your delivery.

ACTIVITIES

1. Go to a library, research a topic that interests you, and take notes on the topic from a journal or magazine article. In class, present a three- to five-minute speech briefing your classmates about the article's content.

2. In groups of three to five students, read an article about a technical concept or mechanism. Prepare a list of questions about the subject matter. Then select one member of the group who will answer questions while the rest of the group members ask them. Do this several times so each group member has an opportunity to answer and ask questions.

3. Prepare an extemporaneous speech of about ten to fifteen minutes. Include an outline, note cards, and visual aids. For your subject matter, focus on a problem at work or school. In your speech, explain how the problem arose, offer alternative solutions, and persuade your audience to pursue a future course of action solving the problem.

4. Prepare a three- to five-minute extemporaneous speech informing your audience of the status of an activity on which you are working. This could be a school project or a work-related activity. Provide visual aids to supplement your presentation.

5. Interview a professional in the career of your choice to determine whether the field will provide you with the challenges, opportunities, and financial remuneration you desire. To perform this interview, set up an interview date, be prepared with questions, and take notes. Then return to class and present a three- to five-minute extemporaneous speech about your interview.

6. Attend a lecture. Assess the speaker's presentation according to the criteria checklist presented in this chapter. Then provide an oral presentation reporting on your findings.

CHAPTER 18

Grammar, Punctuation, Mechanics, and Spelling

It is essential to correctly organize and develop your memo, letter, or report. However, no one will be impressed with the quality of your work, or with you, if your writing is riddled with errors in sentence construction or punctuation. Your written correspondence is often your first contact with business associates. Many people mistakenly believe that only English teachers notice grammatical errors and wield red pens, but businesspeople as well take note of such errors and may see the writer as less competent.

We were working recently with a young executive who is employed by a branch of the federal government. This executive told us that whenever his supervisor found a spelling error in a subordinate's report, this report was paraded around the office. Everyone was shown the mistake and had a good laugh over it, and the report was then returned to the writer for correction. Our acquaintance assured us that all of this was in good-natured fun. However, he also said that employees quickly learned to edit and proof their work to avoid such public displays of their errors. He went on to say that his dictionary was well thumbed and always on his desk.

Your writing at work may not be exposed to such scrutiny by co-workers. Instead, your writing may go directly to another firm, and those readers will see your mistakes. To avoid this problem, you must evaluate your writing for grammar, punctuation, and spelling errors. If you don't, your customers and colleagues will.

This chapter focuses on (a) grammar rules, (b) punctuation, (c) rules for effective mechanics, (d) spelling, and (e) proofreader's marks. In addition, the chapter provides hands-on experience with grammar, punctuation, and spelling in the form of memos and letters needing correction.

GRAMMAR RULES

To understand the fundamentals of grammar, you must first understand the basic components of a sentence.

A correctly constructed sentence consists of a subject and a predicate (some sentences also include a phrase or phrases).

The *meeting*	*began*	*at 4:00 a.m.*
subject	predicate	phrase

Subject: The *doer* of the action; the subject usually precedes the predicate.
Predicate: The *action* in the sentence

He	*ran*	*to the office to avoid being late.*
doer	action	phrases

If the subject and the predicate (a) express a complete thought and (b) can stand alone, you have an *independent clause.*

The meeting began	at 4:00 p.m.
independent clause	phrase

A *phrase* is a group of related words that does not contain a subject and a predicate and cannot stand alone or be punctuated as a sentence. The following are examples of phrases.

at the house

in the box

on the job

during the interview

If a clause is dependent, it cannot stand alone.

Although he tried to hurry,	he was late for the meeting.
dependent clause	independent clause

He was late for the meeting although he tried to hurry.

independent clause   dependent clause

Note: When a dependent clause begins a sentence, use a comma before the independent clause. However, when an independent clause begins a sentence, do not place a comma before the dependent clause.

AGREEMENT BETWEEN PRONOUN AND ANTECEDENT (REFERENT)

A pronoun has to agree in gender and number with its antecedent.

> *Susan* went on *her* vacation yesterday.
>
> The *people* who quit said that *they* deserved raises.

Problems often arise when a singular indefinite pronoun is the antecedent. The following antecedents require singular pronouns: *anybody*, *each*, *everybody*, *everyone*, *somebody*, and *someone*.

> INCORRECT
> *Anyone* can pick up *their* applications at the job placement center.

> CORRECT
> *Anyone* can pick up *his or her* applications at the job placement center.

Problems also arise when the antecedent is separated from the pronoun by numerous words.

> INCORRECT
> Even when the best *employee* is considered for a raise, *they* often do not receive it.

> CORRECT
> Even when the best *employee* is considered for a raise, *she* often does not receive it.

AGREEMENT BETWEEN SUBJECT AND VERB

Writers sometimes create disagreement between subjects and verbs, especially if other words separate the subject from the verb. To ensure agreement, ignore the words that come between the subject and verb.

> INCORRECT
> Her *boss* undoubtedly *think* that all the employees want promotions.

> CORRECT
> Her *boss* undoubtedly *thinks* that all the employees want promotions.

INCORRECT
The *employees* who sell the most equipment *is* going to Hawaii for a week.

CORRECT
The *employees* who sell the most equipment *are* going to Hawaii for a week.

If a sentence contains two subjects (a compound subject) connected by *and*, use a plural verb.

INCORRECT
Joe and Becky *was* both selected employee of the year.

CORRECT
Joe and Becky *were* both selected employee of the year.

INCORRECT
The bench workers and their supervisor *is* going to work closely to complete this project.

CORRECT
The bench workers and their supervisor *are* going to work closely to complete this project.

Add a final *s* or *es* to create most plural subjects or singular verbs, as follows:

Plural Subjects	Singular Verbs
boss*es* hire	a boss hir*es*
employe*es* demand	an employee demand*s*
experiment*s* work	an experiment work*s*
attitud*es* change	the attitude chang*es*

If a sentence has two subjects connected by *either . . . or, neither . . . nor*, or *not only . . . but also*, the verb should agree with the closest subject. This also makes the sentence less awkward.

Either the salespeople or the warehouse worker deserv*es* raises.

Not only the warehouse worker but also the salespeople deserv*e* raises.

Neither the salespeople nor the warehouse worker deserv*es* raises.

Singular verbs are used after most indefinite pronouns such as the following:

another	everything
anybody	neither
anyone	nobody
anything	no one

each	nothing
either	somebody
everybody	someone
everyone	something

Anyone who works here *is* guaranteed maternity leave.

Everybody wants the company to declare a profit this quarter.

Singular verbs often follow collective nouns such as the following:

class	organization
corporation	platoon
department	staff
group	team

The *staff* is sending the boss a bouquet of roses.

COMMA SPLICE

A *comma splice* occurs when two independent clauses are joined by a comma rather than a period or semicolon.

INCORRECT
Sue was an excellent employee, she got a promotion.

Several remedies will correct this error.

1. Separate the two independent clauses with a semicolon.

CORRECT
Sue was an excellent employee; she got a promotion.

2. Separate the two independent clauses with a period.

CORRECT
Sue was an excellent employee. She got a promotion.

3. Separate the two independent clauses with a comma and a *coordinating conjunction (and, but, or, for, so, yet).*

CORRECT
Sue was an excellent employee, *so* she got a promotion.

4. Separate the two independent clauses with a semicolon, a conjunctive adverb, and a comma. *Conjunctive adverbs* include *also, additionally, consequently, furthermore, however, instead, moreover, nevertheless, therefore,* and *thus.*

CORRECT

Sue was an excellent employee; *therefore*, she got a promotion.

5. Use a *subordinating conjunction* to make one of the independent clauses into a dependent clause. Subordinating conjunctions include *after, although, as, because, before, even though, if, once, since, so that, though, unless, until, when, where,* and *whether.*

CORRECT

Because Sue was an excellent employee, she got a promotion.

FAULTY OR VAGUE PRONOUN REFERENCE

A pronoun must refer to a specific noun (its antecedent). Problems arise when (a) there is an excessive number of pronouns (causing vague pronoun reference) and (b) there is no specific noun as an antecedent. Notice that there seems to be an excessive number of pronouns in the following passage, and the antecedents are unclear.

Although Bob had been hired over two years ago, *he* found that his boss did not approve *his* raise. In fact, *he* was also passed over for *his* promotion. The boss appears to have concluded that *he* had not exhibited zeal in *his* endeavors for their business. Instead of being a highly valued employee, *he* was not viewed with pleasure by those in authority. Perhaps it would be best if *he* considered *his* options and moved to some other company where *he* might be considered in a new light.

The excessive and vague use of *he* and *his* causes problems for readers. Do these words refer to Bob or to his boss? You're never completely sure. To avoid this problem, limit pronoun usage, as in the following revision.

Although Bob had been hired over two years ago, he found that his boss, Joe, did not approve his raise. In fact, Bob was also passed over for promotion. Joe appears to have concluded that Bob had not exhibited zeal in his endeavors for their business. Instead of being a highly valued employee, Bob was not viewed with pleasure by those in authority. Perhaps it would be best if Bob considered his options and moved to some other company where he might be considered in a new light.

To make the preceding paragraph more precise, we've replaced vague pronouns (*he* and *his*) with exact names (*Bob* and *Joe*).

FRAGMENT

A *fragment* occurs when a group of words is incorrectly used as an independent clause. Often the group of words begins with a capital letter and has end punctuation but is missing either a subject or a predicate.

INCORRECT
Working with computers.
(lacks a predicate and does not express a complete thought)

The group of words may have a subject and a predicate but be a dependent clause.

INCORRECT
Although he enjoyed working with computers.
(has a subject, *he*, and a predicate, *enjoyed*, but is a dependent clause because it is introduced by the subordinate conjunction *although*)

It is easy to remedy a fragment by doing one of the following:

- Add a subject.
- Add a predicate.
- Add both a subject and a predicate.
- Add an independent clause to a dependent clause.

CORRECT
Joe found that working with computers used his training.
(subject, *Joe*, and predicate, *found*, have been added)

CORRECT
Although he enjoyed working with computers, he could not find a job in a computer-related field.
(independent clause, *he could not find a job*, added to the dependent clause, *Although he enjoyed working with computers*)

FUSED SENTENCE

A *fused sentence* occurs when two independent clauses are connected with no punctuation.

INCORRECT
The company performed well last quarter its stock rose several points.

There are several ways to correct this error.

1. Write two sentences separated by a period.

CORRECT
The company performed well last quarter. Its stock rose several points.

2. Use a comma and a coordinating conjunction to separate the two independent clauses.

CORRECT
The company performed well last quarter, *so* its stock rose several points.

3. Use a subordinating conjunction to create a dependent clause.

CORRECT
Because the company performed well last quarter, its stock rose several points.

4. Use a semicolon to separate the two independent clauses.

CORRECT
The company performed well last quarter; its stock rose several points.

5. Separate the two independent clauses with a semicolon, a conjunctive adverb or a transitional word or phrase, and a comma.

CORRECT
The company performed well last quarter; *therefore*, its stock rose several points.

CORRECT
The company performed well last quarter; *for example*, its stock rose several points.

The following are transitional words and phrases, listed according to their use.

To Add

again	in addition
also	moreover
besides	next
first	second
furthermore	still

To Compare/Contrast

also	nevertheless
but	on the contrary
conversely	still
in contrast	

To Provide Examples

for example	of course
for instance	put another way
in fact	to illustrate

To Show Place

above	here
adjacent to	nearby
below	on the other side
elsewhere	there
further on	

To Reveal Time

afterward	second
first	shortly
meanwhile	subsequently
presently	thereafter

To Summarize

all in all	last
finally	on the whole
in conclusion	therefore
in summary	thus

MODIFICATION

A *modifier* is a word, phrase, or clause that explains or adds details about other words, phrases, or clauses.

Misplaced Modifiers

A *misplaced modifier* is one that is not placed next to the word it modifies.

INCORRECT
He had a heart attack *almost* every time he was reviewed by his supervisor.

CORRECT
He almost had a heart attack every time he was reviewed by his supervisor.

INCORRECT
The worker had to *frequently* miss work.

CORRECT
The worker frequently had to miss work.

Dangling Modifiers

A *dangling modifier* is a modifier that is not placed next to the word or phrase it modifies. To avoid confusing your readers, place modifiers next to the word(s) they refer to. Don't expect your readers to guess at your meaning.

INCORRECT
While working, tiredness overcame them.
(Who was working? Who was overcome by tiredness?)

CORRECT
While working, the staff became tired.

INCORRECT
After soldering for two hours, the equipment was ready for shipping.
(Who had been soldering for two hours? Not the equipment!)

CORRECT
After soldering for two hours, the technicians prepared the equipment for shipping.

PARALLELISM

All items in a list should be parallel in grammatical form.

INCORRECT
We will discuss the following at the department meeting:

1. Entering mileage in logs
2. All employees have to enroll in a training seminar.
3. Purpose of quarterly reviews
4. Some data processors will travel to job sites.

CORRECT
We will discuss the following at the department meeting:

1. Entering mileage in logs
2. Enrolling in training seminars
3. Reviewing employee performance quarterly
4. Traveling to job sites

CORRECT
At the department meeting, you will learn how to

1. Enter mileage in logs
2. Enroll in training seminars
3. Review employee performance quarterly
4. Travel to job sites

PUNCTUATION

APOSTROPHE (')

Place an *apostrophe* before the final *s* in a singular word to indicate possession.

> Jim's tool chest is next to the furnace.

Place the apostrophe after the final *s* if the word is plural.

> The employees' reception will be held next week.

Don't use an apostrophe to make singular abbreviations plural.

> INCORRECT
> The EXT's will be shipped today.

> CORRECT
> The EXTs will be shipped today.

COLON (:)

Use a *colon* after a salutation.

> Dear Mr. Sandhaus:

In addition, use a colon after an emphatic or cautionary word if explanations follow.

> Note: Hand-tighten the nuts.

> Caution: Wash thoroughly if any mixture touches your skin.

Finally, use a colon after an independent clause to precede a quotation, list, or example.

> She said the following: "No comment."

> These supplies for the experiment are on order: plastic hose, two batteries, and several chemicals.

> The problem has two possible solutions: hire four more workers, or simply give everyone a raise.

Note: In the preceding examples, the colon follows an independent clause.

A common mistake is to place a colon after an incomplete sentence. Except for salutations and cautionary notes, whatever precedes a colon *must* be an independent clause.

INCORRECT
The two keys to success are: earning money and spending wisely.

CORRECT
The two keys to success are earning money and spending wisely.

or

The two keys to success are as follows: earning money and spending wisely.

or

The two keys to success are as follows:
1. Earning money
2. Spending wisely

COMMA (,)

Writers often get in trouble with *commas* when they employ one of two common "words of wisdom."

* When in doubt, leave it out.
* Use a comma when there is a pause.

Both rules are inexact. Writers use the first rule to justify the complete avoidance of commas; they use the second rule to sprinkle commas randomly throughout their writing. On the contrary, commas have several specific conventions which determine usage.

1. Place a comma before a coordinating conjunction (*and, but, or, for, so, yet*) linking two independent clauses.

You are the best person for the job, *so* I will hire you.

We spent several hours discussing solutions to the problem, *but* we failed to decide on a course of action.

2. Use commas to set off introductory comments.

First, she soldered the components.

In business, people often have to work long hours.

To work well, you need to get along with your co-workers.

If you want to test equipment, do so by 5:25 P.M.

3. Use commas to set off sentence interrupters.

The company, started by my father, did not survive the last recession.

Mrs. Mittleman, the proprietor of the store, purchased a wide array of merchandise.

4. Set off parenthetical expressions with commas.

A worker, it seems, should be willing to try new techniques.

The highway, by the way, needs repairs.

5. Use commas between items in a series.

Fred, Helene, and Ron were chosen as employees of the year.

We found the following problems: corrosion, excessive machinery break-downs, and power failures.

6. Use commas to set off sections of dates, addresses, and long numbers.

On January 9, 2000, the company opened its new offices.

We live at 11517 Grant, Overland Park, Kansas.

She earns $100,000 before taxes.

Note: Very large numbers are often written as decimals.

Our business netted over $2 million in 2000.

7. Use commas to set off the day and year when they are part of a sentence.

The company hired her on September 7, 2000, to be its bookkeeper.

Note: If the year is used as an adjective, do not follow it with a comma.

The 1999 corporate report came out today.

8. Use commas to set off the city from the state and the state from the rest of the sentence.

The new warehouse in Austin, Texas, will promote increased revenues.

Note: If you omit either the city or the state, you do not need commas.

The new warehouse in Austin will promote increased revenues.

DASH (—)

A *dash*, typed as two consecutive hyphens with no spaces before or after, is a versatile punctuation mark used in the following ways.

1. After a heading and before an explanation.

Forecasting—Joe and Joan will be in charge of researching fourth-quarter production quotas.

2. To indicate an emphatic pause.

You will be fired—unless you obey company rules.

3. To highlight a new idea.

Here's what we can do to improve production quality—provide on-the-job training, salary incentives, and quality controls.

4. Before and after an explanatory or appositive series.

Three people—Sue, Luci, and Tom—are essential to the smooth functioning of our office.

ELLIPSES (. . .)

Ellipses (three spaced periods) indicate omission of words within quoted materials.

"Six years ago, prior to incorporating, the company had to pay extremely high federal taxes."

"Six years ago, . . . the company had to pay extremely high federal taxes."

EXCLAMATION POINT (!)

Use an *exclamation point* after strong statements, commands, or interjections.

You must work harder!

Do not use the machine!

Danger!

HYPHEN (-)

A *hyphen* is used in the following ways.

   1. To indicate the division of a word at the end of a typed line. Remember, this division must occur between syllables.

   2. To create a compound adjective.

He is a well-known engineer.

Until her death in 2000, she was a world-renowned chemist.

Tom is a 24-hour-a-day student.

   3. To join the numerator and denominator of fractions.

Four-fifths of the company want to initiate profit sharing.

4. To write out two-word numbers.

Twenty-six people attended the conference.

PARENTHESES ( )

Parentheses enclose abbreviations, numbers, words, or sentences for the following reasons.

1. To define a term or provide an abbreviation for later use.

We belong to the STC (Society for Technical Communication).

2. To clarify preceding information in a sentence.

The people in attendance (all regional sales managers) were proud of their accomplishments.

3. To number items in a series.

The company should initiate (1) new personnel practices, (2) a probationary review board, and (3) biannual raises.

PERIOD (.)

A *period* must end a declarative sentence (independent clause).

I found the business trip rewarding.

Periods are often used with abbreviations.

D.C.	e.g.	A.M. or a.m.
M.D.	Mr.	P.M. or p.m.
Ph.D.	Mrs.	
B.A.	Ms.	

It is incorrect to use periods with abbreviations for organizations and associations.

INCORRECT
S.T.C. (Society for Technical Communication)

CORRECT
STC (Society for Technical Communication)

State abbreviations no longer use periods.

INCORRECT
KS. (Kansas)
MO. (Missouri)
TX. (Texas)

CORRECT
KS
MO
TX

QUESTION MARK (?)

Use a *question mark* after direct questions.

Do the lab results support your theory?

Will you work at the main office or at the branch?

QUOTATION MARKS (" ")

Quotation marks are used in the following ways.

1. When citing direct quotations.

He said, "Your division sold the most compressors last year."

Note: When you are citing a quotation within a quotation, use double quotation marks (" ") and single quotation marks (' ').

Kim's supervisor, quoting the CEO, said the following to explain the new policy regarding raises: "'Only employees who deserve them will receive merit raises.'"

2. To note the title of an article or a subdivision of a report.

The article "Robotics in Industry Today" was an excellent choice as the basis of your speech.

Section III, "Waste Water in District 9," is pertinent to our discussion.

When using quotation marks, abide by the following punctuation conventions:

* Commas and periods always go *inside* quotation marks.

She said, "Our percentages are fixed."

* Colons and semicolons always go *outside* quotation marks.

He said, "The supervisor hasn't decided yet"; however, he added that the decision would be made soon.

* Exclamation points and question marks go inside the quotation marks if the quoted material is either exclamatory or a question. However, if the quoted material is not exclamatory or a question, then these punctuation marks go outside the quotation marks.

John said, "Don't touch that liquid. It's boiling!"
(Although the sentence isn't exclamatory, the quotation is. Thus, the exclamation point goes inside the quotation marks.)

How could she say, "We haven't purchased the equipment yet"?
(Although the quotation isn't a question, the sentence is. Thus, the question mark goes outside the quotation marks.)

SEMICOLON (;)

Semicolons are used in the following instances.

1. Between two independent clauses *not* joined by a coordinating conjunction.

The light source was unusual; it emanated from a crack in the plastic surrounding the cathode.

2. To separate items in a series containing internal commas.

When the meeting was called to order, all members were present, including Susan Bailey, the president; Ruth Schneider, the vice president; Harold Holbert, the treasurer; and Linda Hamilton, the secretary.

MECHANICS

ABBREVIATIONS

Never use an abbreviation which your reader will not understand. A key to clear technical writing is to write on a level appropriate to your reader. You may use the following familiar abbreviations without explanation: *Mrs., Dr., Mr., Ms., Ph.D.,* and *Jr.*

A common mistake is to abbreviate inappropriately. For example, some writers abbreviate *and* as follows:

I quit my job & planned to retire young.

This is too colloquial for professional technical writing. Spell out *and* when you write.

The majority of abbreviation errors occur when writers incorrectly abbreviate states and technical terms. (We list many computer-related abbreviations in Chapter 10.)

States

Writers often abbreviate the names of states incorrectly. Use the U.S. Postal Service abbreviations in addresses.

Abbreviations for States

AL	Alabama	MT	Montana
AK	Alaska	NC	North Carolina
AZ	Arizona	ND	North Dakota
AR	Arkansas	NE	Nebraska
CA	California	NV	Nevada
CO	Colorado	NH	New Hampshire
CT	Connecticut	NJ	New Jersey
DE	Delaware	NM	New Mexico
FL	Florida	NY	New York
GA	Georgia	OH	Ohio
HI	Hawaii	OK	Oklahoma
IN	Indiana	OR	Oregon
IA	Iowa	PA	Pennsylvania
ID	Idaho	RI	Rhode Island
IL	Illinois	SC	South Carolina
KS	Kansas	SD	South Dakota
KY	Kentucky	TN	Tennessee
LA	Louisiana	TX	Texas
ME	Maine	UT	Utah
MD	Maryland	VT	Vermont
MA	Massachusetts	VA	Virginia
MI	Michigan	WA	Washington
MN	Minnesota	WV	West Virginia
MS	Mississippi	WI	Wisconsin
MO	Missouri	WY	Wyoming

Technical Terms

Units of measurement and scientific terms must be abbreviated accurately to ensure that they will be understood. Writers often use such abbreviations incorrectly. For example, "The unit measured 7.9 cent." is inaccurate. The correct abbreviation for centimeter is *cm*, not *cent*. Use the following abbreviation conventions.

Technical Abbreviations for Units of Measurement and Scientific Terms

absolute	abs
alternating current	AC
American wire gauge	AWG
ampere	amp

ampere-hour	amp-hr
amplitude modulation	AM
angstrom unit	Å
atmosphere	atm
atomic weight	at wt
audio frequency	AF
azimuth	az
barometer	bar.
barrel, barrels	bbl
billion electron volts	BeV
biochemical oxygen demand	BOD
board foot	bd ft
Brinell hardness number	BHN
British thermal unit	Btu
bushel	bu
calorie	cal
candela	cd
Celsius	C
center of gravity	cg
centimeter	cm
circumference	cir
cologarithm	colog
continuous wave	CW
cosine	cos
cotangent	cot
cubic centimeter	cc
cubic foot	cu ft (or ft^2)
cubic feet per second	cfs
cubic inch	cu in. (or $in.^3$)
cubic meter	cu m (or m^3)
cubic yard	cu yd (or yd^3)
current (electric)	I
cycles per second	CPS
decibel	dB
decigram	dg
deciliter	dl
decimeter	dm

degree	deg
dekagram	dkg
dekaliter	dkl
dekameter	dkm
dewpoint	DP
diameter	dia
direct current	DC
dozen	doz (or dz)
dram	dr
electromagnetic force	emf
electron volt	eV
elevation	el (or elev)
equivalent	equiv
Fahrenheit	F
farad	F
faraday	f
feet, foot	ft
feet per second	ft/sec
fluid ounce	fl oz
foot board measure	fbm
foot-candle	ft-c
foot-pound	ft lb
frequency modulation	FM
gallon	gal
gallons per day	GPD
gallons per minute	GPM
grain	gr
grams	g (or gm)
gravitational acceleration	g
hectare	ha
hectoliter	hl
hectometer	hm
henry	H
hertz	Hz
high frequency	HF
horsepower	hp
horsepower-hours	hp-hr

hour	hr
hundredweight	cwt
inch	in.
inch-pounds	in.-lb
infrared	IR
inner diameter or inside dimensions	ID
intermediate frequency	IF
international unit	IU
joule	J
Kelvin	K
kilocalorie	kcal
kilocycle	kc
kilocycles per second	kc/sec
kilogram	kg
kilohertz	kHz
kilojoule	kJ
kiloliter	kl
kilometer	km
kilovolt	kV
kilovolt-amperes	kVa
kilowatt-hours	kWH
lambert	L
latitude	lat
length	l
linear	lin
linear foot	lin ft
liter	l
logarithm	log.
longitude	long.
low frequency	LF
lumen	lm
lumen-hour	lm-hr
maximum	max
megacycle	mc
megahertz	MHz
megawatt	MW
meter	m

microampere	μamp
microinch	μin.
microsecond	μsec
microwatt	μw
miles per gallon	mpg
milliampere	mA
millibar	mb
millifarad	mF
milligram	mg
milliliter	ml
millimeter	mm
millivolt	mV
milliwatt	mW
minute	min
nautical mile	NM
negative	neg or −
number	no.
octance	oct
ounce	oz
outside diameter	OD
parts per billion	ppb
parts per million	ppm
pascal	pas
positive	pos or +
pound	lb
pounds per square inch	psi
pounds per square inch absolute	psia
pounds per square inch gauge	psig
quart	qt
radio frequency	RF
radian	rad
radius	r
resistance	r
revolution	rev
revolutions per minute	rpm
second	s (or sec)
secant	sec

specific gravity	sp gr (or SG)
square foot	ft^2
square inch	in.^{2}
square meter	m^2
square mile	mi^2
tablespoon	tbs (or tbsp)
tangent	tan
teaspoon	tsp
temperature	t
tensile strength	ts
thousand	m
ton	t
ultra high frequency	UHF
vacuum	vac
very high frequency	VHF
volt-ampere	VA
volt	V
volts per meter	V/m
volume	vol
watt-hour	whr
watt	W
wavelength	WL
weight	wt
yards	y (or yd)
years	y (or yr)

CAPITAL LETTERS

Capitalize the following:

1. Proper nouns.

people	cities	countries	companies	schools	buildings
Susan	Houston	Italy	Bendix	Harvard	Oak Park Mall

2. People's titles (only when they *precede* the name).

Governor Sally Renfro

or

Sally Renfro, governor

Technical Supervisor Todd Blackman

or

Wes Schneider, the technical supervisor

3. Titles of books, magazines, plays, movies, television programs, and CDs (excluding the prepositions and all articles after the first article in the title).

Scream
The Catcher in the Rye
The New Yorker
The Taming of the Shrew
Friends

4. Names of organizations.

Girl Scouts
Kansas City Regional Home Care Association
Kansas City Regional Council for Higher Education
Programs for Technical and Scientific Communication
American Civil Liberties Union
Students for a Democratic Society

5. Days of the week, months, and holidays.

Monday
December
Thanksgiving

6. Races, religions, and nationalities.

American Indian
Jewish
Polish

7. Events or eras in history.

the Vietnam War
the Depression

8. North, South, East, and West (when used to indicate geographic locations).

They moved from the North.

People are moving to the Southwest.

Note: Don't capitalize these words when giving directions.

We were told to drive south three blocks and then to turn west.

9. The first word of a sentence.

10. Don't capitalize any of the following:

Seasons—spring, fall, summer, winter

Names of classes—sophomore, senior

General groups—middle management, infielders, surgeons

NUMBERS

Write out numbers one through nine. Use numerals for numbers 10 and above.

10	12
104	2,093
536	5,550,286

Although the preceding rules cover most situations, there are exceptions.

1. Use numerals for all percentages.

2 percent   18 percent   25 percent

2. Use numerals for addresses.

12 Elm   935 W. Harding

3. Use numerals for miles per hour.

5 mph   225 mph

4. Use numerals for time.

3:15 A.M.

5. Use numerals for dates.

May 31, 1948

6. Use numerals for monetary values.

$45 $.95 $2 million

7. Use numerals for units of measurement.

14′   6¾″   16 mm   10 v

8. Do not use numerals to begin sentences.

INCORRECT
568 people were fired last August.

CORRECT
Five hundred sixty-eight people were fired last August.

9. Do not mix numerals and words when writing a series of numbers. The larger number determines the form.

We attended 4 meetings over a 16-day period.

10. Use numerals and words in a compound number adjective to avoid confusion.

The worker needed six 2-inch nails.

SPELLING

The following is a list of commonly misspelled or misused words. You can avoid many common spelling errors if you familiarize yourself with these words.

accept, except	incite, insight
addition, edition	its, it's
access, excess	loose, lose
advise, advice	miner, minor
affect, effect	passed, past
all ready, already	patients, patience
assistants, assistance	personal, personnel
bare, bear	principal, principle
brake, break	quiet, quite
coarse, course	rite, right, write
cite, site, sight	stationery, stationary
council, counsel	their, there, they're
desert, dessert	to, too, two
disburse, disperse	whose, who's
fiscal, physical	your, you're
forth, fourth	

PROOFREADER'S MARKS

Figure 18.1 illustrates proofreader's marks and how to use them.

PROOFREADER'S MARKS

Symbol	Meaning	Mark on Copy	Revision
e	delete	hire better peoples*e*	hire better people
(tr) or ∩	transpose	computer syst∩me	computer system
⌣	close space	we ne⌣ed 40	we need 40
#	insert space	three#mistakes ∧	three mistakes
¶	begin paragraph	¶The first year	The first year
(RUN IN)	no paragraph	financial⁀ ⌐We earned twelve dollars.	financial. We earned twelve dollars.
⊏	move left	⌐ Next year	Next year
⊐	move right	⌐Next year	Next year
⊙	insert period	happy employee⊙	happy employee.
⩜	insert comma	For two years⩜	For two years,
⩘	insert semicolon	We need you⩘ come to work.	We need you; come to work.
⩛	insert colon	dogs⁝brown, black, and white.	dogs: brown, black, and white.
∧	insert hyphen	first∧rate	first-rate
∨	insert apostrophe	Marys car	Mary's car
∨/∨	insert quotation marks	Like a Rolling Stone	"Like a Rolling Stone"
(SP)	spell out	④chips	four chips
(cap) or ≡	capitalize	the meeting	The meeting
(lc)	lower case	the boss	the boss
∧	insert	the left∧margin	the left margin

FIGURE 18.1 Proofreader's Marks

ACTIVITIES

Spelling

In the following sentences, circle the correctly spelled words within the parenthesis.

1. Each of the employees attended the meeting (accept except) the line supervisor, who was out of town for job-related travel.

2. The (advise advice) he gave will help us all do a better job.

3. Management must (affect effect) a change in employees' attitudes toward absenteeism.

4. Let me (site cite sight) this most recent case as an example.

5. (Its It's) too early to tell if our personnel changes will help create a better office environment.

6. If we (lose loose) another good employee to our competitor, our production capabilities will suffer.

7. I'm not (quite quiet) sure what she meant by that comment.

8. (Their There They're) budget has gotten too large to ensure a successful profit margin.

9. We had wanted to attend the conference (to too two), but our tight schedule prevented us from doing so.

10. (You're Your) best chance for landing this contract is to manufacture a better product.

In the letter on page 472, correct the misspelled words.

Fragments and Comma Splices

In the following sentences, correct the fragments and comma splices by inserting the appropriate punctuation or adding any necessary words.

1. She kept her appointment with the salesperson, however, the rest of her staff came late.

2. When the CEO presented his fiscal year projections, he tried to motivate his employees, many were not excited about the proposed cuts.

3. Even though the company's sales were up 25 percent.

4. The supervisor wanted the staff members to make suggestions for improving their work environment, the employees, however, felt that any grievances should be taken directly to their union representatives.

5. Which he decided was an excellent idea.

6. Because their machinery was prone to malfunctions and often caused hazards to the workers.

7. They needed the equipment to complete their job responsibilities, further delays would cause production slowdowns.

8. Their client who was a major distributor of high-tech machinery.

9. Robotics should help us maintain schedules, we'll need to avoid equipment malfunctions, though.

10. The company, careful not to make false promises, advertising their product in media releases.

In the letter on page 473, correct the fragments and comma splices.

March 5, 2000

Joanna Freeman
Personel Director
United Teletype
1111 E. Street
Kansas City, MO 68114

Dear Ms. Freeman:

Your advertizmemt in the Febuary 18, 2000, Kansas City Star is just the opening I have been looking for. I would like to submit my quallifications.

As you will note in the inclosed resum, I recieved an Enginneering degree from the Missouri Institute of Technology in 1995 and have worked in the electronic enginneering department of General Accounts for two years. I have worked a great deal in design electronics for microprocesors, controll systems, ect.

Because your company has invented many extrordinary design projects, working at your company would give me more chances to use my knowlege aquired in school and through my expirences. If you are interested in my quallifications, I would be happy to discuss them futher with you. I look foreward to hearing from you.

Sincerly,

Bob Cottrell

Bob Cottrell

May 12, 2000

Maurene Pierce
Dean of Residence Life
Mann College
Mannsville, NY 10012

Subject: Report on Dormitory Damage Systems

Here is the report you authorized on April 5 for an analysis of
the current dormitory damage system used in this college.

The purpose of the report was to determine the effectiveness of
the system. And to offer any concrete recommendations for
improvement. To do this, I analyzed in detail the damage cost
figures for the past three years, I also did an extensive study of
dormitory conditions. Although I had limited manpower. I gathered information on all seven dormitories, focusing specifically
on the men's athletic dorm, located at 1201 Chester. In this
dorm, bathroom facilities, carpeting, and air conditioning are
most susceptible to damage. Along with closet doors.

Nonetheless, my immediate findings indicate that the system is
functioning well, however, improvements in the physical characteristics of the dormitories, such as new carpeting and paint,
would make the system even more efficient.

I have enjoyed conducting this study, I hope my findings help
you make your final decision. Please contact me. If I can be of
further assistance.

Edie Kreisler

Edie Kreisler

Punctuation

In the following sentences, circle the correct punctuation marks. If no punctuation
is needed, draw a slash mark through both options.

1. John took an hour for lunch (, ;) but Joan stayed at her desk to eat so she could complete the project.

2. Sally wrote the specifications (, ;) Randy was responsible for adding any needed graphics.

3. Manufacturing maintained a 93.5 percent production rating in July (, ;) therefore, the department earned the Golden Circle Award at the quarterly meeting.

4. In their year-end requests to management (, ;) supervisors asked for new office equipment (, ;) and a 10 percent budget increase for staffing.

5. The following employees attended the training session on stress management (, :) Steve Janasz, purchasing agent (, ;) Jeremy Kreisler, personnel director (, ;) and Karen Rochlin, staff supervisor.

6. Promotions were given to all sales personnel (, ;) secretaries, however, received only cost-of-living raises.

7. The technicians voted for better work benefits (, ;) as an incentive to improve morale.

8. Although the salespeople were happy with their salary increases (, ;) the technicians felt slighted.

9. First (, ;) let's remember that meeting schedules should be a priority (, ;) and not an afterthought.

10. The employee (, ;) who achieves the highest rating this month (, ;) will earn 10 bonus points (, ;) therefore (, ;) competition should be intense.

In the letter on page 475, no punctuation has been added. Instead, there are blanks where punctuation might be inserted. First, decide whether any punctuation is needed (not every blank requires punctuation). Then, insert the correct punctuation—a comma, colon, period, semicolon, or question mark.

Agreement (Subject/Verb and Pronoun/Antecedent)
In the following sentences, circle the correct choice to achieve agreement between subject and verb or pronoun and antecedent.

1. The employees, though encouraged by the possibility of increased overtime, (was were) still dissatisfied with their current salaries.

2. The supervisor wants to manufacture better products, but (they he) doesn't know how to motivate the technicians to improve their work habits.

3. The staff (was were) happy when the new manager canceled the proposed meeting.

4. Anyone who wants (his or her their) vote recorded must attend the annual board meeting.

5. According to the printed work schedule, Susan and Tom (work works) today on the manufacturing line.

6. According to the printed work schedule, either Susan or Tom (work works) today on the manufacturing line.

7. Although Larry is responsible for distributing all monthly activity reports, (he they) failed to mail them.

8. Every one of the engineers asked if (he or she they) could be assigned to the project.

9. Either the supervisor or the technicians (is are) at fault.

10. The CEO, known for her generosity to employees and their families, (has have) been nominated for the humanitarian award.

January 8_ 2000

Mr_ Ron Schaefer
1324 Homes
Carbondale_ IL_ 34198

Dear Mr_ Schaefer_

Yesterday_ my partners_ and I read about your invention in the *Herald Tribune_* and we want to congratulate you on this new idea_ and ask you to work with us on a similar project_

We cannot wait to begin our project_ however_ before we can do so_ I would like you to answer the following questions_

- Has your invention been tested in salt water_
- What is the cost of replacement parts_
- What is your fee for consulting_

Once_ I receive your answers to these questions_ my partners and I will contact you regarding a schedule for operations_ We appreciate your design concept_ and know it will help our business tremendously_ We look forward to hearing from you_

Sincerely_

Ron Sandhaus

Ron Sandhaus

In the following memo, find and correct the errors in agreement.

MEMO

Date: October 30, 2000
To: Tammy West
From: Susan Lisk
Subject: Report on Air Handling Unit

There has been several incidents involving the unit which has resulted in water damage to the computer systems located below the air handler.

The occurrences yesterday was caused when a valve was closed creating condensation to be forced through a humidifier element into the supply air duct. Water then leaked from the duct into the room below causing substantial damage to four disc-drive units.

To prevent recurrence of this type of damage, the following actions has been initiated by maintenance supervision:

- Each supervisor must ensure that their subordinates remove condensation valves to avoid unauthorized operation.
- Everyone must be made aware that they are responsible for closing condensation valves.
- The supply air duct, modified to carry away harmful sediments, are to be drained monthly.

Maintenance supervision recommend that air handlers not be installed above critical equipment. This will avoid the possibility of coil failure and water damage.

Capitalization

In the following memo, nothing has been capitalized. Capitalize those words requiring capitalization.

date: december 5, 2000
to: jordan cottrell
from: richard davis
subject: self-contained breathing apparatus (scba) and negative
            pressure respirator evaluation and fit-testing report

the evaluation and fit-testing have been accomplished. the attached list identifies the following:

- supervisors and electronic technicians who have used the scba successfully.
- the negative pressure respirators used for testing in an isoamyl acetate atmosphere.

fit-testing of waste management personnel will be accomplished annually, according to president chuck carlson. new supervisors and technicians will be fit-tested when hired.

all apex corporation personnel located in the new york district (12304 parkview lane) must submit a request form when requesting a respirator or scba for use. any waste management personnel in the north and south facilities not identified on the attached list will be fit-tested when use of scba is required. if you have any questions, contact chuck carlson or me (richard davis, district manager) at ext. 4036.

REFERENCES

CHAPTER 1

Bacon, S. "Want to Get Ahead? Learn to Get Along." *The Kansas City Star*, July 25, 1993: F16.

CHAPTER 3

Adams, Rae, et al. "Ethics and the Internet." *Proceedings: 42nd Annual Technical Communication Conference*. April 23–26, 1995: 328.

Anderson-Hancock, Shirley A. "Society Unveils New Ethical Guidelines." *Intercom* 42 (July/August 1995): 6–7.

Barker, Thomas, et al. "Coming into the Workplace: What Every Technical Communicator Should Know—Besides Writing." *Proceedings: 42nd Annual Technical Communication Conference*. April 23–26, 1995: 38–39.

Bowman, George, and Arthur E. Walzer. "Ethics and Technical Communication." *Proceedings: 34th Annual Technical Communication Conference*. May 10–13, 1987: MPD–93.

Bremer, Otto A., et al. "Ethics and Values in Management Thought." *Business Environment and Business Ethics: The Social, Moral, and Political Dimensions of Management*. Ed. Karen Paul. Cambridge, MA: Ballinger , 1987: 61–86.

Gerson, Steven M., and Sharon J. Gerson. "A Survey of Technical Writing Practitioners and Professors: Are We on the Same Page." *Proceedings: 42nd Annual Technical Communication Conference*. April 23–26, 1995: 44–47.

Girill, T. R. "Technical Communications and Ethics." *Technical Communication* 34 (August 1987): 178–179.

Guy, Mary E. *Ethical Decision Making in Everyday Work Situations*. New York: Quorum Books, 1990.

Hartman, Diane B., and Karen S. Nantz. "Send the Right Messages About E-Mail." *Training & Development* (May 1995): 60–65.

Turner, John R. "Ethics Online: Looking Toward the Future." *Proceedings: 42nd Annual Technical Communication Conference*. April 23–26, 1995: 59–62.

Wilson, Catherine Mason. "Product Liability and User Manuals." *Proceedings: 34th International Technical Communication Conference*. May 10, 1987: WE-68–71.

CHAPTER 4

Horton, William. "The Almost Universal Language: Graphics for International Documents." *Technical Communication* 40 (November 1993): 682–693.

Hussey, Tim, and Mark Homnack. "Foreign Language Software Localization." *Proceedings: 37th International Technical Communication Conference.* May 20–23, 1990: RT-44–47.

King, Janice M. "The Challenge of Communicating in the Global Marketplace." *Intercom* 42.5 (May/June 1995): 1, 28.

Rains, Nancy E. "Prepare Your Documents for Better Translation." *Intercom* 41.5 (December 1994): 12.

Swenson, Lynne V. "How to Make (American) English Documents Easy to Translate." *Proceedings: 34th International Technical Communication Conference.* May 10, 1987: WE-193–195.

Weiss, Edmund H. "Twenty-five Tactics to 'Internationalize' Your English" *Intercom* (May 1998): 11–15.

CHAPTER 7

"Benefits of Being Online." Resumes Online. 25 Mar. 1998 <http://www.kbweb.com/resume/>.

Lorek, L. A. "Searching On-line." *The Kansas City Star.* August 23, 1998: D1.

McNair, Catherine. "New Technologies and Your Résumé." *Intercom* (June 1997): 12–14.

Robart, Kay. "Submitting Résumés via E-Mail." *Intercom* (July/August 1998): 12–14.

Skarzenski, Emily. "Tips for Creating ASCII and HTML Résumés." *Intercom* (June/July 1996): 17–18.

CHAPTER 8

Benson, Philippa J. "Writing Visually: Design Considerations in Technical Publications." *Technical Communication* 32 (Fourth Quarter 1985): 35–39.

Everson, Larry. "Recent Trends in Technical Writing." *Technical Communication* (Fourth Quarter 1990): 396–398.

Keyes, Elizabeth. "Typography, Color, and Information Structures." *Technical Communication* 40 (November 1993): 638–654.

Schriver, Karen A. "Quality in Document Design." *Technical Communication* (Second Quarter 1993): 250–251.

Southard, Sherry G. "Practical Considerations in Formatting Manuals." *Technical Communication* 35 (August 1988): 173–178.

Watzman, Suzanne. "The Approachable Page." *Proceedings: 34th International Technical Communication Conference.* May 10, 1987: VC-85–86.

Watzman, Suzanne. "Visual Literacy and Document Productivity." *Proceedings: 34th International Technical Communication Conference.* May 10, 1987: ATA-48–49.

Wise, Mary R. "Using Graphics in Software Documentation." *Technical Communication* 40 (November 1993): 677–681.

CHAPTER 9

Horton, William. "The Almost Universal Language: Graphics for International Documents." *Technical Communication* (November 1993): 682–693.

Krol, Ed. *The Whole Internet: User's Guide & Catalog*. Sebastopol, CA: O'Reilly & Associates, 1992.

Laquey, Tracy. *The Internet Companion: A Beginner's Guide to Global Networking*. New York: Editorial, 1993.

Legeros, Michael. "Etiquette and Email: Rules for Online Behavior." *Intercom* 42 (July/August 1995): 10–11.

Mazur, Beth. "Coming to Grips with WWW Color." *Intercom*. February 1997: 4–5.

Pike, Mary Ann, et al. *Using Mosaic*. Indianapolis: Que, 1994.

Pratt, Jean A. "Where Is the Instruction in Online Help Systems." *Technical Communication* (February 1998): 33–37.

Rose, Lance. *Netlaw: Your Rights in the Online World*. Berkeley, CA: McGraw-Hill, 1995.

Scott, Michon M. "Learning to Make Web Pages the Hard Way." *Intercom*. January 1998: 16–17.

Shipman, John. "ASCII." January 1, 1995. 4 Apr. 1998 <http://www.nmt.edu/tcc/help/g/ascii.html>.

Stevens, Dawn M. "101 Standards for Online Communication." *Society for Technical Communication: 44th Annual Conference, 1997 Proceedings*. May 1997: 410–412.

Timpone, Donna. "Help! Six Fixes to Improve the Usability of Your Online Help." *Society for Technical Communication: 44th Annual Conference, 1997 Proceedings*. May 1997: 306.

Wagner, Carol A. "Using HCI Skills to Create Online Message Help." *Society for Technical Communication: 44th Annual Conference, 1997 Proceedings*. May 1997: 35.

Wilkinson, Theresa A. "Web Site Planning." *Intercom*. December 1997: 14–15. "World Wide Web." *PC Webopaedia*. 26 Jan. 1998 <http://www.pcwebopedia.com/World_Wide_Web.htm>.

Yeo, Sarah C. "Designing Web Pages That Bring Them Back." *Intercom* (March 1996): 12–14.

Zubak, Chery L. "Choosing a Windows Help Authoring Tool." *Intercom* (January 1996): 10–11, 36.

CHAPTER 12

Daughtery, Shannon. "The Usability Evaluation: A 'Discount' Approach to Usability Testing." *Intercom* (December 1997): 16–20.

CHAPTER 13

Hacker, Diane. *Research and Documentation in the Electronic Age*. Bedford Books, 1997. <http://www.bedfordbooks.com/rd/index.html>.

MLA: *The Modern Language Association of America on the Web*. SilverPlatter International, N.V., 1997. <http://www.mla.org>.

Reynolds, Michael, and Liz Marchetta. "Color for Technical
    Intercom (April 1998): 5–7.

CHAPTER 10

Allen, Douglas. Executive Director of Information Services at Jol
    Community College. Interview. April 7, 1998.
Berst, Jesse. "Seven Deadly Web Site Sins." ZDNet 30
    <http://www.zdnet.com/anchordesk/story/story_1716.html>.
Black & Veatch. 15 Apr. 1998 <http://www.bv.com>.
Byrne, DiAnn. "Marketing on the World Wide Web: How to Get Tl
    to Do." Intercom. (January 1996): 8–9, 43.
Eddings, Earl, Kim Buckley, Sharon Coleman Bock, and Natha
    Software Documentation Specialists at PDA. Interview. Januar
"E-Mail." PC Webopaedia. 24 Apr. 1997. 13 Apr. 1998 <http://w
    dia.com/e_mail.htm>.
"Extranet." What Is? 13 Jan. 1998 <http://whatis.com/extranet.ht
"Firewall." PC Webopaedia. 23 Jan. 1998 <http://www.pcwebo|
    wall.htm>.
Fisher, Sharon, et al. Riding the Internet Highway: Deluxe Editior
    New Riders, 1994.
"FNC Resolution: Definition of 'Internet.'" 30 Oct. 1995.
    <http://www.fnc.gov/Internet_res.html>.
Freeman, Lawrence H., and Terry R. Bacon. Shipley and Associat
    Bountiful, UT: Shipley Associates, 1990.
Gateway 2000. 4 Feb. 1998 <http://www.gw2k.com>.
Goldenbaum, Don, and George Calvert. Applied Communic
    Interview. January 10, 1998.
"From Tech Writer to Web Writer." Applied Communications (
    1998 <http://www.acgtech.com/content/write_to_web.htm>.
Hartman, Diane B., and Karen S. Nantz. "Send the Right Me:
    Mail." Training & Development (May 1995): 60–65.
Henselmann, Mary Anne. "Designing an Online Help System Befc
    Is Ready." Society for Technical Communication: 44th Ann
    1997 Proceedings. (May 1997): 475–476.
"Human-Computer Interaction." Rensselaer Polytechnic Insti
    Programs. 21 Apr. 1998 <http://www.llc.rpi.edu/acad/hcicert.
"Internet." April 24, 1997. PC Webopaedia. 23 Jan. 1998 <http
    pedia.com/Internet.htm>.
"Internet Traffic Growing Quickly." The Kansas City Star April 1
"Intranet." What Is? 8 Jan. 1998 <http://whatis.com/intranet.htn
"Intranet Applications." The Complete Intranet Resource
    <http://intrack.com/intranet/iapp>.
Kim, H. Young. "Looking Toward the Electronic Future in the C
    Electronic Mail." Proceedings: 42nd Annual Internati
    Communication Conference. April 1995: 51–54.

INDEX